Birmingham Archaeology Monograph Series 13

Gorse Stacks – 2000 Years of Quarrying and Waste Disposal in Chester

Richard Cuttler
Sam Hepburn
Chris Hewitson
Kristina Krawiec

BAR British Series 563
2012

Published in 2016 by
BAR Publishing, Oxford

BAR British Series 563

Birmingham Archaeology Monograph Series 13
Gorse Stacks – 2000 Years of Quarrying and Waste Disposal in Chester

ISBN 978 1 4073 1001 5

BAR Publishing is the trading name of British Archaeological Reports (Oxford) Ltd. British Archaeological Reports was first incorporated in 1974 to publish the BAR Series, International and British. In 1992 Hadrian Books Ltd became part of the BAR group. This volume was originally published by Archaeopress in conjunction with British Archaeological Reports (Oxford) Ltd / Hadrian Books Ltd, the Series principal publisher, in 2012. This present volume is published by BAR Publishing, 2016.

Printed in England

BAR
PUBLISHING

BAR titles are available from:

BAR Publishing
122 Banbury Rd, Oxford, OX2 7BP, UK
EMAIL info@barpublishing.com
PHONE +44 (0)1865 310431
FAX +44 (0)1865 316916
www.barpublishing.com

Contents

List of Figures, Plates and Tables

FIGURES

PLATES

TABLES

Summary

The site of Delamere Street lies just outside the North gate of the Roman and medieval Chester and in recent years has been subject to intensive investigation as part of the Gorse Stacks development undertaken by Watkin Jones. (Fig.1.1).This publication represents the culmination of those investigations carried out by Birmingham Archaeology during 2006 and 2008.

The basis for the project was laid out by previous studies including a desk based assessment carried out by Mason Welland Archaeological Consultancy in 2001 (Mason & Frost 2001) and a program of trial trenching carried out by Earthworks Archaeology (Dodd & Clarke 2002). Both studies identified the site as having significant potential for the survival of buried Roman remains in an area that had previously received little attention.

Using the results of the evaluation a total of 5 open areas (A-E) were excavated around the edges of a known post-medieval quarry that dominated the centre of the site (Fig.1.2). The Roman deposits were heavily truncated in the eastern half of the site but remained largely undisturbed to the north. The Roman activity on the site was represented mainly by large sandstone quarry pits which were later reused for the disposal of rubbish from within the fortress. A large quantity of pottery, animal bone and metal artefacts were recovered from the fills of these pits dating to the late 1st-early 2nd centuries, representing a snapshot of daily life within and around the fortress. The results of the pottery analysis showed that a variety of locally made coarsewares as well as imported Gaulish samian were made available to the residents of the fortress. The animal bone also provided an insight into the daily subsistence strategies required to underpin the functioning of the Roman Army within a fortress setting.

In order to understand the nature of the activity outside the main fortress walls environmental samples were analysed for clues as to the function of the area. In the eastern half of the site a well defined Punic-style ditch was excavated and a substantial quantity of hammerscale was recovered from the environmental samples. This, along with a substantial quantity of slag recovered from other features, seems to suggest that metalworking was taking place within the fortress although not at the site itself.

The quarrying which faltered in the Roman period was resumed in the 16th-century and a repeat of the refuse deposition is seen with the quarry being infilled in the subsequent centuries after the sandstone supply had been exhausted. The infilling of this quarry with domestic and industrial waste again provided a snapshot of activity taking place in Chester during the 18th-20th centuries with stratigraphy surviving to depths of over 6m. Due to the size of the quarry it was divided into six areas to make it easier to excavate and record. A mixture of EDM survey with rectified photography, and laser scanning, was used to record the quarry face to provide detailed information regarding toolmarks and quarrying techniques.

SECTION 1 – INTRODUCTION TO THE PROJECT

Background

This volume reports on the series of investigations at the site of the former Bus Station in Delamere Street, Chester, prior to the redevelopment of the site as part of the Gorse Stacks project (Fig 1.1). An evaluation of the site was undertaken by Earthworks Archaeology in 2002 and culminated in open area excavation by Birmingham Archaeology in 2005-2007 and a watching brief in 2008. A total of five open areas (A-E) were excavated in 2005 around the edges of the post-medieval quarry that was known to dominate the centre of the site (Krawiec 2008). The watching brief of the removal of the former quarry and its infill was undertaken in 2008 (Hepburn and Krawiec 2009). This volume describes the results of these investigations with reference to work subsequently undertaken adjacent to Delamere Street.

The initial investigation of the site comprised a desk-based assessment which was undertaken by the Mason Welland Archaeological Consultancy in 2001 (Mason & Frost 2001). This assessment identified the potential for Roman occupation of the site in the form of a ribbon settlement along the road heading from the north into the fortress. A program of test pitting and a borehole survey carried out by Earthworks Archaeology in 2002 identified a high potential for the survival of multi-period archaeological remains (Dodd & Clarke 2002; Fig 1.2). The test pits confirmed the presence of undisturbed Roman deposits containing domestic refuse but were too small to identify the feature types. An initial trench (Trench 1) excavated by Birmingham Archaeology in 2005 further characterised these deposits as extremely deep and extensive (Krawiec 2008).

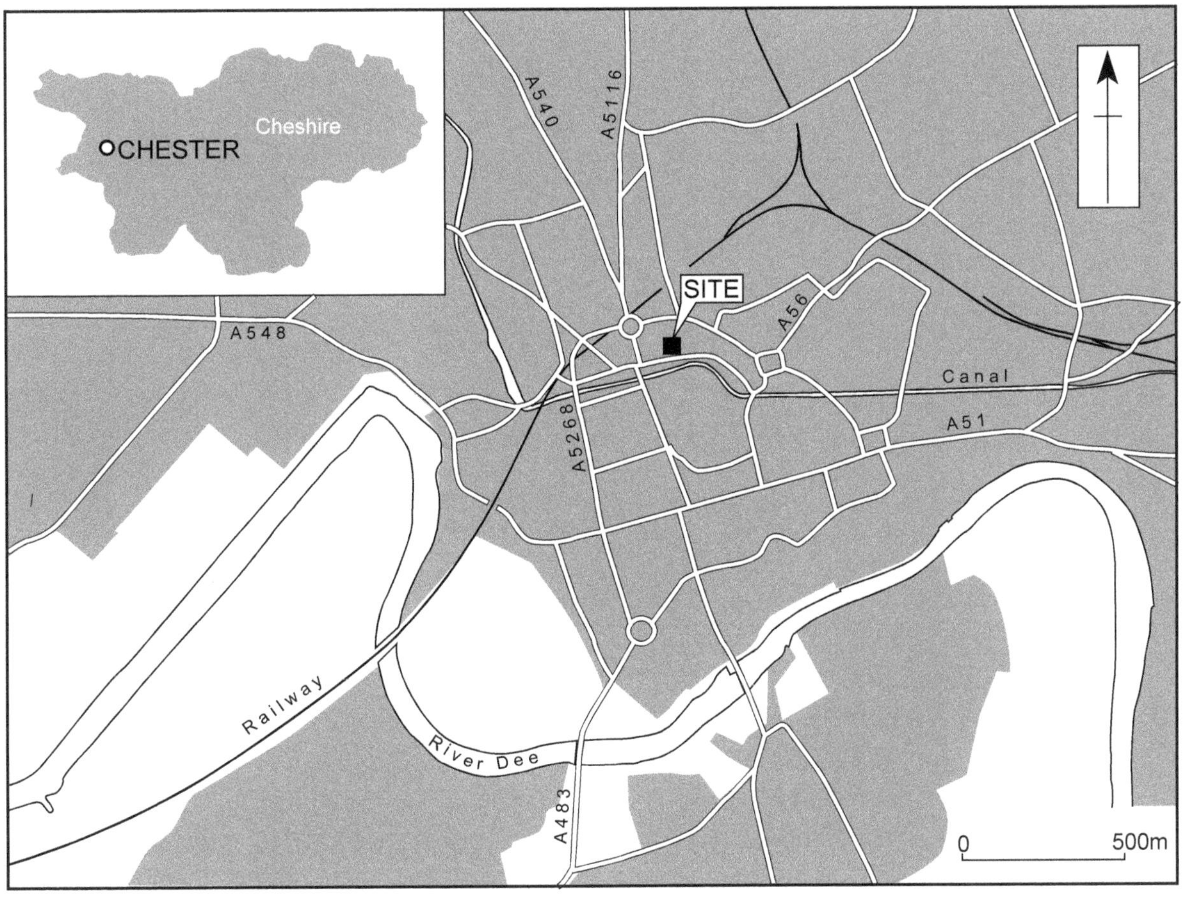

FIGURE 1.1 LOCATION MAP.

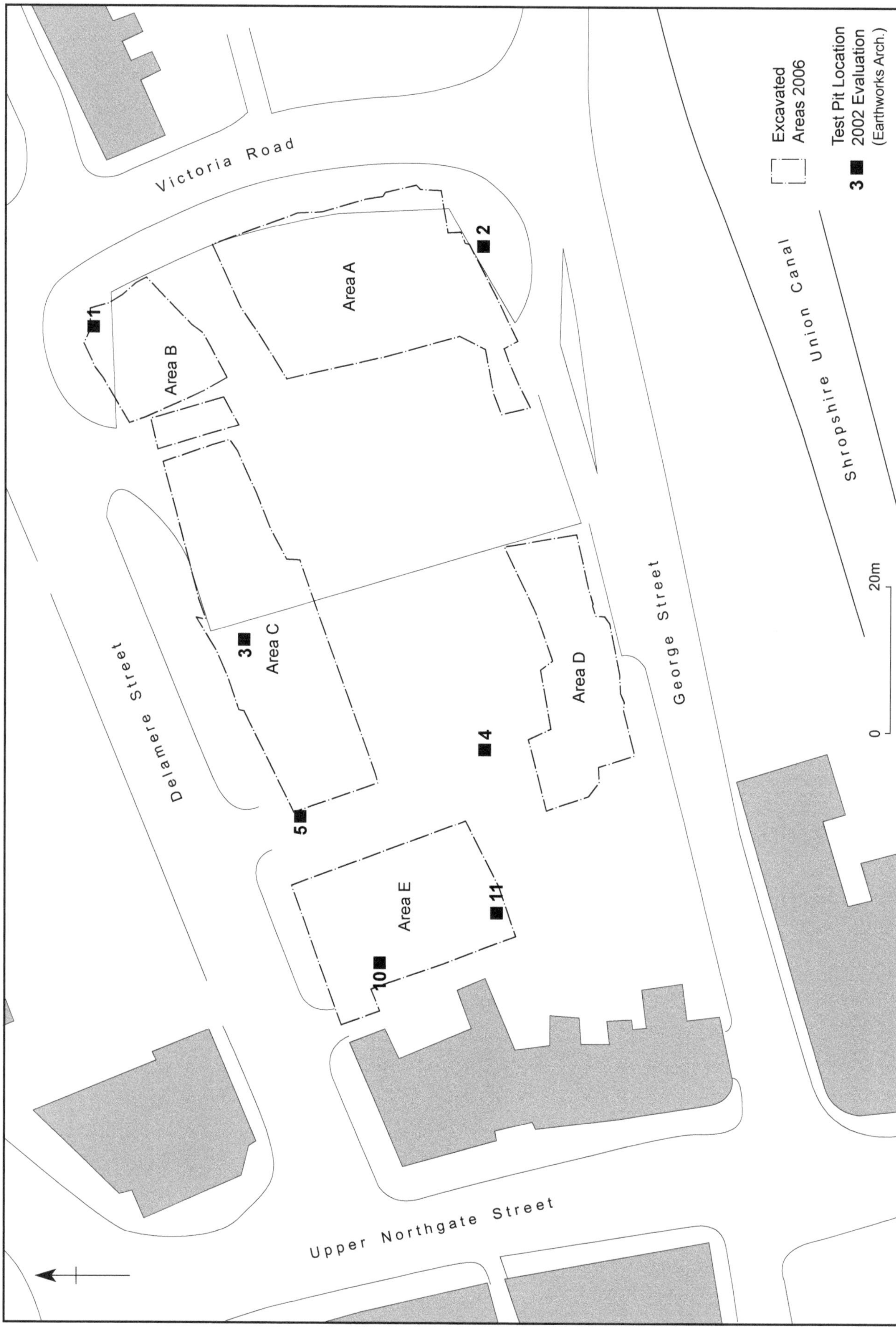

FIGURE 1.2 EXCAVATION AND TEST PITS LOCATION MAP.

Archaeological background

Location and geology

Chester occupies a shallow bowl of the Cheshire plain and its geology is characterised by ridges of underlying sandstone bedrock which outcrops at several locations in and around the area. This is overlain in places by softer glacial deposits of sand and gravel. Chester itself lies at the southern end of this Triassic sandstone ridge that overlooks the River Dee and the surrounding valley.

Archaeology of the region

Prehistoric activity at Chester has yet to be uncovered although as is indicated by several hillforts in the region it may have been occupied in the Iron Age. Excavations at the site of the legionary parade ground to the east have revealed possible pre-Roman Iron Age plough marks and coupled with the 'Celtic' name for the settlement (Deva) all points to a possible pre-existing community at the site (Mason 20001:27). Certainly Chester was firmly located within Cornovii territory, which itself lay between the territories of the Deceangli and Ordovices to the west and the territory of the Brigantes to the east.

The presence of an established early settlement is in part supported by the theory that the path the legionary forces took through the northwest of England was based on areas with sufficient existing infrastructure from the native population to support such a large movement of people (Wells 2003:81). However the bias in archaeological investigation towards the investigation of military sites has precluded the excavation of sites that support this theory (Philpott and Brennard 2005:3). Despite the bias towards military sites most lack detailed chronologies and many investigations have not been carried out to modern standards (Philpott and Brennard 2005:6). There is also a lack of fully published material although this is beginning to be addressed through the work of David Mason and Peter Carrington who have sought to pull together these disparate datasets to form a coherent narrative (Mason 2001:2007, Carrington forthcoming).

Extensive work has been carried out in the *canabae* which has defined the areas to the east, south and west as areas of extra-mural settlement as well as the presence of a port. These areas are relatively well-understood in comparison with the area to the north which has been conjectured to be unoccupied and used as a cemetery. Systematic excavation has yet to be carried out to confirm this and the relationship with the activity to the north and the workings of the fortress and *canabae* are poorly understood.

The post-Roman period in Chester is not well documented in the archaeological record as many of the antiquarian excavations of the early 20th century failed to adequately record layers of activity overlying the Roman remains (Mason 2007:30). It is likely that the area outside the walls was occupied during the medieval period but any trace of this has been erased by subsequent development. In addition the widespread damage to properties along the frontage of George Street during the Civil War meant the site had to be cleared. There are considerable gaps in the development of this area of Chester and documentary and cartographic sources suggest the presence of a sandstone quarry in the early 16th century which would also have severely truncated earlier remains. The development of the frontages along George Street continued from this period onwards and it appears the quarrying ceased around *c* 1870 at which point the quarry which occupied the centre of the Delamere Street site was infilled and developed further. The last clearance of the site occurred in the 1960s just prior to the construction of the bus station. This sequence of events has led to a more compressed timeline for the development of Chester than has been seen at other urban sites such as York, presenting challenges when attempting to place these changes in the regional picture of the post-Roman development of the region.

Aims of the Investigation

The principal aim of the archaeological fieldwork was to identify, excavate and record the archaeological remains, and to preserve those remains by record. The more specific research objectives of the site were:

- To recover evidence of the history and development of the site that may be set against the wider archaeological and historical background of the city of Chester.
- To establish the presence of specific activities which may have been undertaken in the excavation area, such as industrial or manufacturing activities, throughout the Roman, medieval and post-medieval periods.
- To further enhance our understanding of the development, functions and status of the site during the Roman period.
- To establish a more secure site chronology from the earliest evidence of activity onwards, giving particular emphasis to a defined chronology for various phases of activity during the Roman period.
- To provide comparative material to contribute to our understanding of the site in relation to the *canabae.*
- To build up a picture of the history and development of the post-medieval quarry and set it within the framework of local stone extraction activity outside the city walls.
- To engage in a recovery policy of material from the backfill of the quarry in order to provide a finds assemblage for comparative study of material in Chester. The lifespan of the quarry from the 16th to the 19th century is poorly understood and hence the recovery of a securely stratified assemblage of material will aid future study within the city.
- To build up a picture of the history and development of the post-medieval quarry through the thorough recording and examination of the quarry

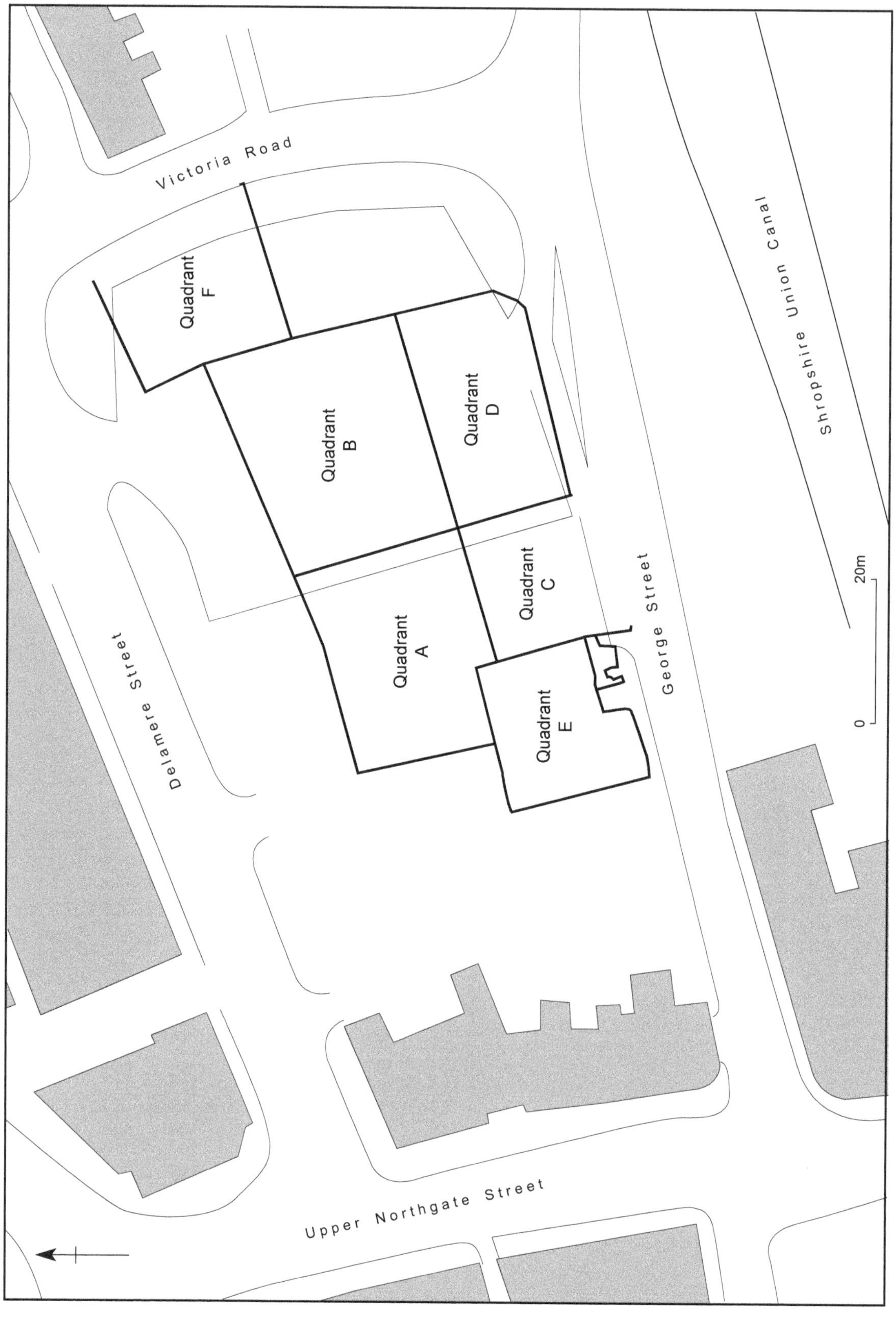

FIGURE 1.3 WATCHING BRIEF QUADRANTS.

fabric. This should be achieved by the accurate recording of the quarry faces, and the recording of longitudinal and cross-sections of the quarry fills.
- To establish quarrying techniques involved with stone extraction through the identification, examination and recording of tool marks within the quarry.

Methodology

The proposed development area covered approximately 5000m². Two phases of work were undertaken as defined by the timetable of the development.

Open Area Excavations

Around the perimeter of the site Roman remains survived the truncation caused by the creation of the post-medieval quarry therefore the site was divided into five areas of excavation (labelled A-E). These were specifically positioned over known archaeological deposits as defined by the test pitting and expanded to ensure all areas with the potential for buried deposits were investigated.

All topsoil and modern overburden was removed using a mechanical excavator with a toothless ditching bucket, under direct archaeological supervision, down to the top of the uppermost archaeological horizon or the subsoil. Subsequent cleaning and excavation was by hand. All spoil heaps and trenches were scanned by an experienced metal detectorist.

All stratigraphic sequences were recorded, even where no archaeology was present. A comprehensive written record was maintained using a continuous numbered context system on *pro-forma* context and feature cards. After excavation the various features were assigned group numbers that defined specific feature types and buildings.

Features were planned at a scale of 1:50, and sections were drawn through all cut features and significant vertical stratigraphy at a scale of 1:20. The excavation area was surveyed using an EDM total station and the results mapped into CAD software. Spot levels were taken where appropriate and all levels were recorded relative to an Ordnance Survey datum. Written records and scale plans were supplemented by photographs using monochrome, digital and colour slide photography.

Forty litre soil samples were taken from datable archaeological features for the recovery of charred plant remains. The environmental sampling policy followed the guidelines contained in the *Birmingham Archaeology Guide to On-Site Environmental Sampling*. Finds were cleaned, marked and remedial conservation work was undertaken as necessary. Treatment of all finds conformed to guidance contained within the Birmingham Archaeology Fieldwork Manual and *First Aid for Finds* (Watkinson and Neal 1998).

The Watching Brief

The large post-medieval quarry-pit covered an area of approximately 2000-2500m² and was investigated after the open area excavations had been completed. A watching brief was undertaken during the process of site clearance and involved the monitoring of all machine excavation including the digging of test-pits for contaminated land samples and the scouring down of the edges of the quarry.

To aid with the recording, the main quarry was divided into quadrants labelled A-D with the two extensions E and F, and each face of the main quarry numbered 1-4 (Fig 1.3). The quadrants were roughly equal in size at each stage in order to create an accurate profile of the quarry fills. This allowed the recording of sections of the quarry fill both longitudinally and in cross-section. Each quadrant was excavated in successive 2m deep strips, until the piling formation level was reached, 8-10m below the surface (18.60m AOD). The surrounding rock face of the quarry was then reduced to the same depth. This process was repeated after each 2m strip had been recorded.

The sections were recorded by means of reflectorless electronic distance measurement (EDM) and photographic record. Digital photographs of the quarry faces were rectified to allow detailed drawing. In addition to the written and photographic record the quarry face was recorded using laser scanning in order to produce an interactive three dimensional model. Where tool marks or particular details of the quarry face were encountered these were drawn by hand at a scale of 1:20. A full photographic survey was undertaken, with scales where it was safe to position them, of all sections through the quarry fill and elevations of the quarry face.

As the quarry deposits were excavated artefacts were recovered by hand by experienced archaeologists. Where large quantities of material were encountered these were removed by machine to a safe location to be allow more time for hand sorting to take place. The sampling policy was subject to on-going review in consultation with Mike Morris, Chester City Archaeologist, regarding a suitable discard policy. In addition, Leigh Dodd acted as an on-site finds specialist where unusual or large quantities of material were recovered.

Phasing of the Site

There were several problems with the phasing of the site which only became clear in the late stages of the post-excavation process. The Roman activity at the site all broadly sits within the late first to mid-second century. The problems of dating the stone cut pits as well as the problems of the midden material that infills them has meant that a less traditional system must be used in order to understand the site. The activity has been broken down into 'Events' rather than phases which are as follows:

- Phase 1: Roman period
    - Event 1: Quarrying of stone
    - Event 2: Deposition of quarry waste
    - Event 3: Rubbish deposition
- Phase 2: 16th century quarry group 110.
- Phase 3: 17th and 18th century deposition.
- Phase 4: 19th and 20th century buildings.
    - Phase 4a: mid-19th century sandstone buildings groups 107, 108,109 & 111.
    - Phase 4b: late-19th century Methodist chapel and Delamere Street dwellings.

Phase 5: 20th century deposition of material

The Roman activity on the site was relatively short-lived, spanning the late 1st to mid-2nd centuries AD and was characterised by large stone extraction pits which were then infilled with quarry waste and refuse from the fort. The dating of the stone extraction is difficult to achieve with any certainty and it is only the later refuse deposits that provide a chronology of activity around the site. The quarry waste deposited at the base of the larger pits may be assumed to date from the end of active quarrying at the site and therefore all that can be known with certainty is that the quarrying must have a short time before this. There are no deposits within the pits that suggest a build up of infill over a lengthy period of time, rather the pits were infilled in two distinct events: quarry waste and refuse.

The Post-medieval activity on the site spans a much longer period, from the resumption of the quarrying at the end of the 16th century to the subsequent infilling and redevelopment of the site during the 18th and 19th centuries. Here three distinct phases of activity have been identified. The first (Phase 2) was the excavation of the quarry; the second was a period of waste deposition within the quarry during the 17th and 18th centuries (Phase 3). The third was during the 19th century when the quarry saw the construction of several buildings (Phase 4). The final phase saw the demolition and backfilling of the site as part of the construction of the bus station in the 1960's. The details of these phases will be discussed in Section 3.

The following chapters will provide the results of these investigations and place then within the context of the development of the fort and *canabae* and post-medieval town.

SECTION 2 – THE ROMAN QUARRY

This section discusses the Roman phase of activity (Phase 1) recorded during the excavations in 2005 (Figs 2.1-2.5), residual material recovered from the watching brief on the quarry in 2008 and the evaluation test pits carried out in 2002. The results of these investigations revealed a short-lived period of activity between the 1st and 2nd centuries AD.

The Excavations

by Kristina Krawiec

Introduction

The natural bedrock that was encountered across the site was a compact red sandstone that sloped gently down to the east and was reached at 29.22m AOD at the western end of the site and 27.26m AOD at the eastern end of the site. The sandstone was extremely weathered to the south-east becoming more compact to the north-west.

Phase 1: The Roman Period: Groups 112 and 113

The Phase 1 activity at the site is confined to the early Roman period (late 1st to mid 2nd century). Due to the nature of the deposits and features at the site this activity has been ascribed 'Event' numbers rather than sub-phases. There are three distinct events represented on site; the activity in Event 1 is characterised by the removal of stone from the site for use elsewhere;. Event 2 is represented by the initial backfilling of these pits with rubble and quarry waste once the quarry had fallen out of use; Event 3 involves the clearance of areas within the fort, the transportation of that waste to the site and the deposition of this waste within the partially infilled quarry pits (Plate 2.1).

There are also several features which are not associated with this activity and are infilled with material which is undated or indistinguishable from the bulk of the activity on site. These comprise ditches and short gully features

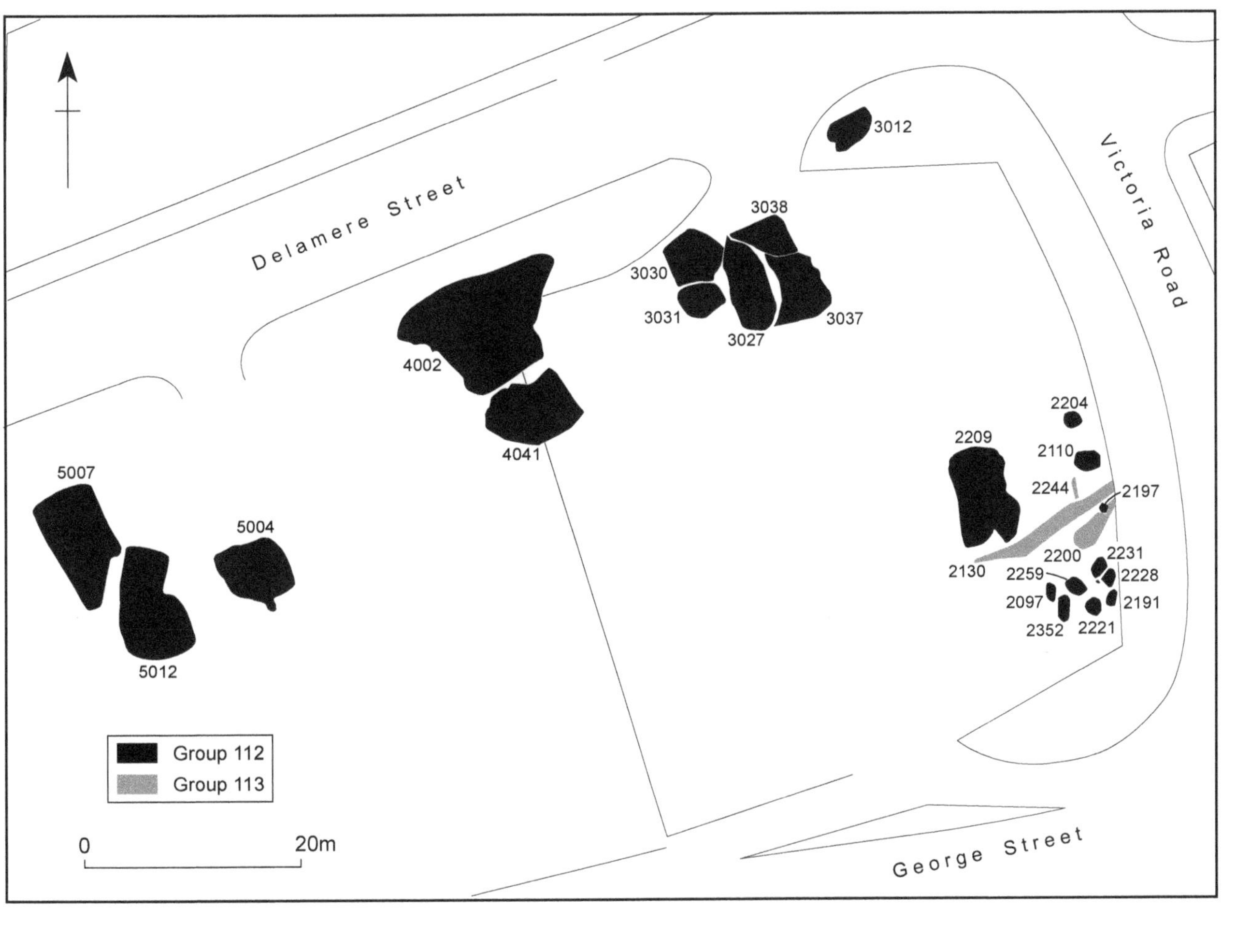

FIGURE 2.1 GROUP 112 AND 113 FEATURES.

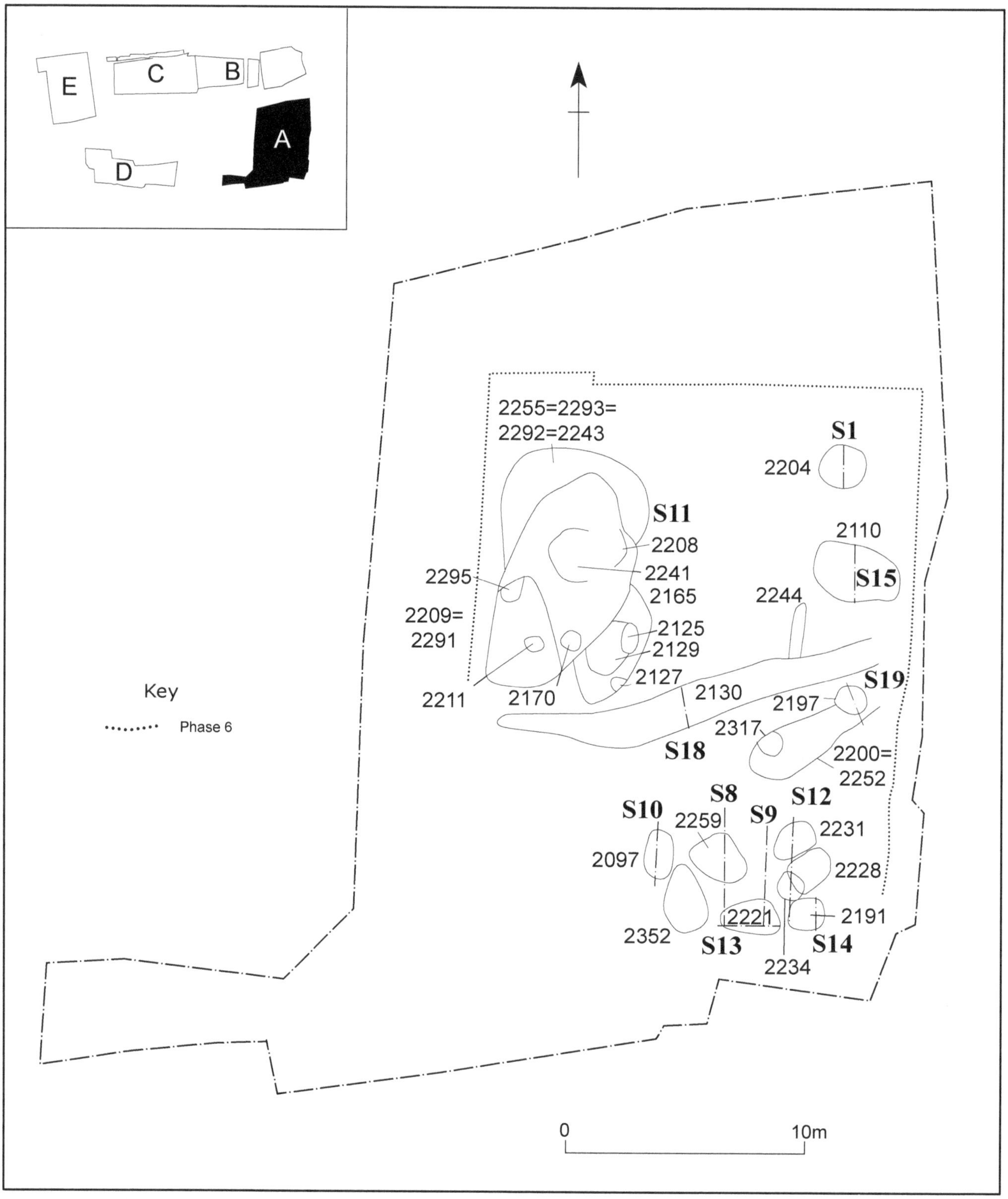

FIGURE 2.2 AREA A FEATURES.

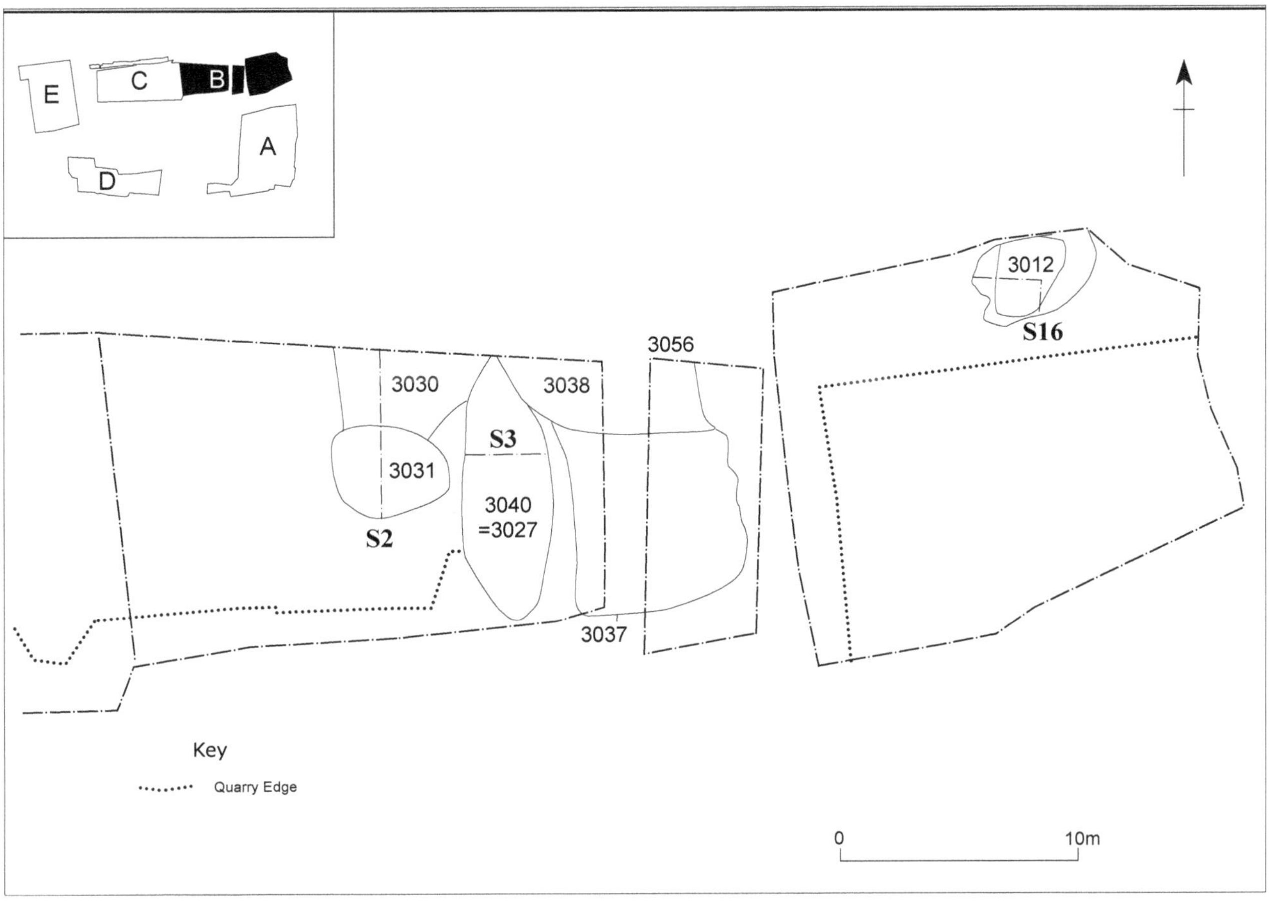

FIGURE 2.3 AREA B FEATURES.

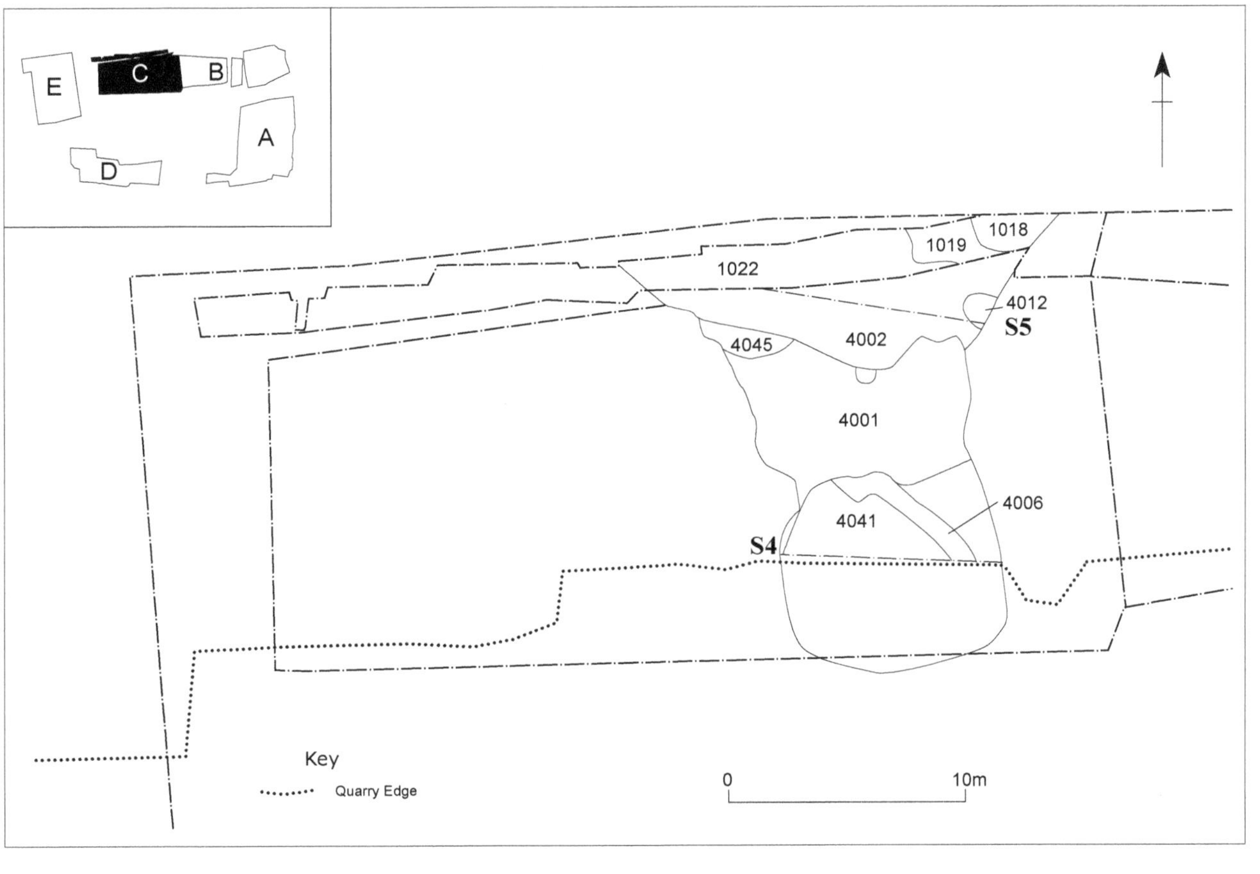

FIGURE 2.4 AREA C FEATURES.

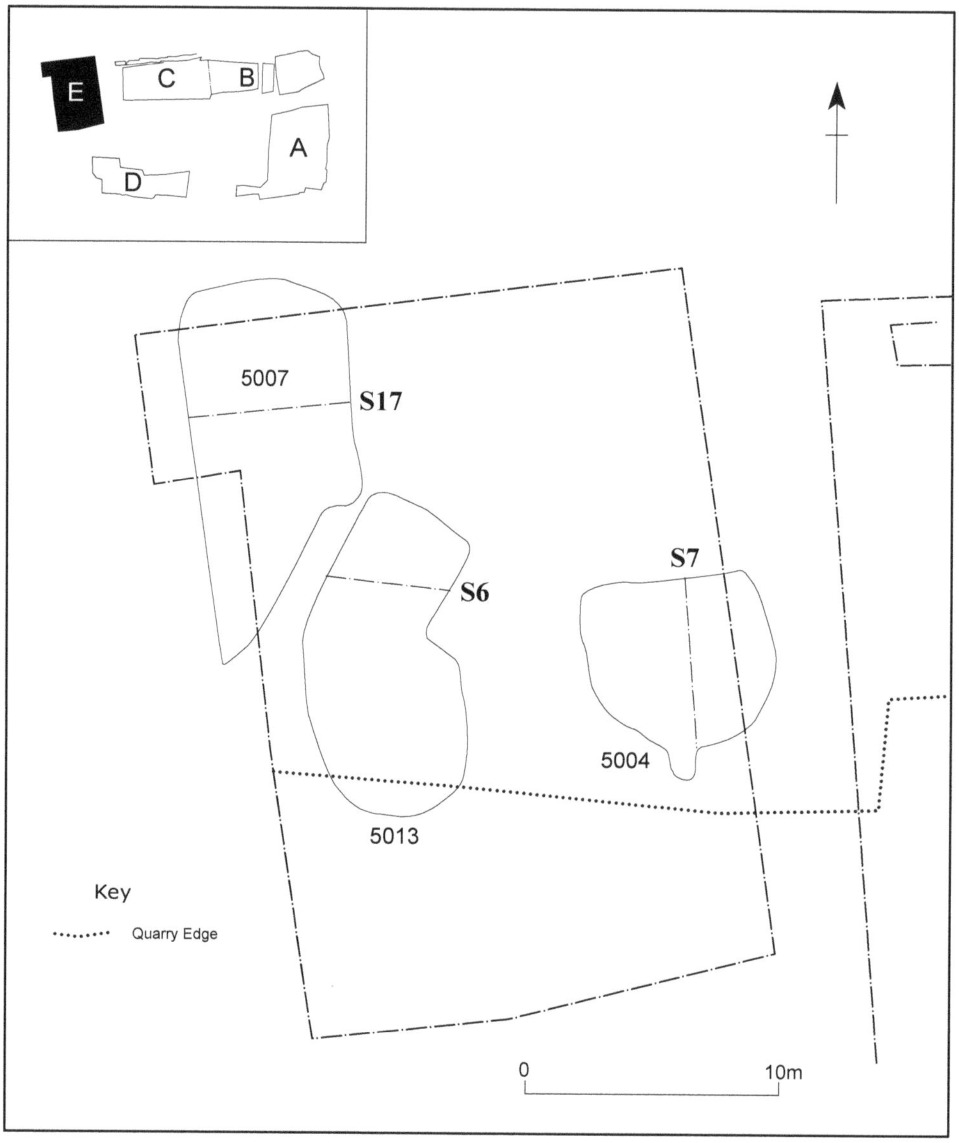

FIGURE 2.5 AREA E FEATURES.

which have no known function and appear not to be related to the quarrying (Plate 2.2). They will be described in a separate section and have not been assigned Event numbers.

Event 1- The Quarry

Across the northern and eastern edges of the site several large quarry pits were excavated (Group 112, Figs. 2.1-2.5). The toolmarks and techniques used to extract the stone are not diagnostic and therefore the exact date of the active life of the Roman quarry cannot be ascertained directly from them. The initial infilling of these pits may at least provide a *terminus ante quem* for the end of the quarrying period.

The stone extraction may have taken place across the whole area but the presence of a post-medieval quarry pit had removed any trace of earlier activity across much of the site, with few residual Roman finds surviving. There was no evidence of any structural remains to accompany the stone extraction, however, structures may have been present within the area truncated by the later activity across the site.

The Roman activity within the area appears to have begun with the removal of stone from the northern, eastern and western areas of the site and consisted of a series of large quarry-pits (Areas C, B, A and E). The cutting and removal of stone must have occurred at some point before the date suggested by the pottery recovered from the primary quarry waste deposits (Event 2). The deposition of the quarry waste within these pits is dated to the Flavian--Trajanic period (AD 80--120) and it is unclear whether this disposal of waste would have taken place immediately after stone extraction was terminated at the site, or whether the site was abandoned and the quarry waste left lying on the surface until the site was re-activated for rubbish disposal (Event 3).

Given the probability that the quarry waste deposits (Event 2) are not primary fills but represent the clearance material from the site it would be realistic to date these pits to the earlier part of this period. The quarry-pits are mainly irregular in shape and the toolmarks suggest that the stone was not removed in discreet blocks and may have been removed as rubble for the core of walls rather than the dressed outer surface.

Area A

As described above the stone extraction at the site occurred in large irregular pits. Within Area A was a wide shallow pit [2209] which was 3.85m wide and 0.60m deep (Plate 2.1). This area of the site had been subject to later building development that in all probability has truncated this type of feature. The building of a chapel in the late 19th century and bus station in the 1960s and their subsequent demolition may have removed a significant depth of material leaving this pit shallower than it originally was.

This pit [2209] was truncated by a series of modern drains and walls so each section was assigned a separate cut number but are all part of the same pit, [2209, 2241, 2291]. It was also located in a part of the site where the sandstone was extremely weathered and of a poor quality having a tendency to crumble easily. Again this would suggest stone may have been extracted as rubble only. The basal fills of this pit contained similar material to the upper fills of the large stone extraction pits recorded in Area C (*see below*) but contained a much higher concentration of hammerscale. This would suggest the waste came from an area of the fortress given over to smithing or metalwork repair.

To the east of pit [2209] a circular pit [2204] directly truncated the bedrock and differed from the stone extraction pit in its regular shape which may indicate some other primary function (Plate 2.3). It was 1.69m in diameter and 0.48m in depth and was infilled in a single episode with a mid grey brown sand clay (2206) which contained pottery dating to the Flavian--Trajanic period including a Dorset Black Burnished flat rim bowl that cannot be earlier than Trajanic (Event 2). It was truncated by a later pit [2321] presumably due to the fact that it was far easier to dig into soft fill than digging another pit into the bedrock (Event 3, Fig. 2.6, S1).

Area B

In Area B several inter-cutting irregularly shaped quarrying pits were also excavated. The earliest of these was a wide, shallow (*c* 0.40m) quarry pit [3030] infilled with a weathered sandstone deposit (3047) which was overlain by a mid-grey brown silt sand (3029) which had sandstone rubble throughout (Fig. 2.6, S2). This was then truncated by a pit which dated to the same period [3031] and was infilled by similar material which also contained sandstone

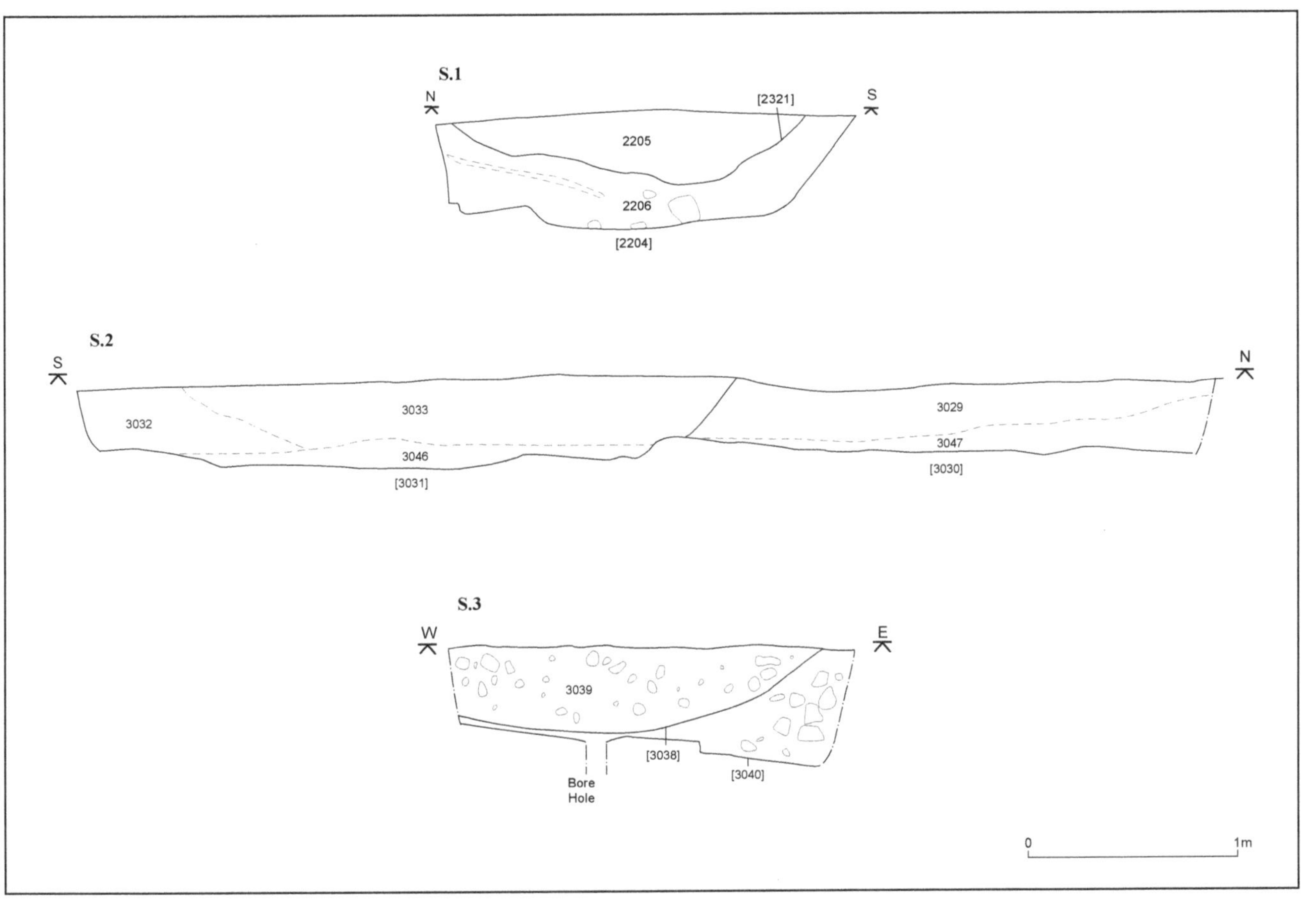

FIGURE 2.6 S1. [2204]WEST FACING SECTION, S3 [3030] AND [3031] EAST FACING SECTION, S3. [3040] AND [3036] SOUTH FACING SECTION.

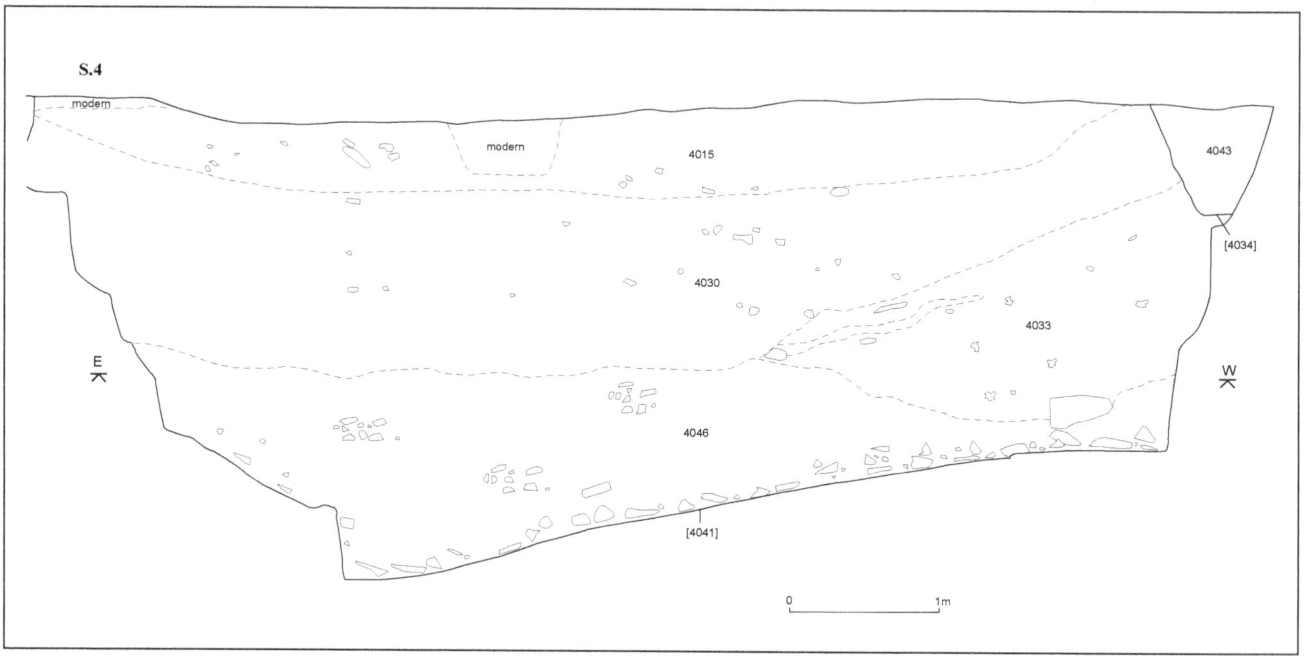

FIGURE 2.7 S4. [4041] NORTH FACING SECTION.

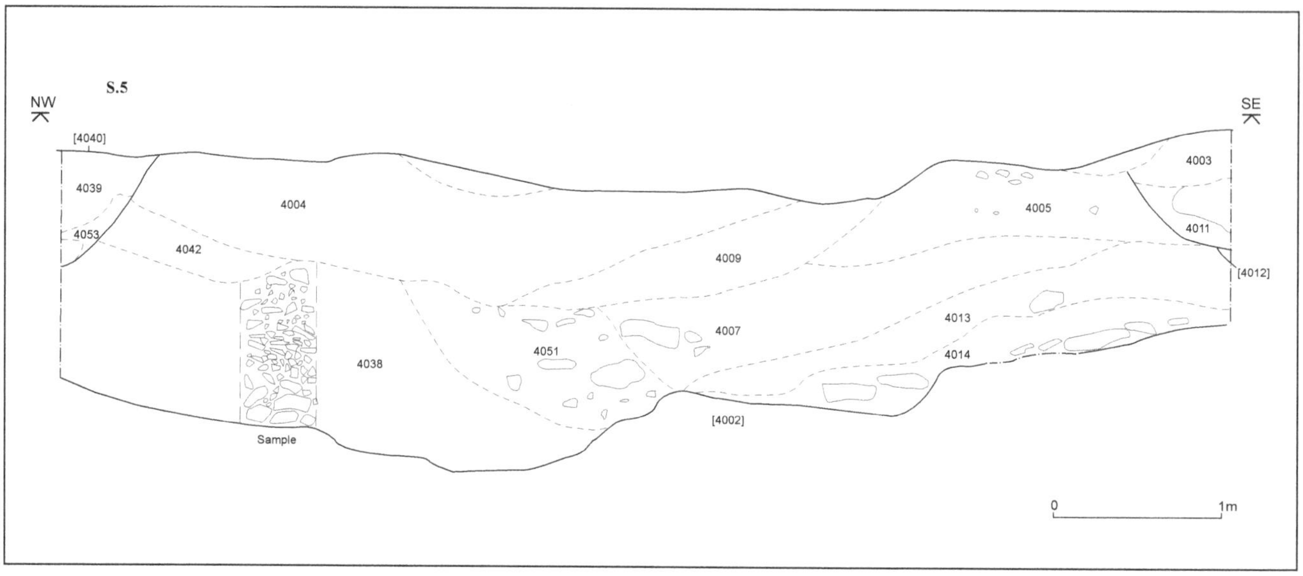

FIGURE 2.8 S5. [4002] SOUTHWEST FACING SECTION.

rubble (3032, 3033). These two pits were some of the only quarry features on site with an intercutting relationship indicating that infilling of some of the pits may have occurred during the active life of the quarry. Perhaps a 'leapfrogging' from pit to pit may have occurred, placing the waste of an active pit into an inactive pit, although this is speculative.

In the north-eastern portion of Area B a large, shallow (*c.*0.52m) square pit [3037/ 3058] and a similar more irregular pit [3027/ 3040] were excavated (Fig. 2.6, S3). No dateable finds were recovered from [3027/3040] although a blade-like flint flake was recovered from the upper fill (3028) which may be Roman although it is not possible to closely date the form (Fig. 2.28.32).This was truncated by pit [3056=3038] which was shallow with the majority of the feature continuing under the baulk.

The most regular pit [3012] was almost circular and was much deeper than the other pits in this area being 1.20m deep. This again has toolmarks along the edges indicative of stone removed for rubble building material. This pit was isolated from the others with no others features surrounding it.

Area C

Within Area C were the main quarry-pits, the largest of which [4041] being 14m wide and 3.5m deep with a stepped profile (Plate 2.4). This contrasted with the smaller stone extraction pits described above. The largest pit [4041] was severely truncated by the large post-medieval stone quarry to the south and therefore its full extent is not known. The faces of the quarry pit had evidence of scabbling toolmarks (rough dressing to remove irregular surfaces) as well as steps that were probably a result of stone removal before the pit was re-used for the disposal of waste (Plate 2.5, Fig. 2.7, S4). The irregular profile of the pit suggested that stone was removed in small blocks or rubble as opposed to large regular ashlar blocks.

To the north of this was a pit of similar size which was principally excavated in Trench 1 (an initial evaluation trench excavated prior to the main work) [1022] and extended into Area C [4002] (Plates 2.6 and 2.7; Fig. 2.8, S5). Again it appeared this was originally a pit created

PLATE 2.1 AREA A PHASE 1 EAST FACING.

PLATE 2.2 DITCH 2130 WEST FACING.

PLATE 2.3 PIT [2204] FULLY EXCAVATED.

PLATE 2.4 PIT [4140] WEST FACING.

PLATE 2.5 SCABBLING MARKS ON FACE OF PIT [4140].

PLATE 2.6 PITS [4002] AND [4140].

PLATE 2.7 FILLS OF PIT [4140].

during the extraction of stone and reused as a refuse pit. This pit also had stepped sides that became vertical towards the centre of the cut. As with [4041] there was evidence of tool marks caused by the stone extraction. The full extent of this pit is also unknown as it continued under the baulk but it may be preserved to the same extent beneath the pavement and road surface of Delamere Street. The upper fills of this pit can be correlated with those deposits recorded in Test Pit 3 of the evaluation (Dodd and Clarke 2002).

Area E

In Area E, to the west of the main quarrying activity, were several other quarry-pits of varying size (Plate 2.8). The irregular shape of these pits, a by-product of the quarrying process, was best represented in pit [5013] (Fig. 2.9, S6) although this was quite shallow at *c* 0.50m in depth.

The largest of the quarry-pits in Area E was [5004] which measured 6.50m in length by 8.60m in width and consisted of steep, vertical sides to the north and south of the feature, with a less steeply sloping side from west to east before again shelving vertically (Fig. 2.9, S7; Plate 2.9). The pit [5004] contained evidence of two parallel grooves in the side of the pit which could be interpreted as the remains of wedge marks from stone extraction or a possible structure associated with a pulley system for the removal of cut stone from the pit (Plate 2.10). As well as the grooves there was also a significant number of 'scabbling' toolmarks on the walls of the pits, made by picks as part of the extraction process as seen in Area C.

Stone extraction may have ceased in some locations at the site by the time these pits were active suggesting this was the last area to be quarried before the site fell out of use. The possibility remains that there may have been quarry pits, and associated structures, located in the area truncated by the later post-medieval quarry. Overall, quarrying activity appears to have declined by the early 2nd century and soon after the site was cleared ready for the reuse of the pits for the deposition.

PLATE 2.8 AREA E NORTH-WEST FACING.

PLATE 2.9 PIT [5004] EAST FACING.

PLATE 2.10 SLOTS ON QUARRY WALL OF [5004].

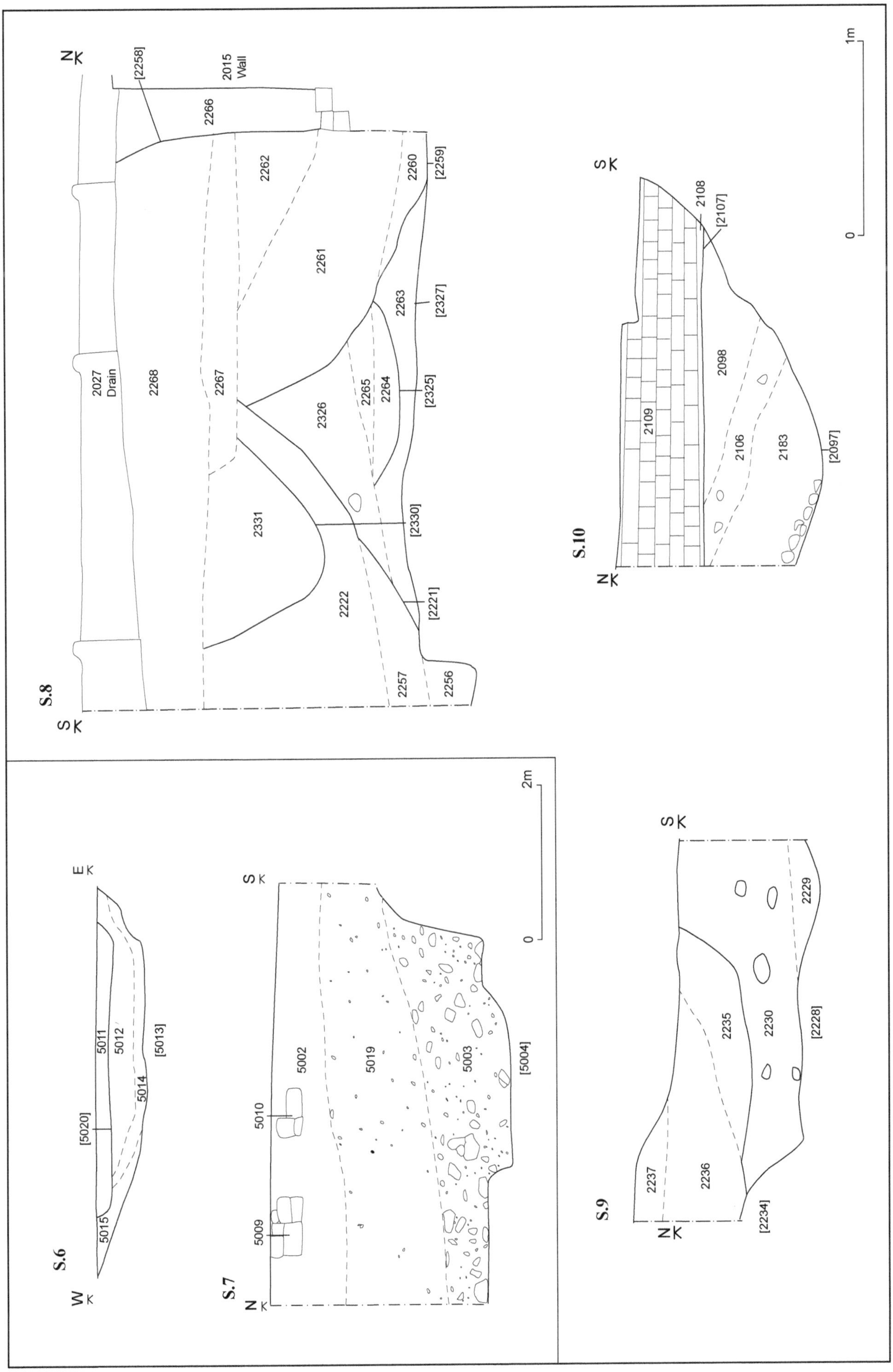

FIGURE 2.9 S6 [5013] AND [5020] SOUTH FACING SECTION, S7. [5004] WEST FACING SECTION, S8. [2327], [2221],[2325],[2259],[2330] EAST FACING SECTION, S9 [2228] AND [2234] EAST FACING SECTION AND S10. [2097] EAST FACING SECTION.

Event 2 - Deposition of quarry waste

After the removal of stone at the site had ceased the remaining waste was cleared and deposited within the pits which represent Event 2. These deposits were characterised by their high rubble and sand content. Whether this clearance occurred as the stoneworkers moved from pit to pit or as a wholesale clearance is unclear. However there was evidence for the gradual deposition of sediment in the base of the pits suggesting they were not open to the elements for very long before the primary backfilling began.

Area A

There does not appear to be any deposition of quarry waste in Area A and most of the pits in this area fall into the category of refuse pits which occur as part of Event 3.

Area B

In Area B the quarry pit [3030] was primarily infilled with a weathered sandstone deposit (3047) which was overlain by a mid-grey brown silt sand (3029) which had sandstone rubble throughout (Fig. 2.6, S2; Plate 2.11). This rubble-rich deposit (3029) contained a pottery assemblage dating predominantly to the Trajanic period with a relatively high amount of Samian (9%). This was truncated by a pit which dated to the same period [3031] and was infilled by similar material which also contained sandstone rubble (3032, 3033). The high sandstone content of the fills suggests that quarry waste lay scattered across the site after the removal of the sandstone and was cleared into the pits before the site was used for the deposition of refuse from the fortress (Event 3). These fills also contained pottery including again a high percentage of Samian (14%) and Dorset Black Burnished ware that may place the pit [3031] slightly later in the sequence towards the mid-2nd century. A copper alloy slide lock bolt and the largest quantity of nails from the site were recovered from fill 3029 ([3030], Fig. 2.28.37). A belt buckle frame was also recovered from pit [3031], (3033, Fig. 2.30.73). A quantity of animal bone, predominantly cattle (*Bos* f. domestic), bone from domestic fowl (*Gallus* f. domestic) was also recovered from both pits. Generally the waste from the pits suggests clearance was occurring across the site with the incorporation of both sandstone waste and small amounts of domestic refuse.

Area C

In Area C, the lower fills of both the large quarry pits [4002, 4041] contained abundant remains of crushed sandstone and the finds assemblages were very similar in composition, with high proportions of amphorae and samian. The proportion of Samian within the fills was consistent with other areas of the site (7--9%) and dated the pits to the Flavian--Trajanic period with a *terminus post quem* in the later period.

PLATE 2.11 PIT [3030] EAST FACING.

Area E

In Area E, the lowest fill of [5013] was a sandy rubble deposit (5015) as was common to most of the stone extraction pits at the site. Personal items recovered from this pit included a Polden Hill brooch (5012, Fig. 2.28.2). Military equipment recovered included a strap fitting (5012) and a copper alloy stud (5015, Fig. 2.31.89).

The primary backfilling of pit [5004] was also waste derived from the quarrying process that contained large irregular pieces of sandstone rubble. The pottery recovered was dominated by Flavian--Trajanic wares but later Hadrianic and early Antonine wares were also present suggesting this activity may relate to the very end of the site clearance. A single fragment of sawn antler was also recovered (5003, Plate 2.20 G).

The pottery from the pits in Area E was consistent with a Late 1st to Early 2nd Century date. Evaluation Test Pits 10 and 11 identified layers of mixed Roman and post-medieval material in the same area which can be interpreted as part of the overall spread of material sealing the bedrock rather than relating to specific features.

Event 3 Rubbish disposal

This event is characterised by the removal of waste from workshops, kitchens and tables that had accumulated within the fortress and deposition in the disused quarry pits. This was presumably to facilitate the rebuilding of structures within the fortress in the late 2nd century. The material deposited in this event is very mixed suggesting they derive from a variety of sources and supporting the theory that there was middening occurring within the fort.

Area A

The south-eastern corner of the site (Area A; Plate 2.1; Fig. 2.2) was extensively developed in the post-medieval period with the presence of structures from Groups 108 (Phase 3), 103 (Phase 4) and 100 (Phase 5), this left the Roman deposits severely truncated and therefore difficult to interpret with any confidence. All that remained of the Roman activity in this area was a confusing array of basal deposits with few indicators of the original extent of the features they infilled (Plate 2.2). However it is likely these deposits represent the remains of episodes of refuse disposal with pits excavated for this purpose.

The earliest refuse pits [2225, 2327, and 2228] in Area A were mostly truncated by successive pits and can only be seen in section (Fig. 2.9, S8 & S9). This left the range of finds recovered fairly limited with no pottery and only very degraded animal bone retrieved from the lower fill of pit [2228], which was a sterile sand deposit (2229) sealed by light brown silt sand (2230). These sandy deposits preserved little in the way of environmental remains and shed no light on what other organic material may have been discarded along with the animal bone.

The south-eastern corner of Area A was the focus for further refuse pitting. Three of these pits, [2097, 2231, 2259], had a quantity of hammerscale present in the upper and lower deposits as well as 2nd century pottery and animal bone (Fig 2.9, S10). These deposits echo the activity seen in the central part of Area A with multiple waste sources within the same deposit.

The upper fills of the central pit [2209] contained an assemblage of pottery consistent with a 2nd century date. Of the wares, jars and tablewares with bowls (19%) and flagons (17%) made up a high proportion of the assemblage. Of the bowls 40% were Dorset Black Burnished wares. These and a belt buckle frame (2175, [2209], Fig. 2.30.74) suggest the waste derives from a military context. The pit also contained *tegulae, opus spicatum* (Herringbone floor) and two fragments of the wall of *clibanus* (portable oven; 2175 [2291]; 2290 [2291]; Fig. 2.26.1). Recovered from these deposits was a small amount of cattle bone (*Bos* f. domestic) and a single fragment of duck bone (*Anas platyrhynchos*). The presence of the portable oven fragments and animal bone may suggest this refuse came from within the fort, possibly from the small kitchens in the barrack block.

The large quarry pit in Area A [2209] was re-cut numerous times along its eastern edge by a series of small refuse pits [2125, 2127, 2129, 2165, 2170, 2208, 2241] (Fig. 2.10, S11). These pits contained less samian but more Hadrianic and Hadrianic--Antonine pieces, a lower proportion of greywares and more coarsewares suggesting a date range towards the middle of the 2nd century. Other finds from this assemblage include a blue-green prismatic bottle fragment which had been retouched to form a blade (from pit [2165], 2164, Fig. 2.29.33). In addition to this were small quantities of slag and flake scale indicative of smithing activities.

A wide shallow pit [2243/ 2255/ 2292/ 2293] truncated the northern edge of pit [2209]. The fill of the pits contained a number of fragments of *pedalis* (brick; 2285 [2292]), a *Chateline* (2283, [2292], Fig. 2.28.23) and a cuirass girdle plate (2311, [2293]). Two smaller pits [2191, 2221] in the south eastern corner of Area A were badly truncated by Phase 3 building activity and were only identifiable in section (Fig. 2.10, S12, S13 & S14). Adjacent to these, pit [2228] was truncated by a later pit [2234] from which a drilled and sawn bone fragment was recovered (2235).

To the east of this pit cluster was a single sub-circular pit [2110] that contained high concentrations of pottery displaying re-deposition from Flavian-Trajanic deposits with a small number of Hadrianic-early Antonine pieces (Fig. 2.10, S15). This pit also contained a number of iron nails and a split pin staple as well as animal bone and hammerscale which again represents the waste from a broad range of activities.

Area B

In Area B, from the large, shallow square pit [3037/ 3058]

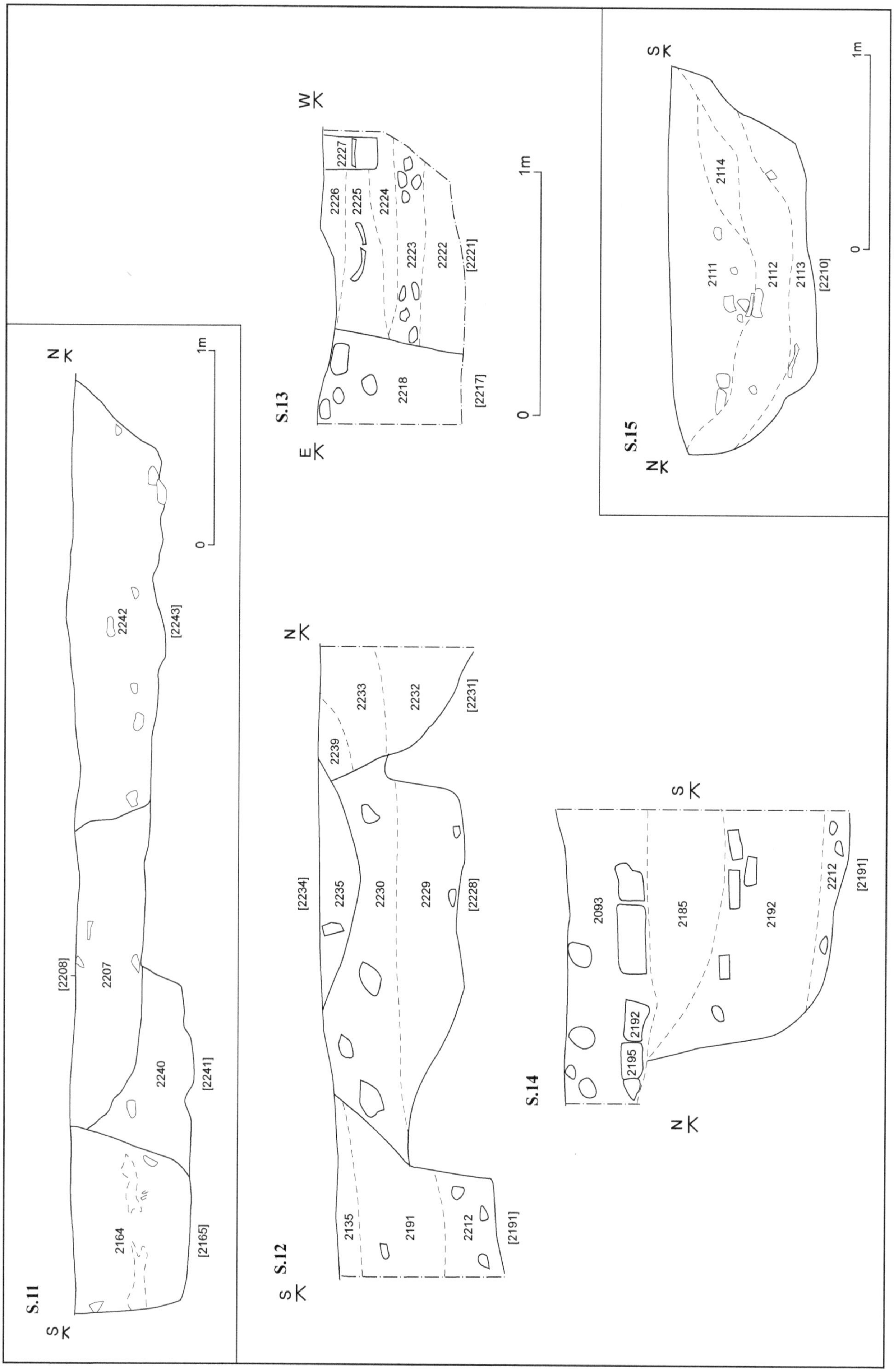

FIGURE 2.10 S11. [2241], [2243], [2165] AND [2208] EAST FACING SECTION, S12. [2191], [2228], [2231] AND [2234] EAST FACING SECTION, S13. [2221] AND [2217] SOUTH FACING SECTION, S14. [2191] WEST FACING SECTION AND S15. [2210] WEST FACING SECTION.

the recovered pottery assemblage suggests that the infilling of this feature dates to the Flavian period, confirmed by the recovery of a coin dating to the consulship of Vespasian (post-AD 71). This pit was probably excavated for the extraction of stone and then subsequently infilled with rubble and refuse. The assemblage recovered also produced *tegulae* cutaways dating to the period *c* AD 240-380 and the corner of a ?box tile although this is probably intrusive, as it is much later in date (?*tubulus*; Fig. 2.26.4).

In Area B these pits were recorded along the Delamere Street frontage where a discreet circular pit [3012] and an irregular ovoid pit [3038/3056] were excavated and both contained similar material to the pits in Area A (Fig.2.11, S16). From pit [3012] two wall fragments of *clibani* (portable oven) were recovered (3010, Fig. 2.28.7; 3011, Fig. 2.28.8) which further supports the hypothesis that this refuse came from within the fort. Both pits appeared to represent the remains of quarry pits that were subsequently back-filled by waste material showing a continuation of the practice at the site throughout its life.

Area C

Later rubbish pit [1019] contained *tegulae* cutaways which dated to the later part of the 2nd century and into the 3rd century. However, the samian again points to high levels of re-deposition with a small number of Hadrianic-early Antonine pieces. This pit also produced a number of nails of Type 1b which show evidence of reuse. Personal ornament items recovered also included a brooch chain (1008) along with other copper alloy items such as a chain and a hinged mount (Fig. 2.30.51). A harness pendant (1008 Fig. 2.31.87) and lead pipe fragment was also recovered from the same context. Interestingly fragments of human bone were also present in (1008) again suggesting either medical waste or clearance which has disturbed burial deposits.

A number of coins were also recovered from the deposits within pit [4041]. From the secondary fill (4033) a coin dating to the reign of Vespasian, AD 69-71 and from upper fill 4016 (equivalent to 4015) a coin dated to Domitian AD86 were recovered. From a cleaning layer over pit [4041] two coins also dating to the reign Domitian, AD 85-96 and AD 88-96 respectively, were recovered. Overall this probably suggests a Trajanic date for the accumulation of the waste with the active stone extraction occuring probably at a slightly earlier date, possibly in the last decades of the 1st century AD.

Large quantities of ceramic building material were recovered from throughout the deposits infilling the pits, the majority of which has been identified as *tegulae*, but also included in the assemblage were ridge tiles, box tiles, *parietales* and brick fragments. The general range of material incorporated into these two quarry pits is consistent with demolition deposits from a range of buildings. Evidence for vitrified mortar adhering to *tegulae* from the lower fill (4007) of [4002] may suggest heat damage prior to deposition.

Personal items recovered from these pits include a copper alloy Nauheim Derivative brooch (4007, [4002]), a Hinged T-shape brooch (4030, [4041] Fig. 2.28.4), a Trumpet brooch (4030, [4041], Fig. 2.28.5), a Beaded Torc in two parts (4030, 4046, [4041], Fig. 2.28.11), a finger ring (4030, [4041]) and melon beads. In addition to this was toilet equipment including a nail cleaner (4016, [4041], Fig. 2.28.24), a *Ligula* (4030, [4041], Fig. 2.29.26) as well as a copper alloy needle (4004, [4002], Fig.2.29.28).

Other metal items included a copper alloy suspension loop (4030, [4041], Fig. 2.29.48); copper alloy studs and nails including a large flat-headed stud (4030, [4041], Fig. 2.30.59); miscellaneous items included lead alloy discs and copper alloy rings and an iron loop (4033, [4041]). A number of items of military equipment were also recovered from the pits including fragments of cuirass armour such as a buckle frame (4004, [4002], Fig. 2.30.75), a hinge plate (4030, [4041], Fig. 2.30.77), a girdle tie plate (4030, [4041]); a strap terminal (4033, [4041], Fig. 2.30.86), a copper alloy collar (4016, [4041], Fig. 2.31.90), and a button and loop fastener (4015, [4041], Fig. 2.31.92). These objects were accompanied by quantities of slag and metallurgical residues from contexts (4016) and (4030) of pit [4041] which are indicative of smithing waste. Taken together there is a strong suggestion that some of this refuse may have originated from a military workshop (*fabrica*) within the fort. A swing handle from a large copper alloy bowl (4033, [4041], Fig. 2.29.29), a single whetstone of palaeozoic sandstone (4030, [4041], Fig. 2.29.30) and fragments of predominantly blue-green vessel glass which included a tubular rimmed-jar (Fig. 2.27.12) were also recovered from pit [4041].

Over 50% of the animal bone recovered from the site came from the two large quarry pits [4002] and [4041], cattle bone (*Bos* f. domestic) making up an overwhelming majority of the assemblage. The animal bone recovered from pit [4041] made up over 40% of the overall bone assemblage recovered from all areas of excavation. Other species represented included significant quantities of pig (*Sus scrofa*), sheep/goat (*Ovis/ Capra* f. domestic) as well as small quantities of domestic fowl (*Gallus* f. domestic), horse (*Equus caballus*), dog (*Canis familiaris*) and red deer (*Cervus elaphus*). This suggests that at least some of the waste may be derived from kitchen area within the fort.

In addition to food waste, evidence for bone craft was also recovered including the by-products of pin manufacture with some of the bone displaying saw marks (4006, 4009 [4002], 4030 [4041], Plates 2.20B-F). A small amount of human bone was also recovered from the quarry pit fill 4046 [4041] and may be from clearance activities or possibly from medical practices.

In Area C the large quarry pits along the northern perimeter of the area were already partially infilled by the time the rubbish deposition was underway and smaller refuse pits were dug into the top of the upper deposits. The large quarry pit [4002] was truncated by a series of smaller refuse pits

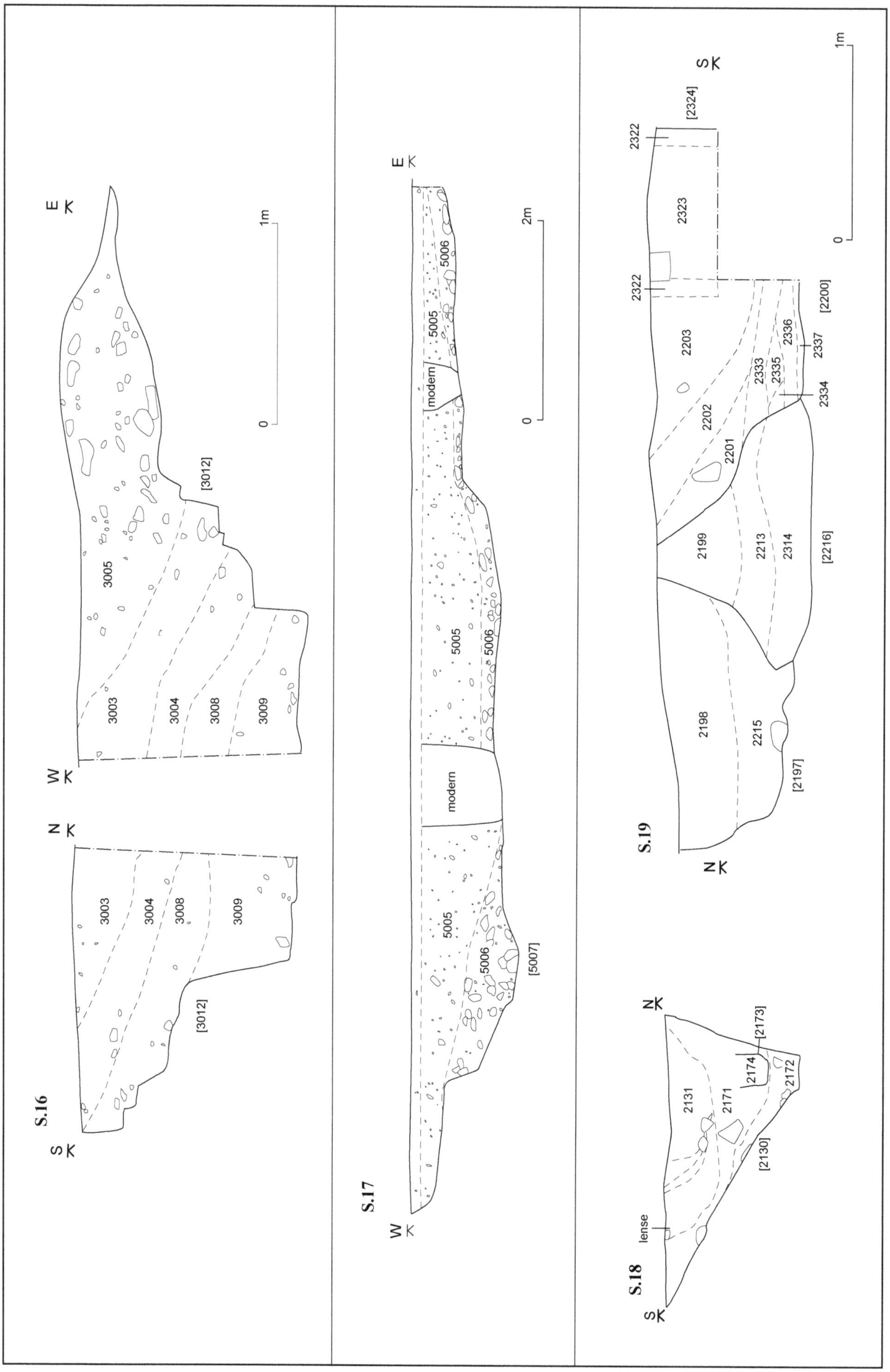

FIGURE 2.11 S16. [3012] EAST FACING SECTION AND SOUTH FACING SECTION, S17. [5007] SOUTH FACING SECTION, S18 [2130] AND [2173] EAST FACING SECTION A DN S19. [2197], [2216], [2200] AND [2324] WEST FACING SECTION.

[4012], [4045], [1018] and [1019] which contained early to mid-2nd century pottery and animal bone. The occurrence of samian within these pit deposits again reflects a high proportion of residual material with smaller numbers of Hadrianic and early Antonine pieces suggesting the material was either redeposited or mixed with the material from the quarry pits. As well as this pottery, a moment in time was captured in a *tegulae* fragment from pit [4012] which produced the faint impression of a dog's paw print. The pit also produced a fragment of *clibani* (portable oven; Fig. 2.26.5). Also within the pit were blue-green utilitarian vessel glass fragments including a square bottle base (Fig. 2.27.40), which suggest the pit may have been associated with deposition from a barrack block kitchen area where cooking would have taken place along the verandas using these small ovens. In addition a number of copper alloy studs and nails were uncovered from this pit.

Area E

From [5013] the pottery assemblage was comparatively large in terms of average sherd weight (49g in comparison with 29.4g from 5004) which is explained by a large number of amphorae sherds, whilst jars make up 75% of the assemblage by EVE (Estimated Vessel Equivalents). There is also a relatively low samian count (1%) and the strong suggestion is that this may relate to disposal associated with waste as opposed to a clearance deposit.

In pit [5004], included within the finds assemblage was a Polden Hill brooch (5002, Fig. 2.28.3), a Chatelaine tool (5002), a number of copper alloy nails, rivets and studs and military equipment included a cuirass tie ring (5002, Fig. 2.30.81) and a cuirass girdle plate (5002).

In Area E a wide shallow stepped quarry pit [5007] was excavated (Fig. 2.11, S17, Plate 2.13). It had similar rubble and sand infill deposits that suggested quarry waste had become intermingled with domestic waste. Functional tableware such as Flagons (24%) and bowls (22%) dominated the pottery assemblage. Two coins were also recovered, the first dating to the reign of Trajan (AD 103-111) and the second slightly later to Hadrian (AD122-8).

A large assemblage of ceramic building material was recovered from [5007] which included the usual range of *tegulae*, *imbrices*/ ridge tiles and brick. A number of small personal items were also recovered including a copper alloy chain (5005, Fig. 2.29.46), a split pin and a single melon bead (5005), an enamelled strip with quatrefoil inlay (5005, Fig. 2.31.93) and a copper alloy ring. Military items included a cuirass hinge plate (5006, Fig. 2.30.76),

PLATE 2.12 PIT [3012] NORTH FACING.

PLATE 2.13 PIT [5007] NORTH-WEST FACING.

PLATE 2.14 PICK-CUT SLOT IN AREA A QUARRY.

and a cuirass girdle tie plate (5005). The presence of a metalworking punch and a polished red deer antler fragment (*cervus elaphus*; 5005, Plate 2.20H) from within these deposits suggests that workshop (*fabrica*) waste was being deposited. Cattle (*Bos* f. domestic) still dominated the animal bone assemblage recovered from the pit, but small quantities of domestic fowl (*Gallus* f. domestic), goose (*Anser/Branta* sp.), roe deer (*Capreolus capreolus*) and flatfish (cf. *Pleuronectes* sp.) were present indicating that the soldiers within the fort enjoyed a varied diet.

The extensive range of the material recovered from these pits supports the hypothesis that a phase of clearance was carried out within the fortress with the removal of overflowing midden deposits to this site which was outside the habitation area. The assemblages recovered are closely dated to the years AD 90—120 which indicates the date of the accumulation within the fort. The later material recovered, dating to the Hadrianic and Antonine periods may indicate the date of the midden removal and therefore the date of its deposition at the Delamere Street site. This would suggest stone removed during the quarrying episode (Event 1) was part of a late-1st century building episode in the fortress or *canabae*.

Group 113 Other features

In Area A there were indications that perhaps prior to stone extraction the site may have had another function. A v-shaped ditch [2130] orientated east--west was excavated that appeared to be of the Punic-profile type which suggests that this may relate to the earliest defensive activities at Chester. The pottery assemblage recovered from the infill of the ditch included DOR BB1 and South Gaulish Flavian samian which is consistent with the early 2nd century date of much of the rest of the site (Fig. 2.11; S18, Plate 2.2). Without further dating evidence and as the ditch was heavily truncated by subsequent phases of post-medieval activity, the suggestion that this ditch may predate the quarry is tenuous at best and has therefore not been assigned to a specific Event.

Immediately to the north of this ditch was a short, shallow gully feature [2244] of similar date. However, its function is unclear and it appeared to respect ditch [2130]. A second ditch [2200/ 2252] perpendicular to this gully and parallel with [2130] was located to the south. This feature was so heavily truncated by post-medieval (Phase 4) and Roman features that the ditch was not visible in plan and could only be identified in section. This ditch appeared have been truncated by as many as four Roman refuse pits (Event 3) [2197], [2301], [2317], [2329] (Figure 2.11, S19). The fragments of pottery recovered from the deposits in the ditch are of the same date as those deposits infilling the pits. However, the ditch clearly predates these pits and the suggestion is that they were all infilled in quick succession during the early 2nd century and may be part of the same Event. These other features indicate that the site was used prior to stone quarrying and the extent of this activity can only be guessed at.

Summary

The excavations at Delamere Street have produced an overwhelming quantity of material culture and represent one of the largest assemblages excavated from Chester in modern times. The following chapters will attempt to elucidate the origins of this material in relation to activities occurring within and around the fortress. Unlike other assemblages the finds from Delamere Street represents the range of industries and activities that occurred within and around the fortress rather than the limited range of processes occurring at the site itself.

THE ARTEFACTS

The excavations of the Roman quarry pits uncovered a range of archaeological artefacts as they were back-filled rapidly with waste. The finds recovered include an extensive range of Roman pottery with both local and continental wares represented; glass; ceramic building material and stonework; metalwork, including domestic accessories and non-domestic material including nails and metallurgical residues. Although chronologically short-lived the assemblage is valuable as it provides a well-preserved cross-section of the activity that reflect both the military occupation of the forts but also the domesticity associated with the occupation of the canabae in the late-1st and 2nd centuries AD. It provides useful indicators of both the local, regional and international activity associated with the Roman fortress and canabae.

THE ROMAN POTTERY

by Jane Timby with a report on the samian by Felicity Wild

Introduction

The evaluation and subsequent excavation at Delamere Street resulted in the recovery of 11,617 sherds of Roman pottery weighing *c* 241 kg with 195.27 estimated vessel equivalents (rim) (EVE). In general terms the assemblage was in very good condition, reflected in the overall average sherd weight of 20.8 g. Most sherds are quite fresh with unabraded edges and there are several examples of multiple sherds from the same vessels. Surface treatments such a slipping, painting or roughcasting have been preserved.

The evaluation recovered a modest assemblage of 199 sherds with the bulk coming from the main excavation. In the latter pottery was recovered from some 65 cuts with an additional material from 42 other contexts along with 147 sherds from unstratified collection. A small quantity of additional material was subsequently recovered from the soil samples.

Following a comment on the methodology used, the assemblage is described by fabric and form. The assemblage is then discussed in terms of the site and its chronology and then placed into its local and regional context.

Methodology

The assemblage was sorted into broad fabric groups based on inclusions present, the frequency and grade of the inclusions and the firing colour. Known regional or traded wares were coded following the system advocated for the National Roman reference collection (Tomber and Dore 1998) (=T&D). Many of the groupings are fairly generic reflecting the general homogeneity of the assemblage and the problems inherent in dealing with material from the Cheshire Plain (see below). A visit was made to the Chester Archaeological unit to look at the fabric series for comparison and some of the Chester codes are referenced in.

The sorted assemblage was quantified by sherd count and weight for each recorded context. Rims were measured for diameter and percentage present, the latter for the calculation of EVE's. Although not specifically targeted where joining sherds from distinctive vessels were observed between layers these were noted. With time, patience and space it is possible that more links might exist amongst the less distinctive sherds but such an exercise would not have proved cost-effective. The pottery data was entered onto an MS Excel spreadsheet to which site phasing and other information was added. A copy of this is deposited with the site archive.

A selection of material has been illustrated demonstrating the range of forms present. These have been arranged following the stratigraphic phasing sequence.

Description of fabrics and forms

The assemblage has been divided into three groups: continental imports, regional imports and local wares. The continental imports are further subdivided into samian; other table wares; mortaria and amphorae. Table 2.1 provides a quantified summary of the complete assemblage.

CONTINENTAL IMPORTS

The samian *by Felicity C Wild*

The site as a whole produced about 1080 sherds of samian ware, much of it fragmented, but otherwise in a good state of preservation. Of these, some 502 sherds were from decorated bowls, mainly of form 37, a strikingly high proportion of just over 46%.

The material forms a remarkably consistent group dating from the Flavian through to the Hadrianic-early Antonine period. Based on a rough count of vessels within each context group, 74% was South Gaulish and 26% Central Gaulish, with a roughly equal percentage of 13% from Les Martres-de-Veyre and 13% from Lezoux or other centres. Here, as elsewhere in Britain, La Graufesenque was the main source of wares during the first century and the Trajanic period, supplemented by Les Martres-de-Veyre from *c*.AD 100. Lezoux did not start to export widely until *c*.AD 120. Les Martres-de-Veyre was a very much smaller producer than either La Graufesenque or Lezoux, and the fact that the quantity of wares from there is almost equal to

TABLE 2.1: QUANTIFIED SUMMARY OF THE POTTERY ASSEMBLAGE

	Code	Description	No	No %	Wt	Wt %	Eve	Eve%
IMPORTS	SG SAM	South Gaulish samian	825	7.2	8138.5	3.4	158	0.8
	CGSAM	Central Gaulish samian	258	2.2	2413.5	1.0	115	0.6
	ARG CC	Argonne colour-coat	8	0.1	28	0.0	40	0.2
	CNG CC	Central Gaulish colour-coat	6	0.1	14	0.0	30	0.2
	CAM WH	Italian whiteware (jug)	2	0.0	44	0.0	12	0.1
	KOL CC	Cologne colour-coat	1	0.0	3	0.0	0	0.0
mortaria	NOG WHM	North Gaulish mortaria	11	0.1	1231	0.5	38	0.2
	GLG OX	Gallia Lugdunensis (Rhone Valley)	2	0.0	141	0.1	12	0.1
amphorae	BAT AM	Baetican amphora (all)	560	4.9	63036	26.3	392	2.0
	BAT AM	Haltern 70 (featured)	4	0.0	1064	0.4	0	0.0
	CAD AM	Cadiz amphorae	24	0.2	1740	0.7	31	0.2
	CAM AM	Campanian fabric ?Dr 2-4	3	0.0	159	0.1	0	0.0
	Dressel 2-4	Dressel 2-4 amphorae (Baetican)	3	0.0	240	0.1	0	0.0
	Dressel 7	Dressel 7? South Spain	1	0.0	294	0.1	0	0.0
	GAL AM	Gallic amphorae	78	0.7	1846	0.8	0	0.0
	RHO AM1	Rhodian	10	0.1	400	0.2	0	0.0
	AMP	unassigned amphorae	21	0.2	1036	0.4	0	0.0
lamp	LAMP	lamp fragment	1	0.0	1	0.0	0	0.0
REGIONAL	DOR BB1	Dorset black burnished ware	755	6.6	10599	4.4	1705	8.7
	LNV CC	Lower Nene Valley colour-coat	1	0.0	4	0.0	0	0.0
	PNK GT	Midlands pink grog-tempered	2	0.0	68	0.0	0	0.0
	SVWOX?	Severn Valley ware	16	0.1	131	0.1	0	0.0
	VER WH	Verulamium white ware	68	0.6	1206	0.5	77	0.4
	VER WH (M)	Verulamium white ware mortaria	26	0.2	2818	1.2	109	0.6
	WIL OX	Wilderspool oxidised	1	0.0	27	0.0	5	0.0
	WRX WH	?Wroxeter whiteware mortaria	7	0.1	907	0.4	71	0.4
Cheshire	GLAZE	British glazed ware	4	0.0	19	0.0	0	0.0
fine	MICOX	mica-slipped oxidised ware	263	2.3	3709	1.5	543	2.8
	WWEGG	white eggshell ware	7	0.1	40	0.0	29	0.1
	WWF	fine white ware	5	0.0	30.5	0.0	20	0.1
	OXIDRC	roughcast oxidised/colour-c ware	12	0.1	71	0.0	20	0.1
coarse	GREY	misc grey /black wares	4604	40.0	61811	25.8	9001	46.1
	BW	black ware	40	0.3	502	0.2	148	0.8
	OXID	oxidised ware	2150	18.7	38547	16.1	3451	17.7
	OXIDWPT	white painted oxidised ware	50	0.4	614	0.3	86	0.4
	GY/OX	grey exterior, oxidised interior	108	0.9	1835	0.8	60	0.3
	WSGY	white-slipped grey ware	2	0.0	14	0.0	0	0.0
	WSOXID	white-slipped oxidised ware	1117	9.7	17194	7.2	2596	13.3
	WW	whiteware	57	0.5	813	0.3	56	0.3
mortaria	HOL OX	Holt oxidised mortaria	20	0.2	2148	0.9	67	0.3
	HOL WS	Holt white-slipped mortaria	2	0.0	188	0.1	10	0.1
	MORTOX	oxidised mortaria no grits visible	10	0.1	572	0.2	24	0.1
	MORTOX1	oxidised mortaria, coarse grits	7	0.1	1827	0.8	69	0.4
	MORTOX2	oxidised mortaria, sparse grits	1	0.0	126	0.1	0	0.0
	MORTOX3	oxidised mortaria, igneous grits	2	0.0	70	0.0	0	0.0
	MORTOX4	oxidised mortaria mixed grits	8	0.1	1071	0.4	43	0.2
	MORTOX5	paler oxidised with iron	4	0.0	259	0.1	0	0.0

	MORTWS	white-slipped mortaria	18	0.2	1053	0.4	46	0.2
	MOWSOX1	white-slipped mortaria	1	0.0	182	0.1	8	0.0
	MOWSOX2	white-slipped mortaria	5	0.0	493	0.2	22	0.1
	MOWSOX3	white-slipped mortaria	41	0.4	2672	1.1	37	0.2
	MOWSOX4	white-slipped mortaria	22	0.2	2543	1.1	103	0.5
LOCAL/	BUFF/PALE	misc buff/ cream wares	36	0.3	383	0.2	81	0.4
UNKNOWN	BUFFMIC	micacous buff ware	1	0.0	4	0.0	0	0.0
	BB1COPY	BB1 copies, black sandy ware	19	0.2	344	0.1	31	0.2
	BWSLIP	black-slipped grey/black ware	6	0.1	90	0.0	6	0.0
	BWF	fine black ware	19	0.2	264	0.1	42	0.2
	CWQZ	quartzite-tempered coarseware	1	0.0	10	0.0	0	0.0
	GROG	grog-tempered	75	0.7	1773	0.7	0	0.0
	GYIGN	grey igneous rock-tempered	1	0.0	8	0.0	0	0.0
	LIMEOX/GY	limestone tempered oxid/grey ware	3	0.0	92	0.0	27	0.1
	GYMIC	micaceous grey ware	3	0.0	147	0.1	7	0.0
	GYMISC	other miscellaneous greywares	4	0.0	57	0.0	6	0.0
	MORTWH	whiteware mortaria	2	0.0	209	0.1	0	0.0
	MORTBUFF	buff mortaria	3	0.0	13	0.0	0	0.0
	OXIDF	oxidised fine ware	7	0.1	27	0.0	41	0.2
	OXIDSY	sandy oxidised ware	8	0.1	329	0.1	45	0.2
	SHELL	shelly ware	6	0.1	56	0.0	0	0.0
	WWBPT	brown painted whiteware	1	0.0	22	0.0	7	0.0
	CRUMBS	small unidentified crumbs	49	0.4	53	0.0	0	0.0
	MISC	miscellaneous unclassified	5	0.0	44	0.0	0	0.0
TOTAL			11503	100.0	239918	100.0	19527	100.0

that from Lezoux suggests that little of the material in the present deposit was manufactured much after *c.*AD 120. Table 2.2 lists the number of vessels (excluding scraps where the form is uncertain) from South Gaul (SG), Les Martres-de-Veyre (MdV) and Lezoux or other Central Gaulish potteries (CG).

Very few pieces are likely to be earlier than the Flavian period. Form 29, production of which ceased *c.*AD 85, is heavily outnumbered by form 37, as is form 15/17 by form 18. In view of this, uncertain base sherds, described as form 15/17 or 18, are perhaps more likely to be of form 18. The purely pre-Flavian forms are absent. Of the decorated ware, the earliest pieces are likely to be D1, D2, D17 and D18, which may well be Neronian.

The material consists of high-status table ware: cups, dishes and, particularly, decorated bowls, dumped into quarry pits just outside the North gate of the fortress. It clearly represents rubbish disposal from the fortress. In view of the exceptionally high proportion of decorated ware, it may not be going too far to suggest that this is the detritus from officers' dining tables.

A crucial question is at what stage were the pits refilled and the material from the fortress redeposited. It is likely that we are indeed dealing with unintentional breakages over a period of time. The lack of complete or almost complete

TABLE 2.2: SAMIAN WARE, BY FORM AND FABRIC

Form	SG	MdV	CG	Total
29	24			24
30	17	1	4	22
37	210	45	30	285
27	72	12	19	103
33	6 (all 33a)	1	3	10
15/17	11			11
15/17 or 18	25			25
18	90	9		99
18 or 18/31	10	5	3	18
18/31		9	17	26
18R	2	1?		3
18 or R form	3			3
18/31 or 18/31R	2		1	3
18/31R	14	3	10	27
31			2	2
35		1		1
35/36		1		1
36	3			3
36 or 42	1			1
42	1			1
81			1	1
Curle 11	16			16
Curle 15	2			2
Beaker	14		3	17
Dish		2		2
Total	523	90	93	706

bowls would militate against the hypothesis that this expensive tableware had simply been dumped *en bloc*. Was the departure of the legion, or much of it, to Hadrian's Wall in the early AD 120s perhaps seen as a good opportunity for clearing out the rubbish, or is it more likely that this took place on their return, with the refurbishment of the fortress? The samian ware cannot answer these questions conclusively, but may provide some pointers.

The material from Areas A, B and C was all remarkably similar in date, with the highest proportion and largest quantity of South Gaulish ware coming from Area C. As might be expected, few sherds came from contexts attributed to the quarrying phase (Event 1): a mere three sherds from Area A and eight from Area B. Although one of the sherds from Area A was among the earliest from the site (D1), another was Central Gaulish and likely to be Hadrianic. Seven of the sherds from Area B were South Gaulish and Flavian or Trajanic, including D16 below, the eighth from Les Martres-de-Veyre and possibly as late as Hadrian. This may suggest that quarrying was taking place up to the time of Hadrian.

There was little, if any, noticeable difference in date between the primary filling of the Areas (Event 2) and the deposition of material from the fortress (Event 3). In both Areas B and C, sherds from the same decorated bowls came from contexts attributed to both Events (D26-D27 in Area B, D46-48 in Area C). Almost all of the material was South Gaulish or from Les Martres-de-Veyre, though in each case with the odd later sherd of *c*.AD 125-145.

In Area A, only three sherds of plain ware were attributed to Event 2, two of which were Central Gaulish and likely to be Hadrianic-early Antonine. The decorated ware from Event 3 was evenly divided between South Gaul and Les Martres-de-Veyre, the latest pieces being D13, *c*.AD 125-145 and a stamp of Reginus ii (S12), *c*.AD 120-150. Of the Event 2 material from Area B, there were several sherds dated *c*.AD 125-145, the latest being D25, which could even be early Antonine rather than Hadrianic, whereas the Event 3 contexts, apart from a sherd of D27 and a few possible plain sherds, contained nothing likely to have been later than *c*.AD 125. The contents of Area C were similar, with two sherds (D44, D45) from Event 2 in the style of Drusus ii, *c*.AD 125-145, and nothing later than the early AD 120s from Event 3.

Area E contained the latest material on the site. A bowl possibly by Sacer (D58), *c*.AD 125-145, was present in the Event 2 material, while the Event 3 groups contained a greater proportion of Lezoux products than the other Areas, including material of early Antonine date (D65-D68, and Stamp S1, of Albus iii, *c*.AD 145-165).

In all the Areas, the odd sherd of Hadrianic material occurs in Event 2, suggesting that the initial filling, even if it started earlier, was still under way in the Hadrianic period. The deposition of material from the fortress presumably took place, at earliest, in the Hadrianic period, though the deposited material, clearly, was mainly of earlier date. The dumping in Area E was unlikely to have taken place before the Antonine period and the subsequent refurbishment of the fortress. This may also have been the case in the other areas, though it could equally have taken place some twenty years earlier, when the garrison left. There is no evidence to suggest that any great length of time elapsed between the quarrying of the pits and their filling in. The scarcity of material dating to later than *c*.AD 145, even in Area E, is worth stressing.

As might be expected in any large group, specific styles and the work of certain potters are particularly well represented. While these, inevitably, tend to be the potters whose work is most common at the period in question, the presence of a number of pieces by the same potter or group of potters may provide evidence for bulk purchase. Among the South Gaulish wares, work in the styles of M. Crestio and Mercator tend to be common on any site occupied during the period *c*.AD 80-110, as do the slightly later styles of the ovolo with left-turned trident tongue (D3, D5, D16) and pieces in the style of those in the Bregenz Cellar deposit of *c*.AD 90.

More convincing, perhaps, is the evidence from Les Martres-de-Veyre, where the majority of pieces are in the styles of Drusus i or of X13 and his associates, though, again, these were probably the most prolific potters of their period. Of the Lezoux wares, a surprisingly high proportion is in the style of the Sacer-Attianus-Drusus ii group, who appear to have been their direct successors. It has been argued (Hartley and Dickinson 2008- , Vol.3, 331-333) that Drusus i may have moved to Lezoux *c*.AD 120, where he produced the work commonly known as 'Drusus ii'. The style of X13 also appears in Lezoux fabric, where this potter or potters appear to have been at work during the decade *c*.AD 120-130. Sacer and Attianus shared many of their types and motifs, though the occurrence of their work at Les Martres-de-Veyre may be due to no more than the sale of bowls or moulds between the centres. A bowl with stamp of Attianus, in classic Martres fabric, as yet unpublished, was excavated at Rocester (excavation no. RON01 [8489]).

A fact rather surprising is that the work of Cinnamus, the most prolific of all the Lezoux potters, is conspicuous by its almost complete absence. There is just one scrap of one of his common ovolos (D68), from Area E. It is currently estimated that his career, along with Cerialis, started *c*.AD 135. It may be merely coincidence that so little of his work found its way into the material deposited in the quarry pits: the latest pieces from the site are certainly contemporary with his early work, though these are indeed few. It does, though, perhaps confirm that the material redeposited in the pits dated to the time before the legion left for Hadrian's Wall.

The decorated ware

In the text below, the significant decorated ware is listed by Area, with cross references to the stamp report, which follows, in alphabetical order, at the end. Figure types are quoted from Oswald 1936-37 (O.) or from the supplementary list in Rogers 1999 (R.), Central Gaulish decorative motifs from Rogers 1974 (Rogers) and parallels from Stanfield and Simpson 1958 (S&S).

Catalogue

Area A

Event 1

D1. Form 29, South Gaulish. Fragment of upper zone with panel containing two small dogs (O.1968) chasing animals, probably stags (O.1737). The types were used too widely to suggest a particular potter, but a similar scene with the two types appears on a form 29 from London by Modestus (Dannell *et al.* 2003, Modestus i, Taf. F1, 0712). Bowls with panels containing similar hunting scenes occur in the Neronian wreck at La Nautique (Fiches *et al.* 1978, fig. 7, 1, 2). *c.*AD 55-70, context (2282). (Fig 2.12).

Event 3

D2. Form 29, South Gaulish, showing a vine scroll in the upper zone, with a small bird pecking the grapes in the lower concavity. The lower zone contains gadroons. The vine-leaf was used by Licinus (Knorr 1919, Taf. 47F), as was the bird (Dannell *et al.* 2003, Licinus, Taf. A2, 0640). Licinus was a Claudio-Neronian potter whose work occurs in the early Neronian wreck at La Nautique and the Boudican burning of AD 61 at Colchester. *c.*AD 50-70. (2112). (Fig 2.12).

D3. Form 37, South Gaulish. Twelve scraps, not joining, but all probably from the same bowl with a distinctive ovolo with the trident tongue turned to the left. Decoration consists of the panthers (O.1437, O.1418) probably to each side of a bush composed of trifid buds, above a basal wreath. The anonymous ovolo occurs fairly frequently on sites founded under Domitian. *c.*AD 90-110. (2206, 2205, 2211). (Fig 2.12).

D4. Form 37, South Gaulish, with zonal decoration, showing a horizontal wreath over a hound and panther opposed across a bush. Beneath the hound is the end of the mould stamp of M. Crestio. The wreath is commonly attested on his work (Mees 1995, Taf. 49, 2), as is the bush motif (*ibid.*, Taf. 45, 3). The animal types are not certainly identifiable from O., but the hound may be as on Mees 1995, Taf. 38, 2, 3. *c.*AD 80-110. (2283). (Fig 2.12).

D5. Form 37, South Gaulish. Three joining sherds of bowl with two rivet-grooves, showing the same anonymous Flavian-Trajanic ovolo with trident tongue as D3 and D16. Decoration shows a stag (as O.1711A but slightly larger) between trees with small, squat birds (as on D42). Trees with the same leaf were used on bowls by Germanus at Banassac (Mees 1995, Taf. 231, 4), though the bird on top is different. Identical decoration, with the stag, bird, tree and probably this ovolo, occurs on a bowl from Holt (Grimes 1930, fig. 37, 42). *c.*AD 90-110. (2283). (Fig 2.12).

D6. Form 37, South Gaulish, with large and distinctive ovolo (Dannell *et al.* 1998, TU), used by the same anonymous group of potters as the ovolo on D5 above. *c.*AD 90-110. (2283)

D7. Form 37, South Gaulish. Sherd of heavy bowl showing the distinctive ovolo of L. Cosius. Little survives of the decoration apart from his triple arcade (Mees 1995, Taf. 28,1). L. Cosius was among the latest of the La Graufesenque potters. The fact that he produced inscribed bowls commemorating Trajan's victories over the Dacians and Parthians shows that he was at work until *c.*AD 120. *c.*AD 100-120. (2111) . (Fig 2.12).

D8. Form 37, Central Gaulish, in the fabric of Les Martres-de-Veyre. Panels show the sphinx (O.855) over a small lion (O.1473?) and a vine-scroll (Rogers M50). The vine-scroll and finely beaded border occur on X13 style, but the two figure types are not recorded for him by Rogers. Oswald (1936-37) notes that the sphinx occurs on a mould at Lezoux, but does not attribute it to any particular potter. *c.*AD 100-120. (2175). (Fig 2.12).

D9. Form 37, Central Gaulish. The ovolo (Rogers B185) was used by X12, who worked at Les Martres-de-Veyre. The urn (Rogers T1) and griffin (O.871A) appear together on bowls in his style from London and Brecon (S&S, pl. 41, 485). *c.*AD 100-120. (2175) . (Fig 2.12).

D10. Form 37, Central Gaulish, probably from Les Martres-de-Veyre. Base sherd, showing decoration of a scroll with acanthus leaves (Rogers K25). The style is that of Drusus i. A bowl from London (S&S, pl. 13, 164) shows a similar scroll with the acanthus the other way up. *c.*AD 100-120. (2111). (Fig 2.12).

D11. Form 37, Central Gaulish, probably from Les Martres-de-Veyre. Five joining sherds of a small bowl in the style of X13, with his characteristic ovolo (Rogers B14). Decoration shows a scroll with his leaf (Rogers J89) and bud (Rogers G137) in the upper concavities and a double medallion with cupid (O.403) over an acanthus (Rogers K10) in the lower. The bead row below the decoration is characteristic of his style. *c.*AD 100-120. (2111, 2112). (Fig 2.13).

D12. Form 37, Central Gaulish in the fabric of Les Martres-de-Veyre. Two joining sherds, showing the ovolo replacement (O.2407A) used by Drusus i (S&S, pl. 11, 133) and his vine scroll (S&S, pl. 11, 134). *c.*AD 100-120. (2285, 2311). (Fig 2.13).

D13. Form 37, Central Gaulish. Three joining sherds of a bowl with scroll decoration in the style of X13 and the Sacer-Attianus group. The ovolo (Rogers B14) was used by X13 and Sacer, the leaf in the upper concavity (Rogers H72) by Attianus. The lower concavity shows a panther

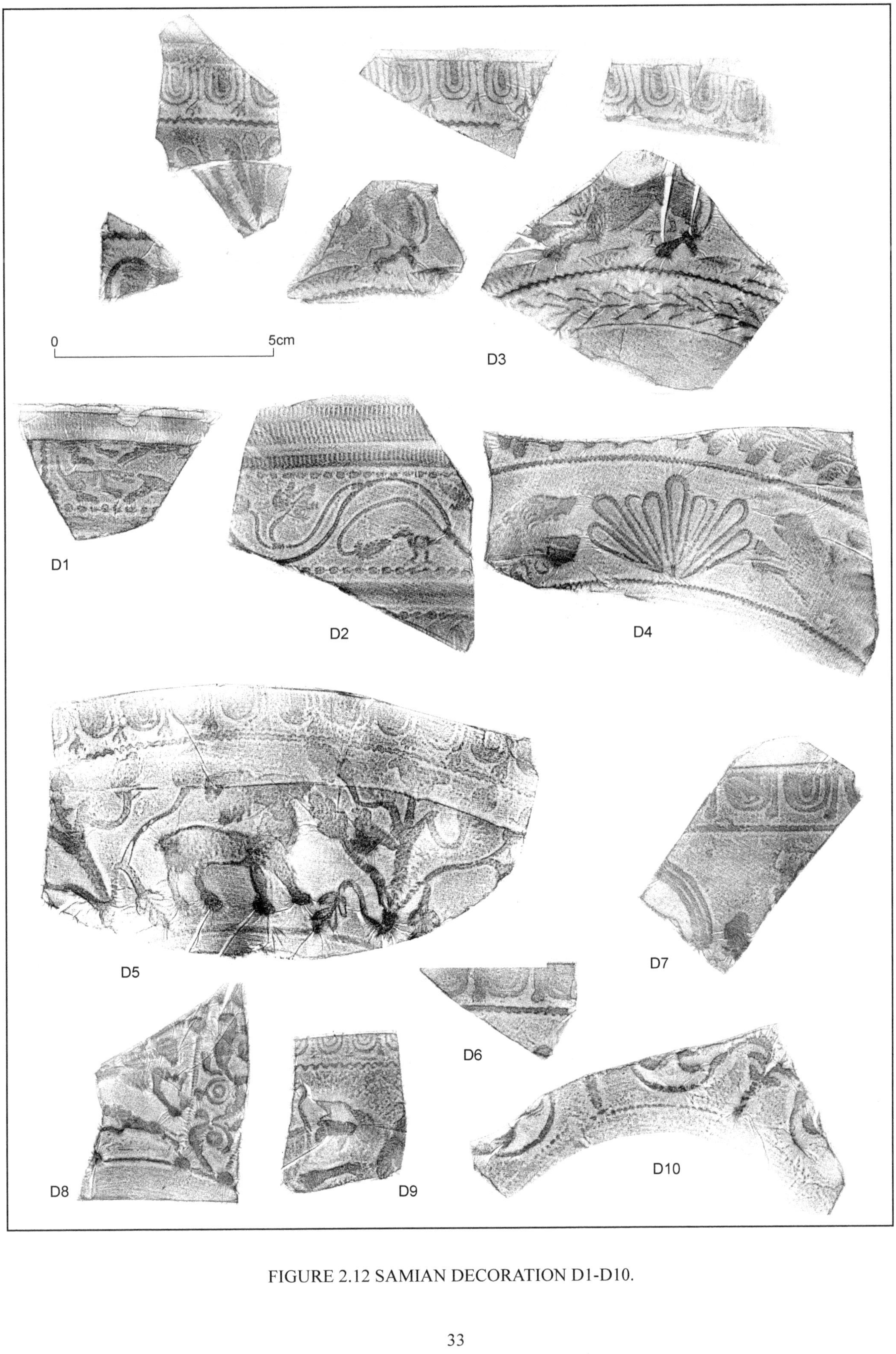

FIGURE 2.12 SAMIAN DECORATION D1-D10.

(O.1518) used by all three potters, over horizontal zones of circles and Rogers G296, not attributed to any named potter. A generally similar decorative scheme occurs on a bowl in X13 style (S&S, pl. 45, 528). The fabric is likely to be from Lezoux, suggesting a date *c.*AD 120-145. (2283). (Fig 2.13).

Also from these contexts: stamps of Masculus (S7), *c.*AD 80-110, and Reginus ii (S12), *c.*AD 120-150.

Fill of ditch [2130]

D14. Form 37, South Gaulish, showing panel decoration with hound (probably O.1922) and triple medallion with archer (O.268). The medallion with archer and a similar leaf tendril was used by C. Valerius Albanus (Mees 1995, Taf. 2, 6). *c.*AD 80-100. (2172). (Fig 2.13).

Later context overlying Roman pit

D15. Form 29, South Gaulish. Three joining sherds, stamped internally with the stamp MANDVILMA, die 6a of Manduillus (S6 below), as on two bowls in the Pompeii Hoard (Dannell *et al.* 2003, Manduillus, Taf. D1, F1; Atkinson 1914, 32, 31). The decoration of scrolls with stirrup leaves in the lower zone is not paralleled on any of his stamped bowls illustrated in Dannell *et al.* 2003. *c.*AD 60-80. (2005, 2009). (Fig 2.13).

Area B

Event 1

D16. Form 37, South Gaulish, showing a zone of festoons containing birds (O.2226, O.2262) with poppy-head pendant over a leaf scroll with the same bird (O.2262). The trident tongue of the ovolo, turned to the left, is just visible at the top of the sherd, apparently the same anonymous ovolo as on D3 above. O.2262 also occurs on another bowl with this ovolo (D5 above). The general style here, with zonal decoration, seems earlier than that of D3 and D5. This is likely to have been among the earlier work of this potter. *c.*AD 85-110. (3040). (Fig 2.13).

Event 2

D17. Form 30, South Gaulish. Two joining sherds showing a single-bordered ovolo and panel with saltire. The ovolo (Dannell *et al.* 1998, FA; Dannell 2006, ovolo A) was used by Lupus ii. A stamped form 30 mould from La Graufesenque (Dannell 2006, fig. 40, 2) shows the ovolo with wavy line above it and a saltire with similar trifid bud. A stamped form 11 from the Cluzel 15 deposit at La Graufesenque shows a saltire with both motifs (Mees 1995, Taf. 99, 6). The tendril ending in three rosettes was a characteristic of Masclus (Mees 1995, Taf. 110, 1 etc.). *c.*AD 50-65. (3029). (Fig 2.13).

D18. Form 29, South Gaulish. Lower zone with panel containing a large eight-petalled rosette (not in Hermet 1934) in a corded medallion. The style is that of a mould-maker, referred to as T-1, who worked at La Graufesenque for a group of Neronian potters including Albus i and Niger ii (Dannell 1996). Two unstamped bowls of form 29 in this style (*ibid.*, Fig. 15.1, 5, 7) show similar medallions in the upper zone with a different rosette, while two with stamps of Albus i (Dannell *et al.* 2003, Albus i, Taf. A2, 1359; Taf. D4, 0044b) show them in the lower zone with two more, different, rosettes. No parallels to the rosette here are forthcoming. *c.*AD 55-70. (3049). (Fig 2.13).

D19. Form 37, South Gaulish, showing two zones of slightly abraded scroll decoration, with the Nile goose (O.2244) in the upper zone. There are few distinguishing features to suggest a potter, but the decoration is typical of the period *c.*AD 70-90. (3009). (Fig 2.14).

D20. Form 37, South Gaulish. Six joining sherds of rim and ovolo. The ovolo, with long straight tongue ending in a tuft, was used by Sabinus iv (Mees 1995, Taf. 176, 2). *c.*AD 90-110. (3049). (Fig 2.14).

D21. Form 37, Central Gaulish, in the fabric of Les Martres-de-Veyre. Three sherds, two joining, of a bowl in the style of Drusus i, with his ovolo replacement motif (Rogers G395) and scroll (S&S, pl. 15, 182). *c.*AD 100-120. (3033). (Fig 2 .14).

D22. Form 37, Central Gaulish, in the fabric of Les Martres-de-Veyre, showing a festoon (Rogers F40) with rosette (Rogers C87) in the panel corner, as on X13 style (S&S, pl. 43, 499). *c.*AD 100-120. (3029). (Fig 2.14).

D23. Form 37, Central Gaulish, in the fabric of Les Martres-de-Veyre and the style of X2. The ovolo (Rogers B28) and small crown (Rogers U61) are typical of his style, and the cupid (O.425/6), Venus (O.324), Bacchus (O.569) and column (Rogers P85) are all attested on his work. *c.*AD 100-120. (3049). (Fig 2.14).

D24. Form 37, Central Gaulish. Two joining sherds showing panel containing double medallion with cock (O.2350). The style is probably that attributed by Rogers to his potter P23. The ovolo (Rogers B12) and cock occur on this style (Rogers 1999, pl. 130, 1), as does the distinctive four-petalled rosette (Rogers C23) masking the border junction (Rogers 1999, pl. 130, 4), also used by Attianus (S&S, pl. 86, 17). The beaded motif is probably P23's cornucopia (Rogers U262), though partially underlying the border here, it looks almost straight. Two bowls in the style of P23 were found at Manchester (Wild 2007, D15, D36). It is argued, from the similarities in style of one to Geminus iv and of the other to Attianus ii, that the dating suggested by Rogers may be too late, and that the potter was likely to have been at work *c.*AD 125-145. (3039). (Fig 2.14).

D25. Form 37, Central Gaulish, showing beaded panels containing the bear (O.1588) and possibly a hare. The bear was used by potters such as Sacer, Docilis, Criciro

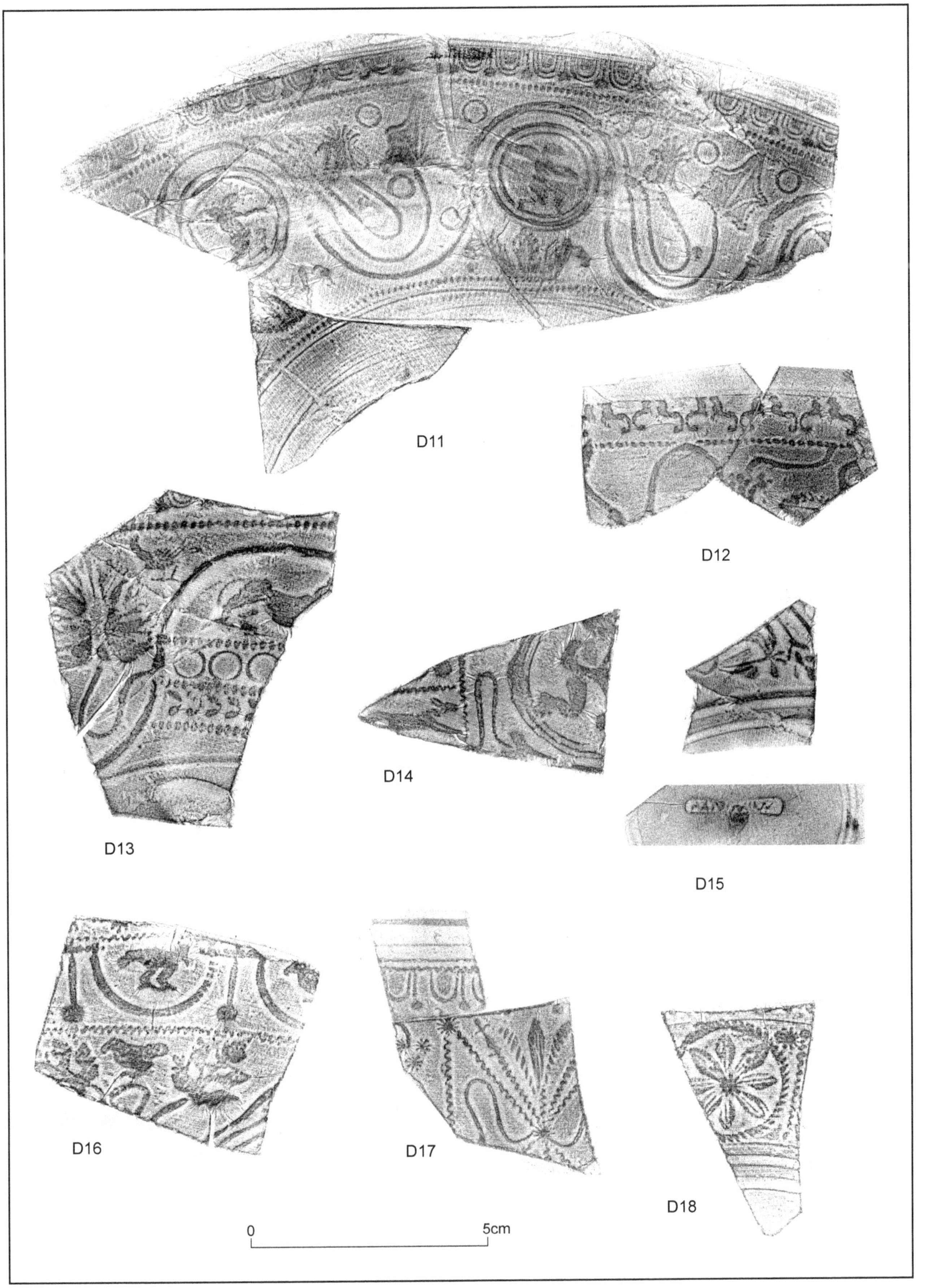

FIGURE 2.13 SAMIAN DECORATION D11-D18.

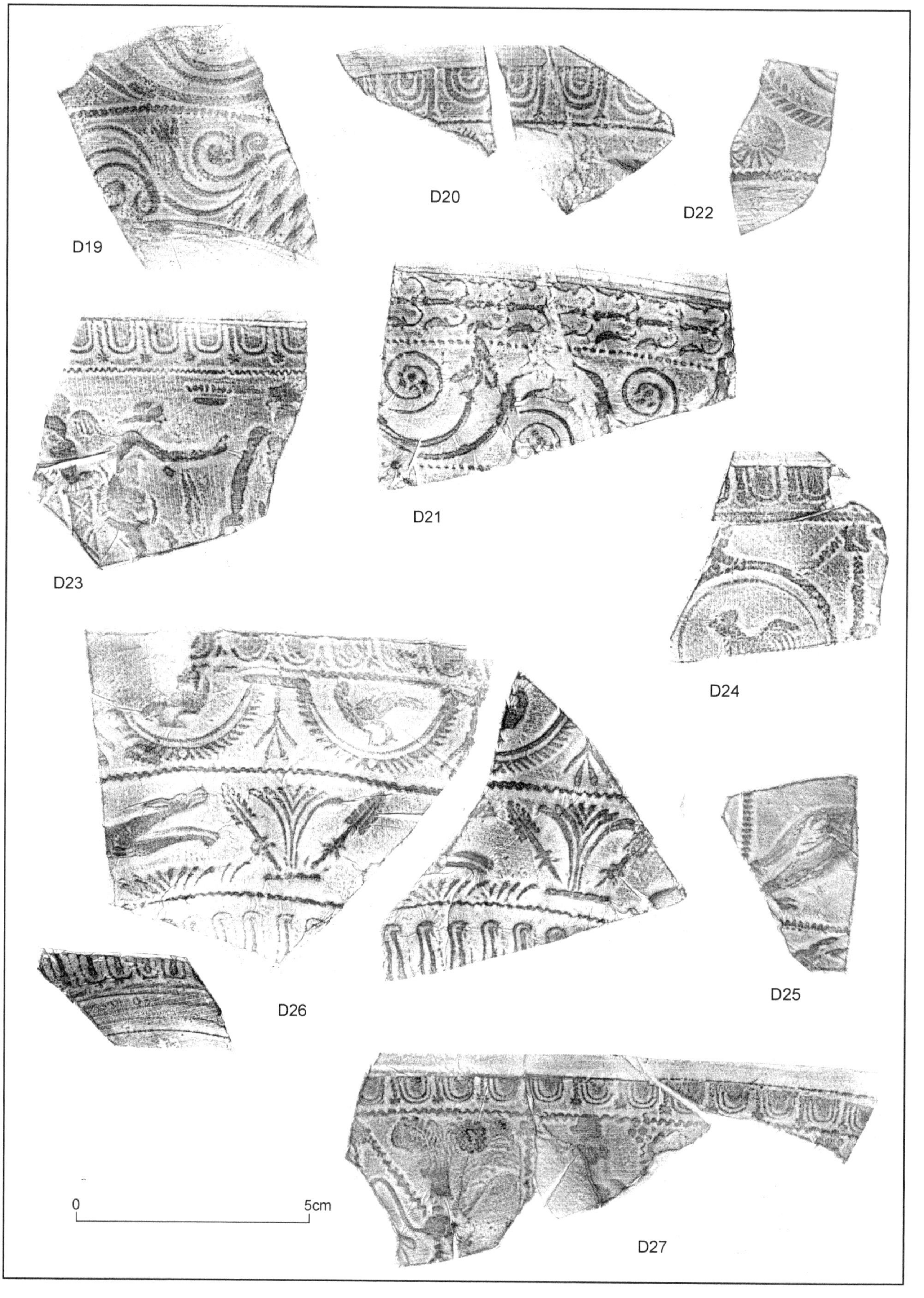

FIGURE 2.14 SAMIAN DECORATION D19-D27.

and Cinnamus, and the general appearance of the piece suggests a Hadrianic or Antonine date. (3033). (Fig 2.14).

Also from these contexts: stamp of Primulus i (S11), *c.*AD 60-85.

Event 2/3

Sherds of the following bowls came from contexts assigned to Events 2 and 3:

D26. Form 37, South Gaulish. Three sherds, non-joining, of a bowl with zonal decoration. The upper zone contains festoons with birds (O.2247, O.2290); the lower contains a bush motif with the hound (O.1925) and stag (O.1738). The types and motifs were all common in the Flavian-Trajanic period, when few potters signed their work. The trident-tongued ovolo is associated with work stamped by Sulpicius, as is the four-pronged pendant (Mees 1995, Taf. 193, 2). The hound, stag, bush and grass motifs were all used by M. Crestio (*ibid.*, Taf. 48, 1), while the squat gadroons around the base were a characteristic of Frontinus (*ibid.*, Taf. 60, 1, 2). *c.*AD 80-100. (3007, 3009). (Fig 2.14).

D27. Form 37, Central Gaulish. Three joining sherds of bowl with panels containing a saltire with ram's horns (Rogers G345) at the top and the dancer (O.363). The style is that of Drusus ii, who used the ovolo (Rogers B15), the wavy-line borders with dot rosette (Rogers C280) at the junctions and both types. A bowl in his style (S&S, pl. 89, 13) shows a similar saltire with the ram's horns and possibly the side leaf (Rogers G200), of which the edge appears here. *c.*AD 125-145. (3003, 3011, 3018). (Fig 2.14).

Event 3

D28. Form 37, South Gaulish. Two joining fragments showing trees with the boar (O.1671) to left and an animal, probably a stag, to right, behind which is a hunter (not in O. or Hermet 1934) holding a rope or leash, with which he has presumably caught the animal (c.f. O.1131 for an East Gaulish type with a similar scene). The tree occurs on work stamped GERMANI.SER (Mees 1995, Taf. 88, 8; 89, 1; 90, 1, which also shows the boar). *c.*AD 90-110. (3003, 3006). (Fig 2.15).

D29. Form 37, Central Gaulish, from Les Martres-de-Veyre. Three joining sherds and another of ovolo, probably from the same bowl, in the style of X13. The ovolo (Rogers B14) and finely-beaded borders are typical of his style. Decoration shows panels, with corded corner tendrils surrounding a double medallion with panther (O.1518) and possibly the lion (O.1422) without its tail, and a saltire, with acanthus (Rogers K23) and leaf (Rogers J127), all attested on his style. A bowl from Brecon (S&S, pl. 49, 588) shows a saltire with the same side leaves, one from London (S&S, pl. 43, 493) shows the large rosette (Rogers C244) used at the junction of border and basal bead row. *c.*AD 100-120. (3005, 3006). (Fig 2.15).

D30. Form 37, Central Gaulish, probably from Les Martres-de-Veyre. Decoration shows a wide panel with beaded borders containing the dancer (O.353), Bacchus (O.566) and cupid (O.403), with the same acanthus as D29 (Rogers K23) and trifid bud (Rogers G171). All the types and details are attested for X13 except the dancer, which was used on the related style of X12. *c.*AD 100-120. (3006). (Fig 2.15).

D31. Form 37, Central Gaulish. Two joining sherds of a small bowl in the fabric of Les Martres-de-Veyre. Beaded panels, without a border junction motif, show Rogers F69 used as an arcade containing the head of a panther (O.1519) over what appears to be a hare (O.2116 or 2117, without its head, to fit into the panel?), festoon with spiral (Rogers F70) over medallion with lion's head (possibly O.1404) and the panther (O.1519) over a crane (O.2196) between two pigmies (one not in O.?, the other probably O.699). The style is that of Stanfield and Simpson's Donnaucus-Ioenalis group (Rogers' potters X11-X13 and P10). The ovolo may be Rogers B14, used by X13, though more widely spaced than on D29 above. The crane and O.699 (S&S, pl. 45, 519), O.1404 and the basal bead row all occur on X13 style. The panther (O.1519) was used by X11, the festoon (Rogers F69) by P10. Although no very close parallels are forthcoming, the date is clearly *c.*AD 100-125. (3005). (Fig 2.15).

D32. Form 37, Central Gaulish. A sherd of similar size and style to the last, but from a different bowl with finely beaded vertical borders. One panel contains the pedestal motif (Rogers Q48) used by P10 (Rogers 1999, pl. 125, 6), with an extra cup motif stamped sideways across it; the other contains a warrior (not in O.; perhaps closest to R.3079), a small mask to left and a seven-dot rosette (Rogers C280) which is common to all the potters in this group. *c.*AD 100-125. (3005). (Fig 2.15).

Also from these contexts: stamp of Sacer Vasil- (S13), *c.*AD 100-125.

Area C

Event 2

D33. Form 37, South Gaulish, showing a single-bordered ovolo with tongue ending in a blob (Dannell *et al.* 1998, JA), bordered by a bead row and with a straight wreath of trifid buds. The ovolo is well-known, though not associated with any named potters. It occurs with the same wreath on a bowl from Ribchester (Wild 1988, fig. 3, 40), in Period 1B-C at Fishbourne (Dannell 1971, fig. 128, 19) and at Verulamium in a pit filled by *c.*AD 75 (Hartley 1972, fig. 86, 41). The bowls are all of form 37, suggesting a date *c.*AD 70-90. (4030). (Fig 2.15).

D34. Form 37, South Gaulish. Two non-joining sherds of a zonal bowl with a scroll with leaf-tips in the lower concavity, a narrow horizontal chevron wreath, a zone of straight gadroons and a repeat of the chevron wreath.

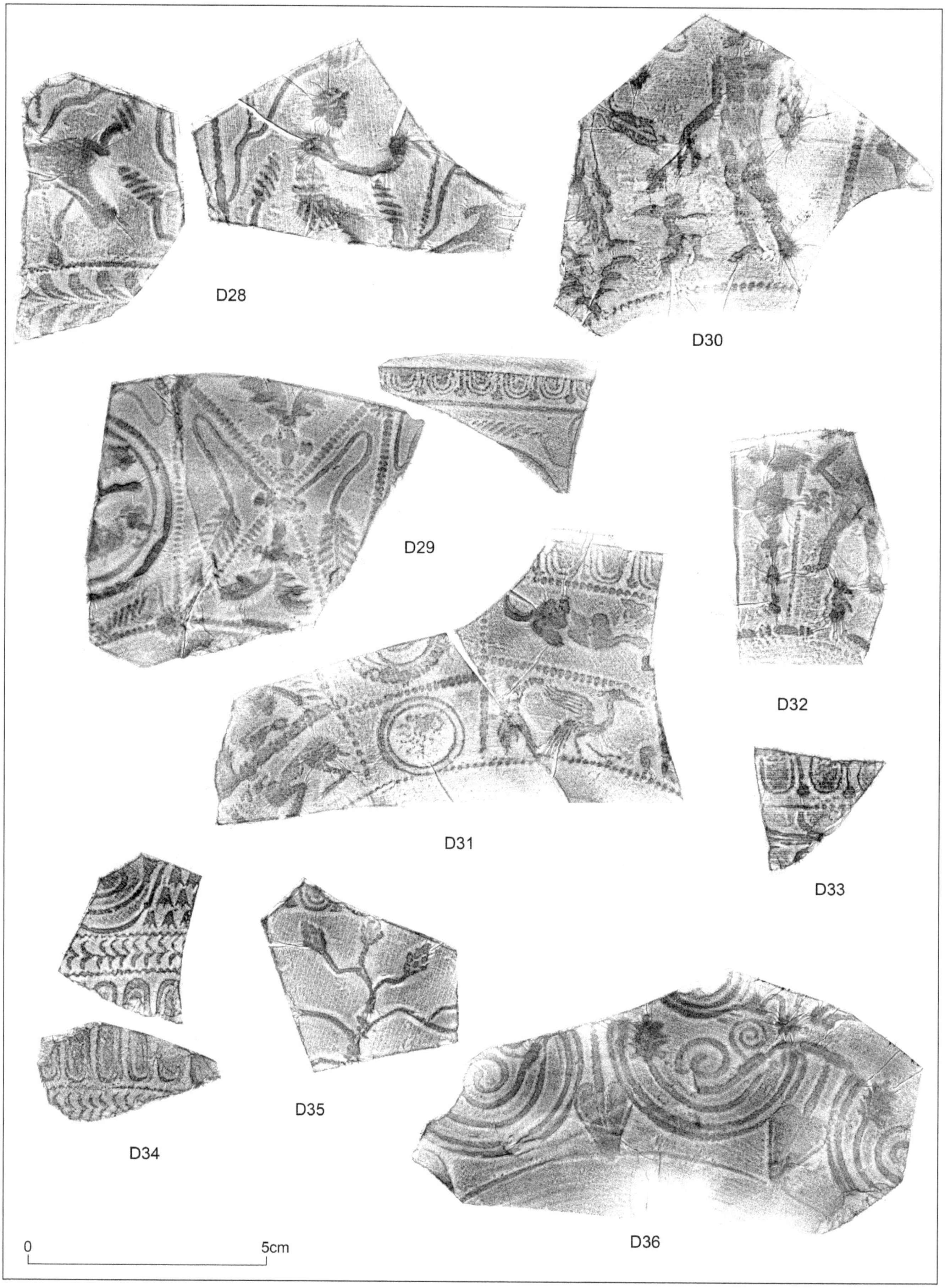

FIGURE 2.15 SAMIAN DECORATION D28-D36.

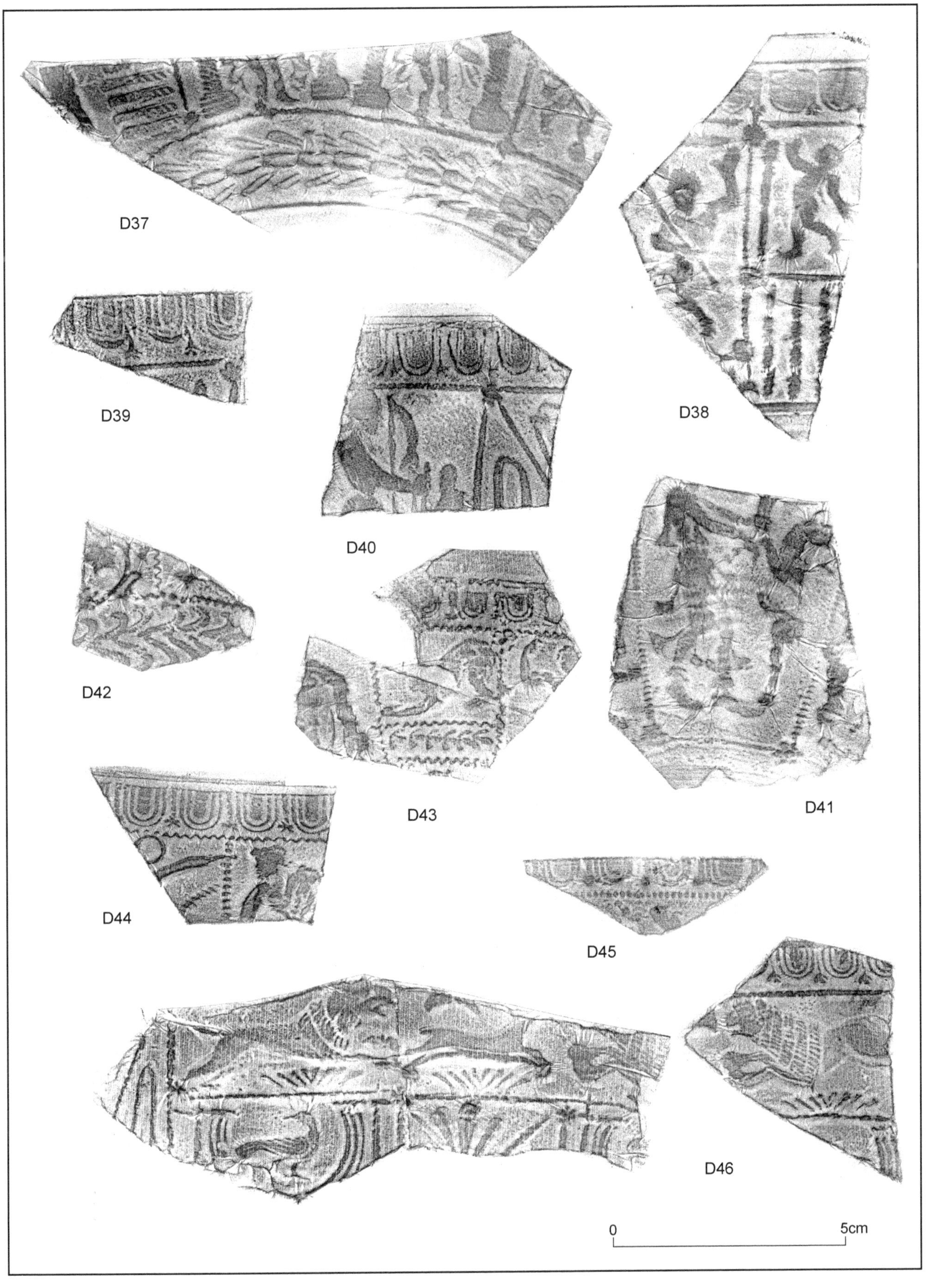

FIGURE 2.16 SAMIAN DECORATION D37-D46.

The decoration is typical of the potters accustomed to producing form 29 who first made form 37, and similar designs occur in the Pompeii Hoard of AD 79 (Atkinson 1914, 40, 44). The chevron wreath and gadroons appear on two bowls from London showing the large rosette ovolo used by Calvus i (Museum of London (ex Guildhall) nos. 4399G, 5860 G). *c.*AD 70-90. (4030, 4046). (Fig 2.15).

D35. Form 37, South Gaulish, with the distinctive ovolo of Germanus and his tree (Mees 1995, Taf. 78, 10). *c.*AD 75-100. (4046). (Fig 2.15).

D36. Form 37, South Gaulish, with an upper zone showing panels containing a festoon or medallion and the hind legs of an animal, and a lower zone of festoons containing leaves and spirals separated by a heart-shaped pendent leaf. The smaller leaf in the lower zone was used by M. Crestio (Mees 1995, Taf. 44, 1). A similar, though smaller, cordate leaf occurs on a bowl with M. Crestio's ovolo from London (Museum of London (ex Guildhall) no. 4620G), though no parallels for a leaf of this size are forthcoming. *c.*AD 80-100. (4030). (Fig 2.15).

D37. Form 37, South Gaulish, showing the lower part of panel decoration, with two arcades containing a figure and the satyr with grapes (O.597) above a basal wreath. The figure in the arcade (not in O. or Hermet 1934) appears at Brive (Moser, Tilhard 1987, fig. 5, 101) and Espalion. It wears a Phrygian cap and is discussed by Tilhard (2009, 89-91), who identifies it as Attis. Although not noted by Hermet, it also appears to have been used at La Graufesenque, where this bowl is likely to have been made. At both sites, the figure tends to appear in an arcade, but the arcade here is similar to those in use at La Graufesenque rather than Espalion, for example on a bowl stamped by Sulpicius (Mees 1995, Taf. 193, 1). *c.*AD 80-110. (4030). (Fig 2.16).

D38. Form 30, South Gaulish, with trident-tongued ovolo and panels containing Pan (O.714) and satyr (O.646) over a panel of 'maggots', short corded bars. The ovolo is probably that used by Mercator, though the general style looks later than that of most of his stamped pieces. Both figure types have been recorded on pieces in the styles of Mercator and Masculus, and occur in the Bregenz Cellar deposit (Jacobs 1913, Taf. 3). *c.*AD 85-110. (4030). (Fig 2.16).

D39. Form 37, South Gaulish, showing the large ovolo with four-pronged, tufted tongue, used by Sabinus iv at La Graufesenque (Mees 1995, Taf. 176, 1, 2) and at Espalion. The tip of the tongue has clearly been applied separately, as it overlaps the core. (See D20 above for a more precise example.) *c.*AD 90-110. (4030). (Fig 2.16).

D40. Form 37, South Gaulish, with panels containing Diana (O.104) and saltire. The ovolo, with trident tongue turned to the right (Dannell *et al.* 1998, SJ) appears on bowls with the stamp of Sulpicius. A stamped bowl from Straubing (Mees 1995, Taf. 193, 1) shows both ovolo and Diana. Both that and the bowl from Wilderspool (*ibid.*, Taf. 193, 2) show saltires, though not identical to the one here. The presence of his work at late first century foundations such as Wilderspool suggests a date *c.*AD 90-110. (1014). (Fig 2.16).

D41. Form 37, Central Gaulish, probably in the fabric of Les Martres-de-Veyre, showing Diana (O.106) in a panel with beaded borders ending in a rosette (Rogers C280). The features all occur on a bowl from Brecon in X13 style (S&S, pl.48, 572). *c.*AD 100-120.

(4030). (Fig 2.16).

D42. Form 37, Central Gaulish, in the fabric of Les Martres-de-Veyre, showing panels with wavy-line borders ending in a rosette (Rogers C280), one containing a vine scroll (Rogers M25), over a chevron basal wreath (Rogers G370). A bowl in X13 style shows the wreath and borders (S&S, pl, 45, 517). The vine scroll also occurs on this style. *c.*AD 100-120. (4033). (Fig 2.16).

D43. Form 37, Central Gaulish. Two joining sherds in the style of X13, with his ovolo (Rogers B14) and panels with Bacchus (O.571) and panther (O.1566) over wreath of the small trifid (Rogers G113). The types are all attested on X13 style except the trifid, which was used similarly as a wreath on a bowl in the style of X8 (S&S, pl. 31, 367). *c.*AD 100-120. (4046). (Fig 2.16).

D44. Form 30, Central Gaulish, showing panel decoration with Vulcan (O.66). The style is likely to be that of Drusus ii, who used the ovolo (Rogers B36), Vulcan and a similar combination of wavy-line and beaded borders (S&S, pl. 88, 1), though here, unusually, the border lacks a masking motif. The corded bar in the left-hand panel is a typical feature of his work (S&S, pl. 88, 1, 8), though the figure type or motif is uncertain. *c.*AD 125-145. (1016). (Fig 2.16).

D45. Form 37, Central Gaulish, in Lezoux fabric and the style of Drusus ii. The ovolo is the same as that on D44, with bead row border and the edge of his acanthus (Rogers K2) (S&S, pl. 88, 5). *c.*AD 125-145. (4030). (Fig 2.16).

Also from these contexts: stamp of Memor (S9), *c.*AD 80-110.

Event 2/3

Sherds of the following bowls were assigned to Events 2 and 3:

D46. Form 30, South Gaulish. Three sherds of panel bowl with saltire, and stag (slightly reduced version of O.1738) between lions (O.1400, O.1454) over smaller saltire between triple festoons with birds (O.2247 and presumably its reverse). The ovolo, types and motifs were used by Mercator. *c.*AD 80-100. (4029, 4046). (Fig 2.16).

D47. Form 37, South Gaulish. Seven sherds, of which four rim sherds join. There is a lead rivet in the rim, and traces

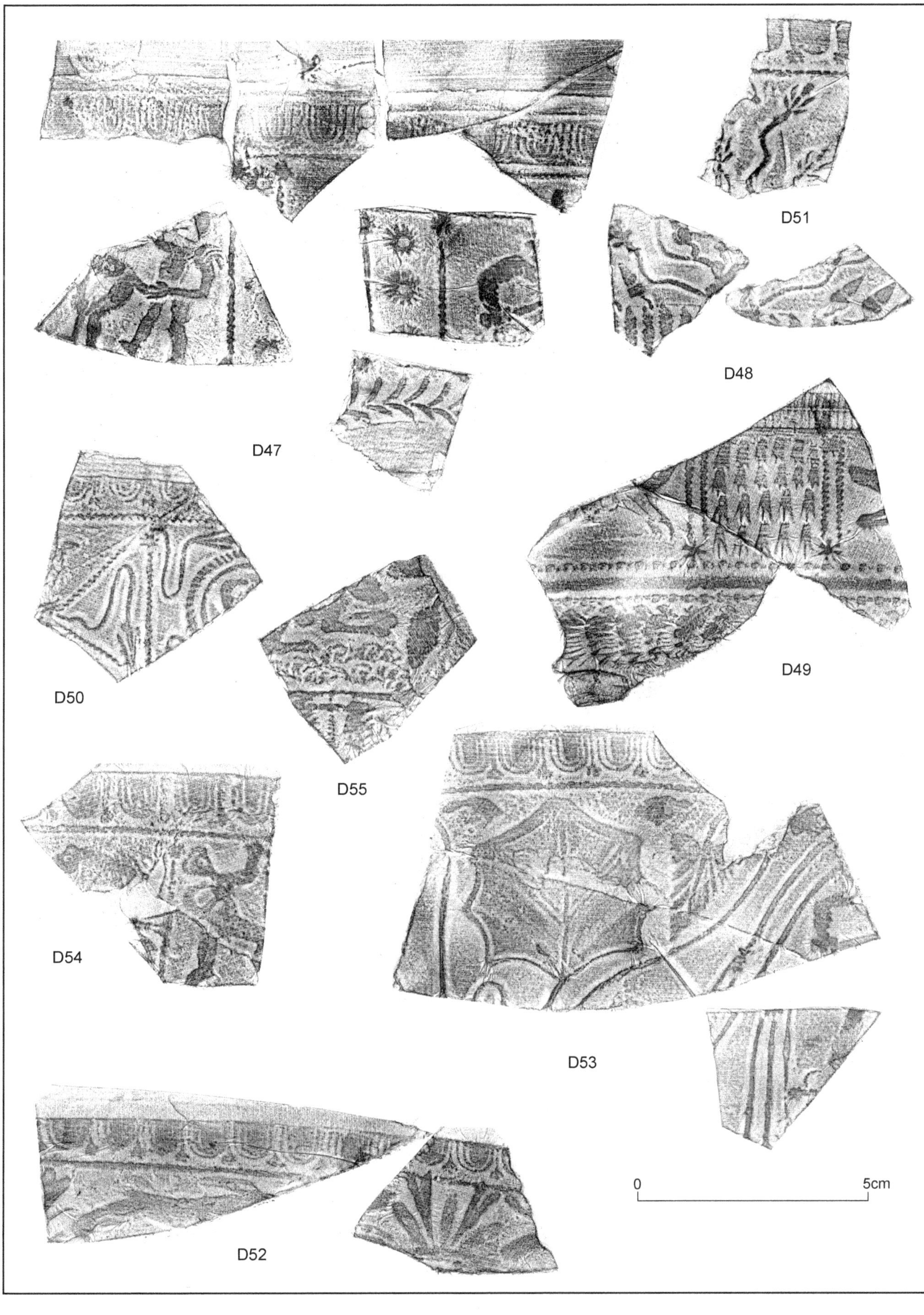

FIGURE 2.17 SAMIAN DECORATION D47-D55.

of a rivet-hole on a sherd showing the basal wreath suggest that they are from the same bowl. Panels show a couple (not in O. or Hermet 1934), a vertical row of rosettes, and a gladiator (probably O.999). The couple appear on a sherd from La Graufesenque (G91 L70) with a distinctive, but anonymous, ovolo with a narrow core and four-pronged tongue. Another sherd with this ovolo from G91 L70 shows a horizontal row of rosettes, though perhaps slightly larger than the rosettes here. The ovolo on these pieces has been recorded widely on Flavian sites in Scotland, such as Camelon and Elginhaugh, and in the north of England. The ovolo on the present piece, poorly moulded and apparently doubly impressed, is probably the same as that on the La Graufesenque pieces. It is interesting that Hartley records double or triple impressions of the ovolo on a bowl from Elginhaugh (Hartley 2007, 385, D34). The potter was clearly at work in the Flavian period, though the style of this particular bowl appears later than some of his earlier, zonal pieces. *c.*AD 80-100. (4016, 4030). (Fig 2.17).

D48. Form 37, South Gaulish. Two non-joining sherds of a thick bowl showing a tree with six-fold buds and acorns, and a row of acorns above a horizontal wreath of similar 'maggots' to D38 above. A tree with similar buds and acorns was used by Sabinus (Mees 1995, Taf. 178, 4, Taf. 179, 1; his styles II and III). A bowl by L. Cosius also shows both motifs (*ibid.*, Taf. 23, 1), though not together, as on the Sabinus bowls. The connections suggest a date *c.*AD 90-110. (4004, 4030) . (Fig 2.17).

Event 3

D49. Form 29, South Gaulish. Two joining sherds showing upper zone with panels probably containing the boar (O.1671) and hound (O.2013) separated by a panel of leaf tips. Both types appear on a bowl from Poitiers stamped by Vitalis (Dannell *et al.* 2003, Vitalis ii, Taf. A4, 3144). A similar arrangement of animals and leaf tip panels occurs with a similar horizontal wreath around the carination on a bowl with the same stamp (*ibid.*, Taf. B2, 3168). The motif of which the wreath is made may be the same as that, used differently, on another bowl with Vitalis' stamp (*ibid.*, Taf. F1, 3156). As Vitalis acquired his moulds from various potters in different styles there is no proof of connection with him, but the style is typical of mould-makers of the early Flavian period. *c.*AD 70-85. (1008). (Fig 2.17).

D50. Form 37, South Gaulish. Two joining sherds of bowl with panels containing a saltire and a medallion with backwards-facing bird to left (O.2289 or similar). The ovolo appears to be a slightly blurred version of the ovolo with four-pronged tongue used by M. Crestio and Crucuro. The medallion with bird appears on a bowl stamped by M. Crestio (Mees 1995, Taf. 39, 7) who also used the pointed leaf in the saltire (*ibid.*, Taf. 43, 1). *c.*AD 80-100. (4004). (Fig 2.17).

D51. Form 37, South Gaulish. Two joining sherds showing an ovolo with rosette tongue of the type used by Calvus and Mercator. A similar tree with bird was used by Mercator (Mees 1995, Taf. 133, 5). *c.*AD 80-100. (4004). (Fig 2.17).

D52. Form 37, South Gaulish. Two non-joining fragments with zonal decoration showing a chase with hare (O.2074), hound (O.1925) and bush motif. The style is that of Mercator, who used the trident-tongued ovolo, bush motif (Mees 1995, Taf. 136, 2) and types (*ibid.*, Taf. 137, 13). *c.*AD 80-100. (4007, 4009). (Fig 2.17).

D53. Form 37, South Gaulish. Four sherds, three joining, of a bowl with leaf-scroll in the style of Mercator, with the gladiator (O.999) in the lower concavity. A stamped bowl from Caerleon (Mees 1995, Taf. 132, 4) shows a scroll with the same leaves and Nile geese (O.2286, O.2244), the gladiator and probably the same trident-tongued ovolo. *c.*AD 80-100. (4015). (Fig 2.17).

D54. Form 37, South Gaulish. Three joining sherds showing an ovolo with tongue ending in a blob and panels with boar (O.1695A?) and a swordsman (not in O. or Hermet 1934). The style is typical of the latest products of La Graufesenque. *c.*AD 85-110. (4004). (Fig 2.17).

D55. Form 37, South Gaulish, showing hare (O.2056) and vine with grapes over a zone of festoons. Identical decoration occurs on a bowl stamped by Masculus (Mees 1995, Taf. 119, 3). *c.*AD 90-110. (4009). (Fig 2.17).

Also from these contexts: stamp of Coelus (S5), *c.*AD 65-85.

Area E

Event 2

D56. Form 29, South Gaulish. Fragment of upper zone, with panel containing the Nile goose to left (O.2286), draped woman (O.938A) and Nile goose to right (O2244) above wavy lines (and probably leaf tips). The connections appear to be with Calvus i. The woman appears on form 37 from La Graufesenque with the mould signature CALVOS and a stamp of Patricius (Mees 1995, Taf. 17, 1). The Nile geese, woman and panel of wavy lines and leaf tips all occur on a form 29 stamped by Q. Iulius Habilis or Habitus from Gloucester (Wild 1998, fig. 1, 1), also possibly from a mould by Calvus. *c.*AD 70-85. (5012). (Fig 2.18).

D57. Form 37, South Gaulish, showing panels with the chariot (O.1161) over satyr with grapes (O.597) and sacrificer (O.977), and probably Victory (O.808B). The style, types and corner leaf tuft are all typical of the potters whose work appears in the Bregenz Cellar deposit of *c.*AD 90 (Jacobs 1913, Taf. 2, 10, 13), including Masculus (Mees 1995, Taf. 120, 2). .AD 90-110. (5003). (Fig 2.18).

D58. Form 37, Central Gaulish. Three joining sherds of bowl showing beaded panels ending in a rosette (Rogers C175) containing a beaded medallion (Rogers E2), a pedestal, probably the base of the caryatid (O.1206), feet of an uncertain type, and another pedestal base. The

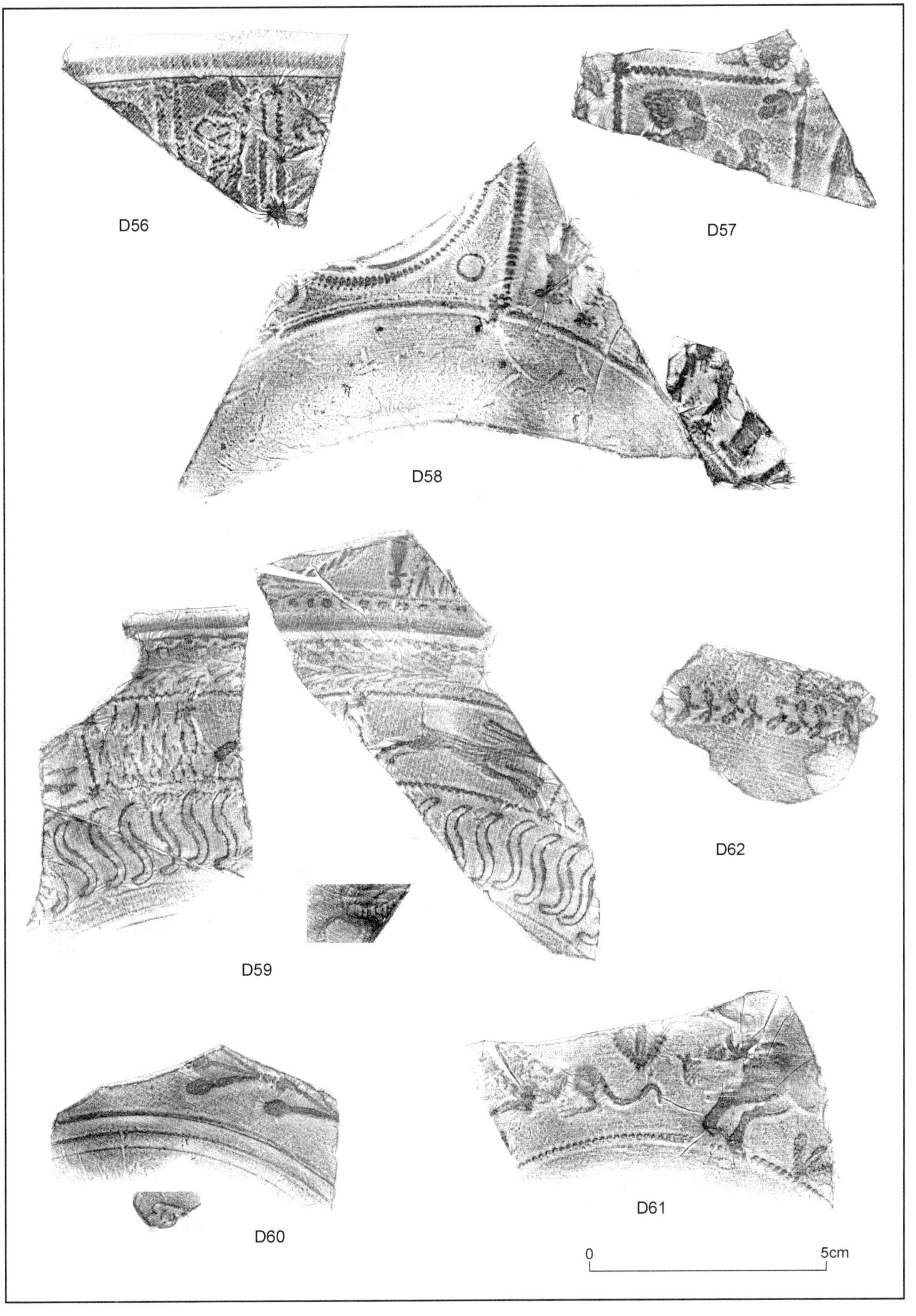

FIGURE 2.18 SAMIAN DECORATION D56-D62.

style is probably that of Sacer, who used C175, E2 and the caryatid. Very faint traces of a mould signature, almost entirely obliterated and covered with scratch marks are just visible below the decoration. Traces of an A and R suggest that the signature may indeed be part of Sacer's. *c*.AD 125-145. (5003). (Fig 2.18).

Event 3

D59. Form 29, South Gaulish. Three sherds, two joining, of a bowl with heavily worn footring. Decoration shows, in the upper zone, a festoon with pendent bottle bud and leaf tips and in the lower, panels with the hare (O.2072) and more leaf tips over a wreath of S-shaped gadroons. The potter's stamp in the interior, badly smudged, appears to read OFCA[, probably one of many dies of Calvus i (S4 below). Most of the motifs here occur on bowls with his stamps: the festoon and bottle bud (Dannell *et al.* 2003, Calvus i, Taf. D2, 1038), the hare (*ibid.*, Taf. F1, 0250a), and gadroons (*ibid.*, Taf. G9, 1684). The general finish of the bowl is poor and its form suggests an early Flavian rather than a Neronian date. *c*.AD 65-85. (5002, 5005, 5014). (Fig 2.18).

D60. Form 29, South Gaulish, showing basal wreath of triple poppy-heads. The edge of the potter's stamp, O[in a swallow-tail frame, survives inside the base (S18 below). The wreath appears on a bowl from Caerleon, with the stamp OFVITA, of Vitalis ii, in a similar frame (Dannell *et al.* 2003, Vitalis ii, Taf. E1, 1004). Although far too little of the stamp survives for certain identification, this may also be the work of Vitalis. *c*.AD 70-85. (5002). (Fig 2.18).

D61. Form 37, Central Gaulish, in the fabric of Les Martres-de-Veyre, showing freestyle decoration with lion (O.1424), boar (slightly larger than O.1642) and trifid bud (Rogers G24). The basal bead row is a particular feature of X13 style, to which the bowl should probably be attributed. The bud occurs on this style, as does a similar lion. *c*.AD 100-120. (5002). (Fig 2.18).

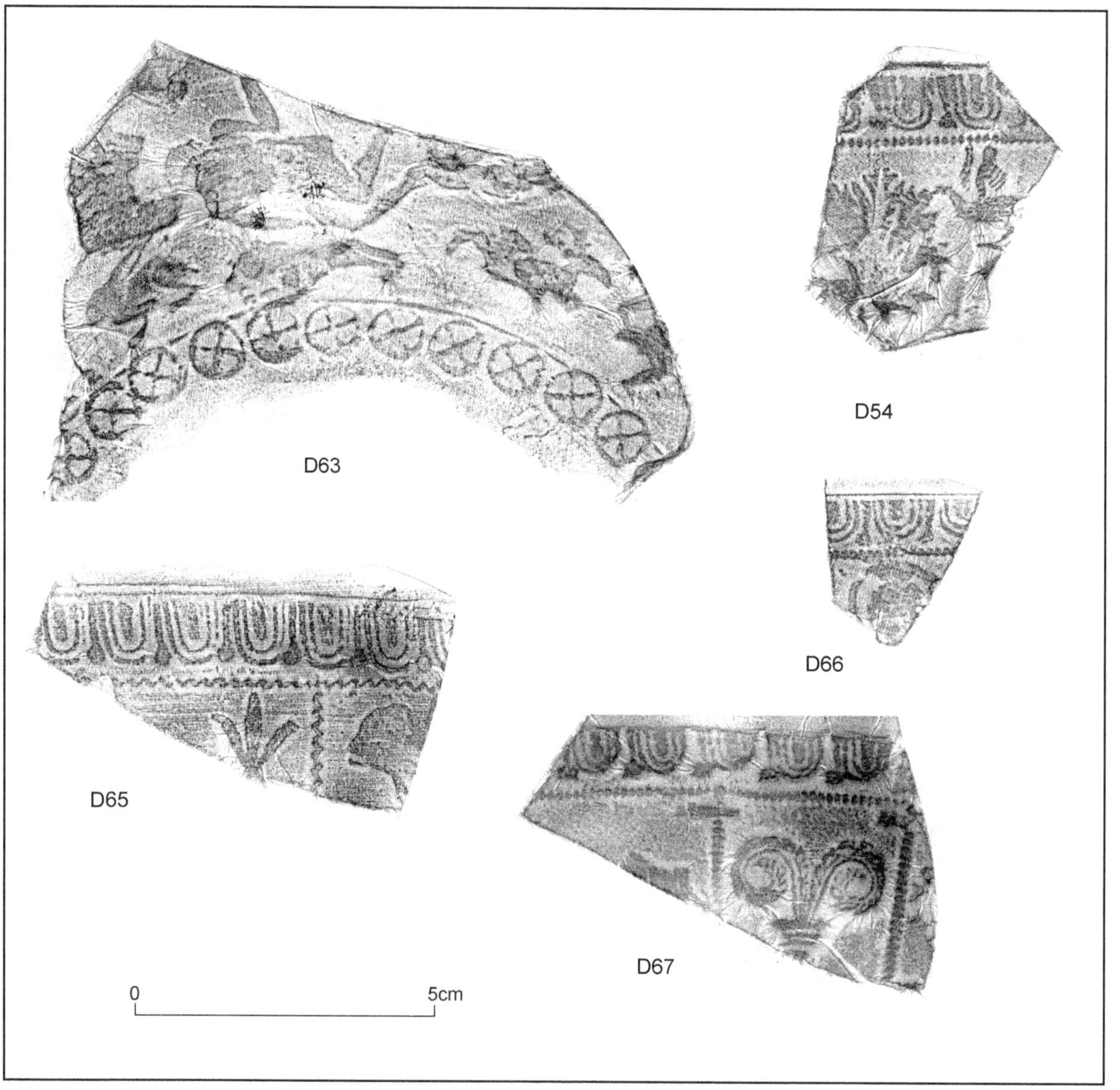

FIGURE 2.19 SAMIAN DECORATION D63-D67.

D62. Form 37, Central Gaulish, in the fabric of Lezoux, showing a basal wreath of the bifid bud (Rogers G311). This appears to be a feature of Rogers' potters P6A and P6B (Rogers 1999, pl. 123, 25; 124, 7, 10). The sherds published by Rogers are all from Lezoux, where this ware was clearly being produced. Rogers dates it on stylistic grounds to *c.*AD 100-120, though as large-scale export from Lezoux did not start until *c.*AD 120, and the fabric of the piece is not typical of the pre-export period, a Hadrianic date perhaps seems more probable. (5005). (Fig 2.18).

D63. Form 37, Central Gaulish. Base sherd, with worn footring, of a very thick bowl in orangey fabric with a black streak beneath the slip in the centre base, presumably caused by incomplete oxidation during firing. Decoration is in freestyle above the most distinctive feature: a row of 'wheels' around the base. Figure types include the hare (O.2116) and bear (O.1628), both slightly reduced in size; the stag (O.1703), gladiator (O.1056) with leg broken, and a large horizontal figure, torso facing up, legs facing down (not in O.) with a large splodge of clay on top of it. The wheel motif, not among Rogers' motifs, appears similar to that on Apollo's chariot (O.102). No close parallels to this piece are forthcoming. Oswald (though not Rogers) record the hare, bear, gladiator and O.102 on pieces by or in the style of Butrio. The stag was used by Acaunissa. The basal wreath of wheels occurs on a small sherd from Chesters, on Hadrian's Wall, rubbed by the late Grace Simpson (I am grateful to Joanna Bird for bringing this to my notice). It shows a wavy-line panel border, but no features attributable to a particular potter. The clumsy nature of the bowl, and its decoration with broken types and types presumably copied by *surmoulage* suggests the work of an amateur. The general connections suggest a Hadrianic date. (5005). (Fig 2.19).

D64. Form 37, Central Gaulish, showing ovolo (Rogers B18), leaf (Rogers H82) and bird (O.2326). The ovolo and types were all used by Austrus (Rogers 1999, pl. 12, 9, 10). *c.*AD 125-150. (5005). (Fig 2.19).

D65. Form 37, Central Gaulish, showing an ovolo (Rogers B49) and wavy-line panels with no junction motif, containing a trifid bud (Rogers G32) and probably the dolphin (O.2394). The ovolo and borders without junctions are characteristic features of Rogers' potter P14 (Rogers 1999, pl. 127, 3), recently shown by Brenda Dickinson (2005) to be equated with Pugnus and to have links with X6. The trifid G32 was used by X6, and occurs on a bowl with ovolo B233, the other Pugnus ovolo discussed (Dickinson 2005, 103, no. 15) as does the dolphin (*ibid.*, fig. 4, 14). The presence of the pieces with ovolo B49 on sites on Hadrian's Wall and its surrounds (Housesteads, Birdoswald and Cardurnock) suggests a Hadrianic date, towards the start of the period *c.*AD 135-165 normally assigned to Pugnus. Another bowl with this ovolo discussed by Dickinson (2005, 102, no. 8) is from Chester. (5002). (Fig 2.19).

D66. Form 30, Central Gaulish, showing ovolo (Rogers B52) and caryatid (O.1207A). The ovolo and type were used by both Criciro and Divixtus (S&S, pl.115, 1). The work of both potters is common on the Antonine Wall. Criciro is now thought to have started work *c.*AD 135, Divixtus slightly later, *c.*AD 140-160. (5002). (Fig 2.19).

D67. Form 37, Central Gaulish. The ovolo (Rogers B24) and bead rows with an astragalus placed just below the junction are characteristic of the late style of Docilis, which shows similarities to that of Casurius (Rogers 1999, pl. 39,1). The ram's horns (Rogers G350) were used by both potters. Below them is the top of Docilis' column (Rogers P62). The edge of the leaf (Rogers J5), also used by both potters, appears in the left-hand panel. Work in this style occurs on the Antonine Wall, at Balmuildy (Miller 1922, nos. 32, 42). *c.*AD 140-160. (5005). (Fig 2.19).

D68. Form 37, Central Gaulish. Small scrap showing the ovolo (Rogers B231) used by Cinnamus and Pugnus. *c.*AD 145-170. (5005. Not illustrated)

Also from these contexts: stamps of Albus iii (S1), *c.*AD 145-165, Balbinus (S2), *c.*AD 105-120, Calvus i (S4), *c.*AD 65-85, Maturitus (S8), *c.*AD 100-130, Primulus i (S10), *c.*AD 60-85, and Tertius (S14), *c.*AD 55-75.

The Potters' Stamps

The information on stamps S1-S5 is based on the first three volumes of *Names on Terra Sigillata* (Hartley and Dickinson 2008), which were available at the time of writing. Details of the other stamps were gleaned from Polak 2000 and other sources, though later revision offered the opportunity to include details from volumes four to seven. Superscript 'a' denotes that the die has been found at the pottery in question; superscript 'b' that other dies of the same potter have been found there and 'c' that the die has been attributed to the pottery solely on features such as fabric and distribution. Underlining denotes ligatured letters.

S1. [A]LBIM (A with a vertical stroke). Form 31, Central Gaulish. Die 3a of Albus iii of Lezoux[b], who may also have worked at Les Martres-de-Veyre[b]. Examples of this die have already been recorded at Chester, on form 31 and 27. The occurrence on form 27 and on sites in the Rhineland suggests a start date before AD 150. The die is likely to have been in use *c.*AD 145-165. (5005; Area E, Event 3)

S2. BΛLB[INI.M]. Form 18/31, Central Gaulish. Die 2a of Balbinus of Les Martres-de-Veyre[a]. This die was heavily used and has been recorded widely on sites in Britain, including Chester. As it deteriorated, its reading appeared more like IINIIIINI.M (first N reversed), the form in which it appears here. The use of this die on form 15/17 suggests a Trajanic date, though it also appears in the Hadrianic fire in London. *c.*AD 105-120. (5002; Area E, Event 3)

S3. BELINI[CCVSF]. Form 27, Central Gaulish. Die 3a of Beliniccus i of Les Martres-de-Veyre[a]. The die regularly occurs on form 27. It has been recorded on form 33 from

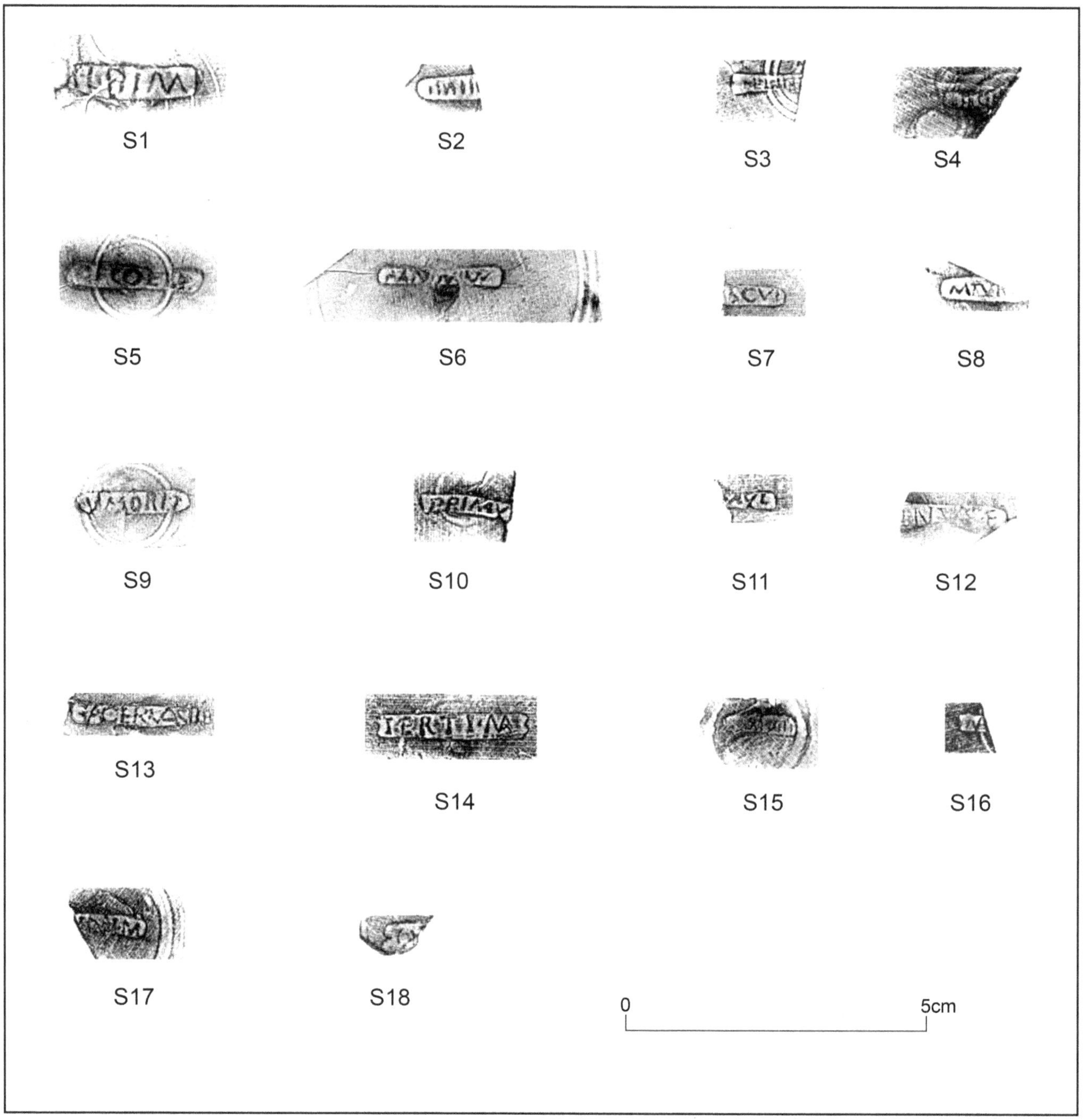

FIGURE 2.20 SAMIAN STAMPS S1-218.

the extra-mural settlement at Chester, but in no closely datable contexts. *c*.AD 100-120. (4006; Area C, cleaning layer)

S4. OFCA[. Form 29, South Gaulish (D58 above). From the style of the decoration, this is likely to be one of the many dies of Calvus i, though it is not possible, from the incomplete and smudged impression, to be certain which. *c*.AD 65-85. (5002, 5005, 5014; Area E, Event 3)

S5. OFCOELI. Form 27g, South Gaulish. Die 1b of Coelus ii of La Graufesenque[a] . This die, too, has been recorded from the extra-mural settlement at Chester. It is suggested, on the basis of his stamped decorated ware, that he began work in the later AD 60s. *c*.AD 65-85. (4007; Area C, Event 3)

S6. <u>MA</u>NDVIL<u>MA</u>. Form 29, South Gaulish (D15 above). Die 6a of Manduillus of La Graufesenque[a] (Polak 2000, M19). The die has been noted at Chester, Ribchester and York as well as in the Pompeii Hoard, and was clearly in use during the early Flavian period. *c*.AD 60-80. (2005, 2009; Area A)

S7. [OFMA]SCVL. Form 18 or 15/17, South Gaulish, showing the end of a stamp recorded by Hartley and Dickinson (2008-) under L. Tr- Masculus (Masclus iii) of La Graufesenque[a] (Polak 2000, M49). Although the final letter looks like I, good impressions from La Graufesenque show it originally to have been L. Polak notes that the die was found at La Graufesenque in the waste surrounding a kiln fired *c*.AD 80-120/130. The site evidence for it includes sites such as Wilderspool, founded under Domitian. *c*.AD 80-110. (2111; Area A, Event 3)

S8. MATVRITM[.S]. Form 18/31, Central Gaulish. Two joining sherds showing die 1a of Maturitus of Les Martres-de-Veyre[c]. This is an uncommon stamp, of which Hartley and Dickinson (2008-) note only three examples, all on form 27. Although neither Terrisse (1968) nor Romeuf (2001) record a potter of that name at Les Martres-de-Veyre, the form of the stamp and fabric of two of the pieces suggest origin there, as does the fabric of the present piece. The stamp is likely to date to the first half of the second century, probably *c*.AD 100-130. (5005; Area E, Event 3 and Area B, unstratified)

S9. MMORM. Form 27, South Gaulish, showing the damaged and worn version of a die of Memor of La Graufesenque[a], die 3a'(Polak 2000, M63*), originally reading MEMORISM. The damaged version occurs on cups (27, 27g, 33) at Domitianic sites such as Holt and Corbridge. It is estimated that the damage to the die took place *c*.AD 85. *c*.AD 80-110. (4030; Area C, Event 2)

S10. PRIMV[L]. Form 18, South Gaulish. Die 9c of Primulus i of La Graufesenque[a] (Polak 2000, P88). Although there is evidence that Primulus started work in the pre-Flavian period, this die has been recorded at both Chester and York (Dickinson and Hartley 1993, 767, fig. 283, 2739), and is likely to be Flavian in date. *c*.AD 60-85. (5014; Area E, Event 3)

S11. [PRI]MVL. Form 18 or 15/17, South Gaulish. Die as on S10. *c*.AD 60-85. (3049; Area B, Event 2)

S12. [REG]INVS.F. Form 18/31, Central Gaulish. Die 2a of Reginus ii of Les Martres-de-Veyre[a] (Romeuf 2001, pl. 35, 136). The site record for the stamp includes Binchester, Catterick and Corbridge, and it appears in the Hadrianic fire in London. *c*.AD 120-150. (2111; Area A, Event 3)

S13. SACER.VASILF. Form 18/31, Central Gaulish. This die, perhaps the larger of the two with this reading illustrated by Terrisse (1968, pl. 54) is thought to represent the partnership of two potters, Sacer and Vasil-, who worked at Les Martres-de-Veyre[a]. The stamps are not closely dated, but a high proportion appears on form 18/31, suggesting a date *c*.AD 100-125. (3006; Area B, Event 3)

S14. TERTI.MA. Dish, possibly form 18 (R?), South Gaulish, with a stamp of Tertius of La Graufesenque[b] with swallow-tail ends (Polak 2000, T9). As the die is known from Caerleon, Heddernheim and Rottweil, it must have been in use into the Flavian period. *c*.AD 55-75. (5014; Area E, Event 3)

Broken and illegible stamps

S15. Form 27g, South Gaulish, with stamp crookedly impressed and smudged. Illegible. Probably Flavian (5002; Area E, Event 3)

S16. MA[. Form 42, South Gaulish. A stamp with similar lettering to that of Manduillus (no. 65 above), but in a swallow-tail frame. The form is not usually stamped. Flavian or Trajanic. (2285; Area A, Event 3)

S17.]INI.M or MA. Form 15/17 or 18, South Gaulish. Probably Flavian. (3018; Area B)

S18. O[, with a swallow-tail frame. Form 29, South Gaulish. A parallel to both frame and basal wreath (D59 above) suggests that this might be a stamp of Vitalis ii, though too little of the stamp survives for certain identification. The form indicates a date *c*.AD 70-85. (5002; Area E, Event 3)

Other table wares

Argonne colour-coated ware (ARG CC) (T&D, 47). A small group of eight sherds all from cornice-rim, roughcast beakers (Fig. 2.23. 80).

Central Gaulish colour-coat (CNG CC) (T&D, 52). Six small beaker sherds with roughcast decoration.

Cologne colour-coated ware (KOL CC) (T&D, 57). A single small sherd.

Campanian tableware (CAM WH). One whiteware rim, (Fig. 2.24 .108), and a bodysherd from a different vessel contain the black-sand temper characteristic of material from South Italy and are likely to be imported jugs or flagons.

Mortaria

North Gaulish whiteware (NOG WH4) (T&D, 77). Eleven sherds of North Gaulish mortaria are present

Gallia Lugdunensis, Rhone Valley (GLG OX) (T&D, 69) A single rim was recovered from pit [1008].

Amphorae

Baetican amphorae (BAT AM) (T&D, 84-5). This fabric originating from Baetica, Southern Spain is by far the commonest amphora fabric present in the assemblage. Most of the sherds appear to be from Dressel 20 globular olive oil forms, but there are in addition, several fragments from Haltern type 70 (Peacock and Williams 1986, class 15). In total Baetican sherds account for 4.8% of the assemblage by count, 26.6% by weight. Sherds of note include a stamped handle which has been reused along the long axis as a rubber or grinder (Fig. 2.21.12). One bodysherd from pit [4041] has a post-firing graffito (Fig. 2.25.G2). Another handle from [1018] has been ground down to a stump. The rim typology of the recorded Dressel 20's is consistent with a late 1st to 2nd-century date. Several sherds show signs of internal burning suggesting reuse, possibly as ovens.

Baetican Dressel 2-4. A double handle fragment from pit [5002] is typical of the Dressel 2-4 amphorae but is in a Baetican fabric. There are several production areas known to have produced this form including Southern Spain (Peacock and Williams 1986, 106).

Gallic amphorae (GAL AM) (T&D, 93-5). Gallic wine amphorae, represented almost exclusively by bodysherds, account for 0.7% of the total assemblage by count. 0.8% by weight.

Cadiz amphorae (CAD AM) (T&D, 87). Several sherds of South Spanish fish-sauce amphorae are present, probably all from Camulodunum type 186C (Peacock and Williams 1986, class 18) (Fig. 2.24.139). These amphorae were in circulation from around the Flavian period through to the early 2nd century.

Dressel 7-11 (P&W AM16) (T&D, 106). A single wide handle with two grooves is probably from a Dressel 7 or similar amphora. The fabric is quite coarse textured, white in colour with a greenish tinge. The paste contains a moderate to common frequency of rounded to sub-angular quartz and rare black grains. The vessel is likely to have come from a South Spanish source and probably used to transport fish-sauce.

Campanian amphorae (CAM AM) (T&D, 88). Two sherds with a classic black sand fabric are probably from Dressel 2-4 wine amphorae imported from Southern Italy. Vessels were imported into Britain from the later 1st century BC through to the mid 2nd century AD.

Rhodian amphorae (RHO AM1) (T&D, 112). Several sherds of Rhodian amphorae (Peacock 1977, fabric 1) are present including part of a peaked handle fragment from pit [4012].

North African amphora (NAF AM2) (T&D, 101-2). A single bodysherd with a brick-red sandy fabric containing a very sparse frequency of white limestone. The exterior surface has a white slipped surface with distinct vertical streaking. The sherd was recovered from pit [1019], phase 1B. Unfortunately it is not possible to identify which of the many types emanating from North Africa this piece belongs but, on the basis of its 2nd-century date, it could be from a Mana C dating from late 3rd century to 1st century AD (Williams and Carreras 1995, 233) or an Africano I or 'Piccolo'. A small number of North African vessels were imported into Britain during the Hadrianic-early Antonine period largely to sites with military connections such as Exeter, Caerleon, York and London (ibid). A handle from a Mana C type has already been documented at Chester (ibid) along with other odd bodysherds of North African types (ibid, appendix no. 7).

Regional imports

South-east Dorset black burnished ware (DOR BB1) (T&D 127). This is by far the commonest of the regional imports accounting for 6.5 % by count, 4.4% by weight of the assemblage. In terms of composition 57.7% of the vessels by EVE are jars (Fig. 2.23. 76-8; 2.24.125-6, 130), 36.6% flat-rim dishes and bowls (Fig. 2.22. 67, 90-1; Fig. 2.24.121, 131) and 2.5% are lids (Fig. 2.21. 16; Fig. 2.24. 135). One flat-rim bowl recovered from cleaning level (4006) has a name scratched into the upper surface after firing (Fig. 2.25 G3). The remaining 3.2% comprise plain-rim or beaded-rim dishes (Fig. 2.24. 132) with single examples of a grooved rim (Fig. 2.22. 66), a flanged-rim conical bowl and a miniature jar or mug (Fig. 2.24. 127). Of the jars 11.6% are short rim or beaded rim forms similar to Gillam (1976) types G30/G31. The chronological bias of the group thus very much lies in the early to middle 2nd century.

Lower Nene Valley colour-coated ware (LNV CC) (T&D, 118). A single sherd from an indented beaker was recovered from pit [2202] which may be intrusive. Such wares are not likely to date before the later 2nd century.

Pink grog-tempered ware (PNK GT) (T&D, 210). Limited to two storage jar sherds.

Severn Valley ware (SVW OX) (T&D, 148). A small group of 16 sherds, the only rim being from a flared rim jar. These wares have been documented from Chester appearing from the end of the 1st century through to the 4th century (Carrington 1977).

Verulamium-region white ware (VER WH) (T&D, 154). A moderately large group of 64 sherds mainly from flagons (eg Fig. 2.21. 15) and jars. One probable jar from pit [4041] has a sooted exterior (Fig. 2.21.11).

Verulamium-region white ware mortaria (VER WH) (T&D, 154). Accompanying the tablewares are a few sherds of mortaria including at least three stamped vessels. One vessel from pit [5020] is stamped with the name SOLLVS (Fig. 2.22. 35; M1). At least four dies are known belonging to this potter with products widely distributed over Britain. This one is probably the same die as one from Verulamium (Hartley 1972, fig 146. no 38), typologically dated to the period AD 70-100/110. The stamped vessel from pit [5004] carries the retrograde stamp BRVCIF, made by Bruccius or Brucius (Fig. 2.22. 61; M4). The die can be paralleled with an example from Verulamium (ibid. fig. 145, no 13). Several examples of his mortaria have been found across England and Wales including the kilns at Brockley Hill where he is presumed to have worked. Typologically his working period is dated to the period AD 80-100. A third stamped mortarium from pit [1019]

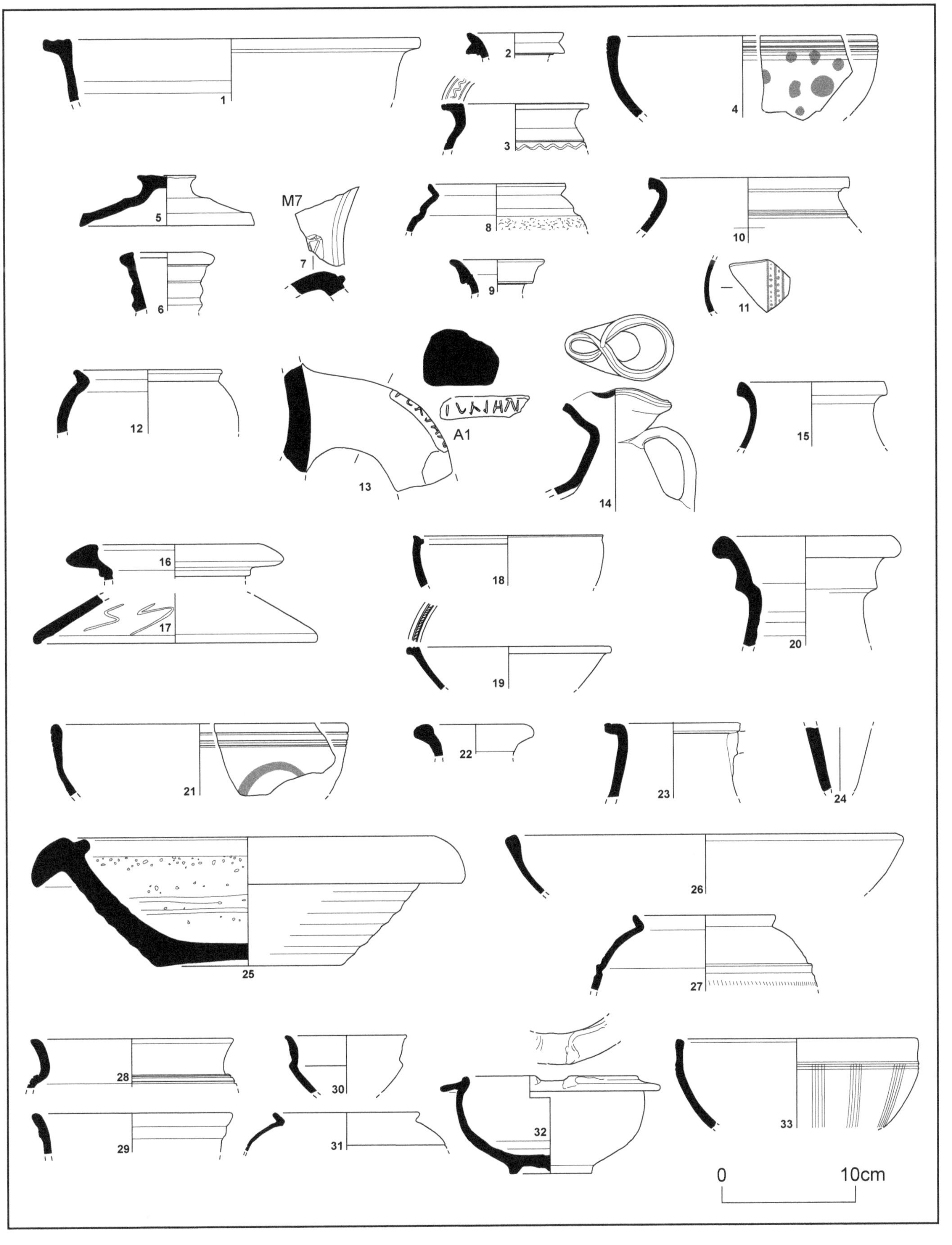

FIGURE 2.21 POTTERY FORMS 1-33.

is broken showing part of a single letter and a herring-bone border (Fig. 2.23. 84; M5) favoured by several of the Brockley Hill potters.

Wroxeter white ware mortaria (WRX WH) (T&D, 179). Three joining sherds from a creamish white ware mortaria (Fig. 2.22. 45; M6) stamped twice with the same stamp across the flange either side of the spout were recovered from pits [4002], [4012] and [4041].

CHESHIRE PLAIN (WILDERSPOOL, HOLT AND CHESTER)

The bulk of the assemblage is remarkably uniform and is typical of the range of wares traditionally linked with the known legionary production centre at Holt (Grimes 1930). However, it is clear that such wares were not exclusively from this single centre and that several known or suspected production centres exist in the Cheshire Plain producing very similar material not easy to distinguish at present either macroscopically or indeed petrologically (Williams 1982). Known kiln sites dating to the Trajanic-Hadrianic period have been documented at Wilderspool, Warrington (Hartley and Webster 1973), Northwich (Curzon and Hanson 1971) and from Chester itself (Swan 1984, fiche 1.239-41; Carrington 1977; Carrington and Ward 1982) (discussed below). As the products of these individual centres cannot for the most part be confidently distinguished, except with one or two exceptions, the wares are grouped generically by firing colour and/ or surface finish.

The wares discussed below have been divided into fine wares (glazed; mica-dusted; white eggshell ware; red painted ware; roughcast beaker); coarse wares; serving, cooking and table ware (oxidised, reduced and white-slipped ware) and mortaria.

Cheshire Plain fine wares

Glazed ware (Chester fabric 204). Four glazed sherds were recorded, one from pit [3012]; one from pit [3031]; one from quarry [3058] and one from cleaning (4006). The sherd from [3012] is probably from a beaker and has impressed oval dimples on the surface (Fig. 2.23. 83). The piece from [3031] is a much bubbled, over-fired sherd, perhaps a waster. The other pieces are plain but also from closed vessels. The sherds have a very clean fine fabric, dark grey or red-brown in colour, with a sparse frequency of fine, rounded quartz, visible at x20 magnification. The glazes are either pale greenish-brown, or almost whitish opaque. Lead glazed wares were produced at Holt (Grimes 1930, 175-8, fig. 76) in the period *c* AD 90/100-130, some imitating cut glass ware, others samian vessels. The sherd from [3012] may be copying the cut-glass style.

Mica-dusted oxidised ware (MICOX) (Chester fabrics 466-7). A large number of mica-dusted oxidised wares were recorded and the ware accounts for 2.3% by count of the total assemblage. The fabric is similar to the main oxidised class of wares (see below) but is distinguished by a thin golden micaceous slip. The vessel range is quite diverse but dominated by bowls and beakers. The former include hemispherical bowls (Fig. 2.22. 41) several with raised cordons and/or beaded rims (Fig. 2.22. 44; Fig. 2.24.106) and one with ridges (Fig. 2.24. 122); flat-rim bowls (Fig. 2.21.18). The beakers include folded and globular examples and at least one bobble-beaker (Fig. 2.22. 53; Fig. 2.23.79). Amongst the less common forms is a wine strainer (Fig. 2.23. 102), a tazza (Fig. 2.22. 47), jars with everted squared rims, a cup-mouthed flagon, a flanged-wall bowl, bowls copying samian form Drag 30, a lid knob and a spout or small tapered flask (Fig. 2.21. 23).

White eggshell ware (WWEGG), (Chester fabric 475). A small group of just seven sherds including at least two small bowls or cups (Fig. 2.23. 88). A similar fine white pottery was made at Holt mainly as small conical cups with rouletted decoration (Grimes 1930, 163; fig. 71.174).

Fine white ware (WWF). A small group of five sherds including a short everted rim globular beaker (Fig. 00. 30) and a necked bowl (Fig. 2.24.108).

Wilderspool oxidised red-painted ware (WIL OX) (T&D 122; Hartley and Webster 1973; Hartley 1981). A single bowl rim has been singled out as a possible Wilderspool product on the basis of the diagonal dark red-painted, burnished lines on the outer rim face (Fig. 2.24. 136). The fabric is quite fine and if undecorated would fall into the main oxidised group of wares. A close parallel can be found in Hartley (1981, fig. 29.1.10) from material in Warrington Museum, and from excavations at Wilderspool (Hinchcliffe and Williams 1992, fig. 421, 394).

Roughcast beaker (OXIDRC). A small group of 12 beaker bodysherds, some with a matt colour-coat and all with clay particles adhering to the outer surface to give a roughcast finish. The fabric is a fine sandy oxidised ware. The sherds are all likely to come from cornice-rim or squat, everted rim, globular or indented beakers. Such beakers are known to have been made at Wilderspool (Hartley and Webster 1973, 83 and fig. 4) (Chester fabric 450) and also figure amongst the material from Holt (Grimes 1930, fig. 71).

Cheshire Plain coarse wares

F*Grey (reduced) wares* (GREY) (Chester fabric 193). A large group of well-fired grey wares of slightly variable texture ranging from fine, smooth to slightly sandy. This category accounts for 39.7% of the assemblage by count, 25.6% by weight. The range of forms and decoration is diverse but dominated by jars and bowls/dishes. Jars effectively account for around 56.5% of the grey ware assemblage by EVE. There is a similarity in form between some of the larger beakers and the jars particularly sharply everted neckless forms and beaded rim forms making classification slightly arbitrary. The jar rims include rounded, beaded (Fig. 2.22. 59; Fig. 2.24. 140) examples, necked with everted squared or rounded simple rims (Fig.

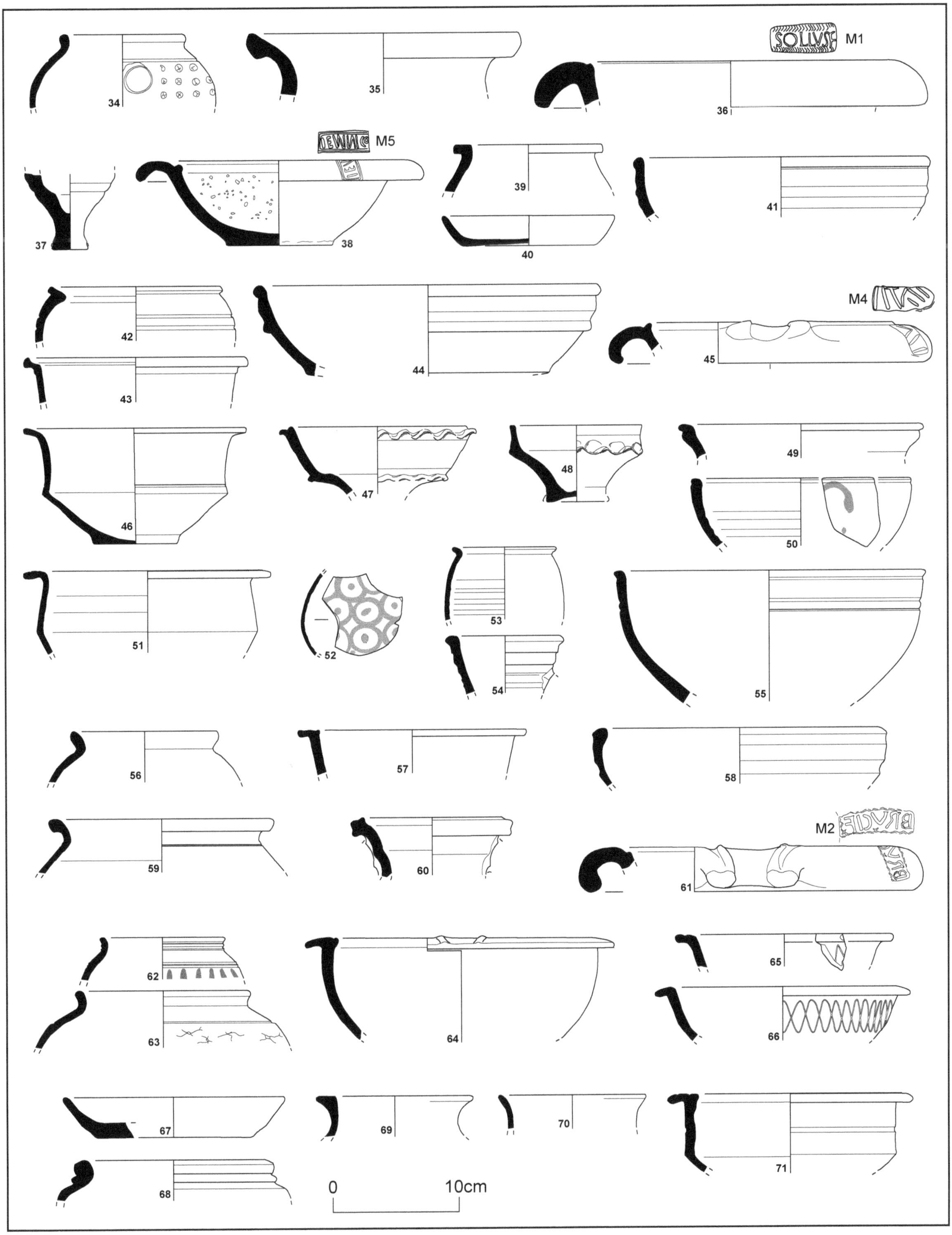

FIGURE 2.22 POTTERY FORMS 34-71.

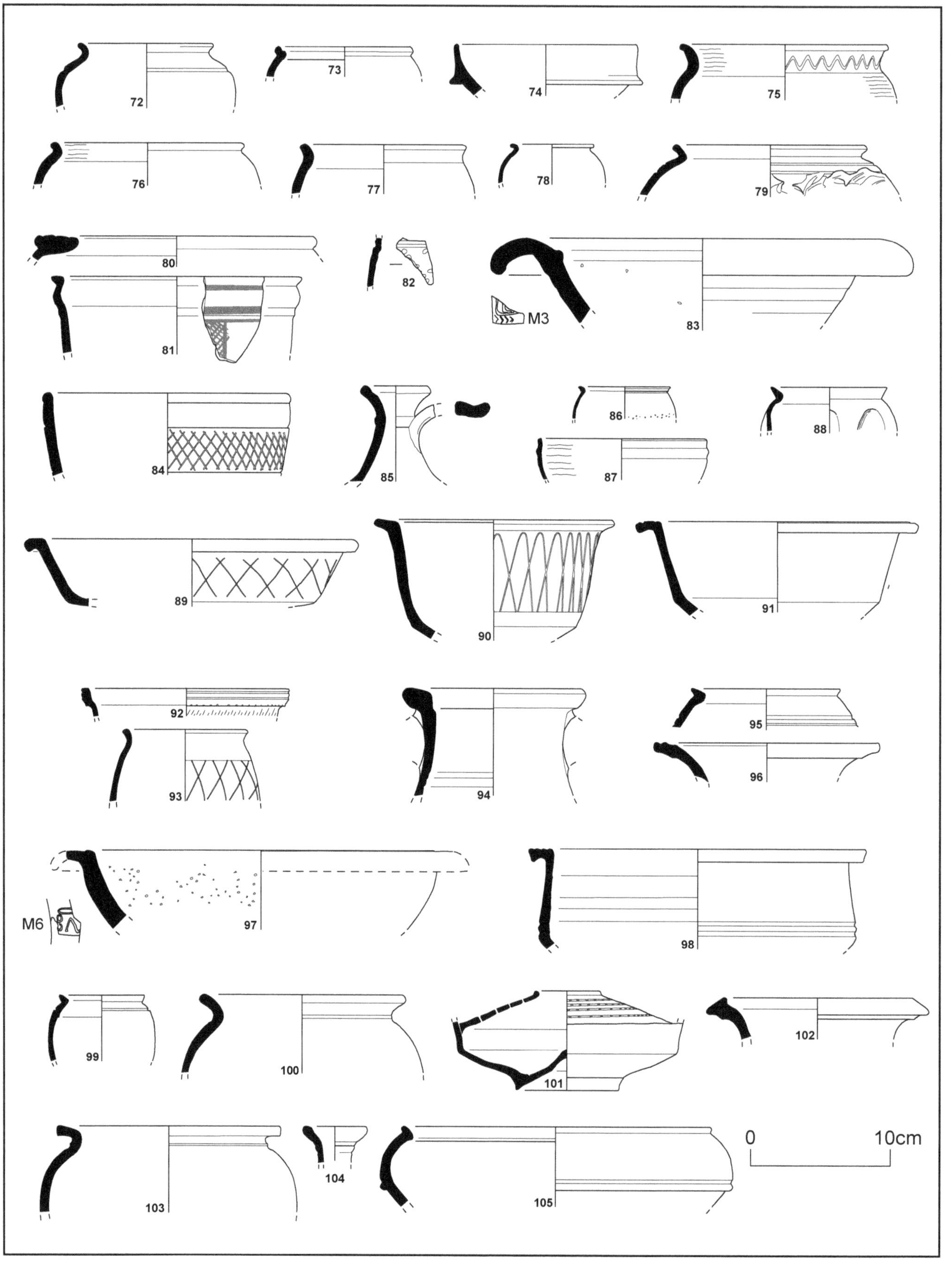

FIGURE 2.23 POTTERY FORMS 72-105.

2.22. 56; Fig 2.24. 116), necked with expanded rims (Fig. 2.24. 128), triangular rims, lid-seated, flared rim and flat-rim examples. Decoration includes cordons, rouletting, barbotine stripes, burnishing and rustication (Fig. 2.22. 64, Fig 2.23. 80; Fig 2.24 129). The latter form of decoration is very extreme and crude in some examples with large lumps of smeared clay, particularly, for example from [4041] which would suggest these are from seconds or wasters. In addition a few sherds are spalled either in production or from use. Bowls, and to a lesser extent dishes, account for 19.2% of the grey ware assemblage by EVE. Most of the bowls are either hemispherical in form (Fig. 2.22. 58), or with flat rims. The former are plain or ribbed, and there are examples with vertical, combed decoration (Fig. 2.21. 32) and rouletted decoration. The flat-rim forms are reeded or plain and where there is sufficient profile, either carinated (Fig. 2.22. 51, Fig. 2.23. 72, 99) or hemispherical (Fig. 2.24. 118). Other less common types include lid-seated (Fig. 00.43), hooked rim and flared rim forms (Fig. 2.21. 28) decorated with examples decorated with incised lattice (Fig. 2.23. 85) and spaced, horizontal, burnished lines. A single carinated bowl with stamp-impressed decoration featured amongst the unstratified material (Fig. 2.24.142). Similar, but not identical stamp decorated vessels featured in the Holt assemblage (Grimes 1930, fig.77). Beakers account for 11.6% by EVE to which can be added 1.6% jar/beaker. Rims are mainly sharply everted, beaded, slightly devolved cornice rim, squat triangular or lid-seated (Fig. 2.21. 7, Fig 2.22 42, 62, Fig 2.23. 73-4, Fig. 2.24. 99-100). Decoration includes barbotine stripes, crescents, rings or dots, vertical rusticated stripes, roughcasting and white painted stripes. Also included is a much finer grey ware globular jar/beaker decorated with grey barbotine rings and dots (Fig. 2.21. 33). The fourth commonest grey ware form is the lid which accounts for 8.6% EVE (Fig. 2.21. 4, Fig. 2.22. 63; Fig. 2.24. 133). Several flat-topped lid knobs are also present. The remaining 2.5% comprise flasks, flagon and a cup (Fig. 2.21. 29). The flagons include ring-necked (Fig. 2.22. 54) and expanded inverted triangular-rim forms.

Black wares (BW). This is a minority group separated out from the grey wares but probably belonging to the same production. The black wares account for 0.3% by count of the total assemblage. Vessels include beakers with rouletted decoration (Fig. 2.21. 26), lids, everted rim necked jars and a bowl. Also present is a detached omphalos base probably from a hemispherical bowl.

Oxidised wares (OXID). This group is the oxidised equivalent of the grey wares with most of the vessels fired to a bright orange colour. As with the reduced wares the texture varies from fine through to slightly sandy. Although not as prolific as the grey wares this group still accounts for a significant proportion of the assemblage making up 19.2% by count, 16.2% by weight. The breakdown of vessels differs from the grey wares in that there are proportionately less jars and beakers but more bowls and flagons. Bowls account for 34.2% EVE with a broadly similar range of types to those in the grey wares. Hemispherical plain rim (Fig. 2.22 .55, Fig. 2.24. 141) and flat-rim carinated or hemispherical bowls (Fig. 2.22. 40, 57, Fig. 2.23. 92-3) are the commonest, some of the former with raised cordons or ridging. Similar forms are known from Holt (eg Grimes 1930, fig.74.226). Two vessels have shallow applied ridges to form spouts (Fig. 2.21. 31, Fig 2.22. 65). Less common types include hooked rim, flared rim (Fig. 2.22. 46), campanulate (Fig. 2.22. 49) and lid-seated examples (Fig. 2.21. 17). Jars form the second commonest category accounting for 28.5% EVE with everted rim, triangular-rim, flat-rim and short everted, neckless forms (Figs 2.21.9, 14, 27, Fig. 2.22. 34, 38, Fig. 2.23. 69-70, 81, 101 and 104). Decoration includes rare examples with rustication (Fig. 2.24. 138), orange barbotine, roughcast stripes, incised wavy line (Fig. 2.21. 2) and burnished lattice (Fig. 2.23. 94). Flagons form 21.9% EVE and are dominated by ring-necked forms, usually with an expanded upper ring. Less common are pinch-mouthed (Fig. 00.13), trefoil-mouthed and flared rim forms (Fig. 2.23. 97). A large double-handled flagon with a slightly bifid rim came from pit [5007] (Fig. 2.24. 117). Beakers account for 6.7% EVE with examples of globular and folded vessels (Fig. 2.23. 89, Fig. 2.43. 112, 123) and one with rusticated decoration (Fig. 2.24. 137). Lids account for 5.3% and include some examples with sooted rim edges from use (Fig. 2.24. 134). Also present is a lid knob waster from [1019]. The remaining 3.4% comprise tazzae (Fig. 2.22. 48), cups (Fig. 2.23.75), dishes (Fig. 2.21. 25, Fig. 2.22. 39, 68), a flask and jugs (Fig. 2.21. 22), two with pulled spouts. Other unusual items include a nozzle (Fig. 2.24. 109), a blind spout from [5007] and a base with a graffiti incised before firing (Fig. 2.25. G4) (see Tomlin below)

White painted oxidised ware (OXIDWPT). This small group of wares were distinguished on the basis of deliberate white painted decoration but in terms of fabric belong to the oxidised class. Nearly all the decorated vessels are hemispherical bowls and the decoration is in the form of blobs or roughly painted rings (Fig. 2.21. 3, 20, Fig. 2.22. 50). One vessel from [4041] is decorated with both red and white paint. Also present is a small jar or beaker with ring-and-dot decoration over the surface (Fig. 2.22. 52). A similar style can be observed on material from Wilderspool (Hartley 1981, fig.29.1.21).

White-slipped wares (WSOXID/WSGY). In this group of wares, mainly oxidised, the upper visible surfaces are covered with a usually quite thin, white slip. Just two white-slipped grey ware sherds were present which may be firing mistakes. White-slipped wares account for 9.7% by count, 7.1% by weight. Flagons make up 88.4% of the vessel range in this ware by EVE. Whilst the bulk of these are ring-necked types (Fig. 2.21. 5, Fig 2.24. 119-20), also present are examples of pulley-wheel (Fig. 2.21. 1), beaded (Fig. 2.24.113), collared (Fig. 2.21. 8, 19), reeded (Fig. 2.22. 60), conical- or cup-mouthed (Fig. 2.23.86, Fig 2.24. 105) and triangular-rimmed types. Both single (Fig. 2.24. 110) and double-handled forms are present (Fig. 2.23. 95, Fig. 2.24. 107). One rim from [2283] is a much

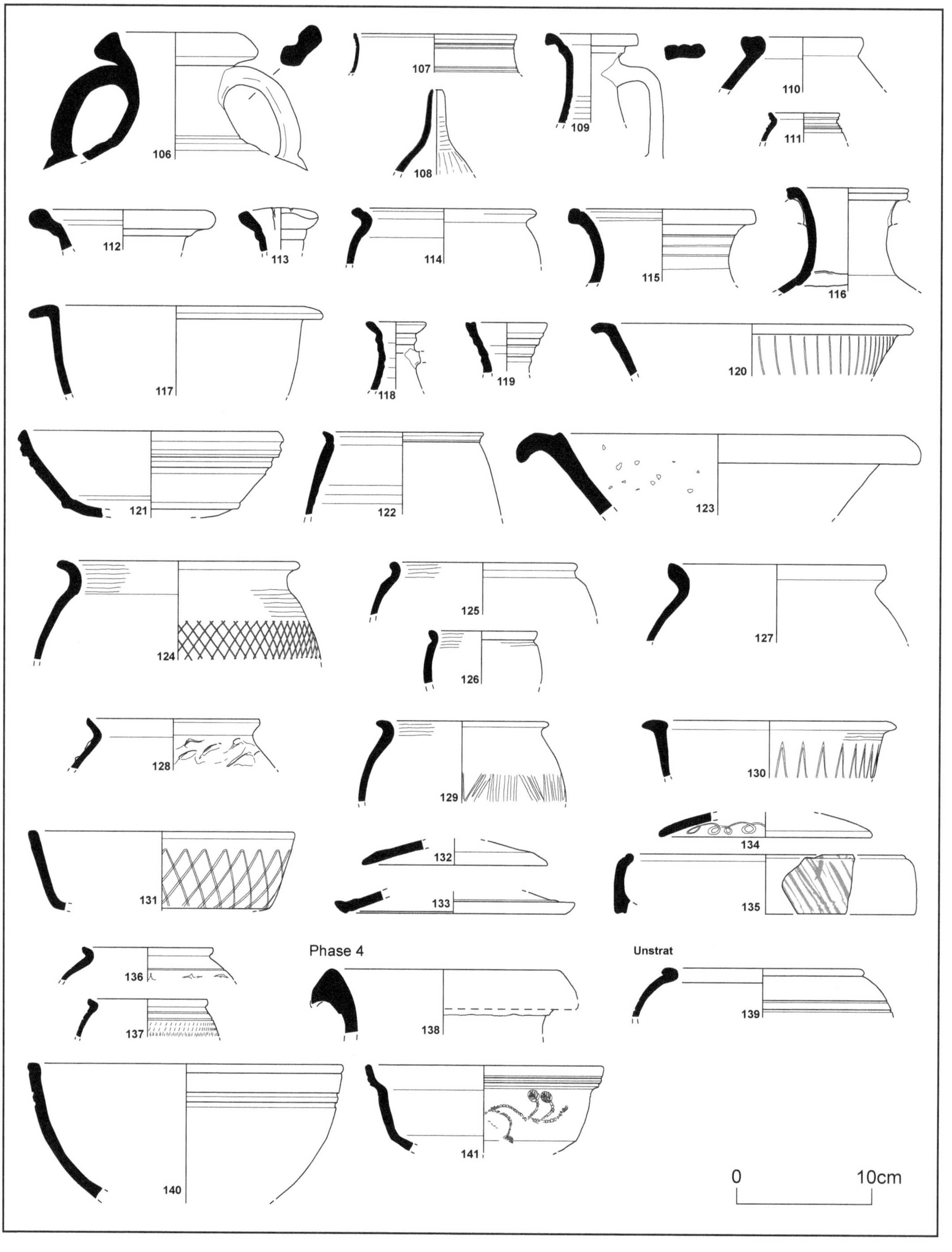

FIGURE 2.24 POTTERY FORMS 106-141.

distorted waster (Fig. 2.24. 114). Jars contribute a further 5.2% and bowls 4.4% by EVE to the group. Other vessels include a beaker with rouletted decoration (Fig. 2.24. 138), and single examples of a tazza, cup and lid. Other than the rouletted beaker the only other decorated pieces are from a vessel with combing through the slip.

White wares (WW). A small group of plain white wares may also belong to this Cheshire Plain group of material representing a continuum from the finer white wares. This group makes up only 0.5% by count of the assemblage and includes rims from a hemispherical bowl, a lid-seated jar (Fig. 2.24. 115) and two flagons (Fig. 2.21. 21, Fig 2.24. 103). The footring of a very large flagon came from [4002]. A single beaker sherd is decorated with red-orange vertical tramline and dot decoration (Fig. 2.21. 10).

Cheshire Plain mortaria

Holt oxidised and white-slipped mortaria (HOL OX/ HOL WS) (T&D, 207-8) (Chester fabrics 139, 530-2). Twenty oxidised sherds and two white-slipped oxidised mortaria sherds have been identified as probable Holt products. These are distinguished by fine, sandy fabric, very slightly micaceous, with a sparse scatter of coarse, quartzite, trituration grits which places them apart from the other mortaria found at the site. All the featured vessels have a hooked flange and a shallow bead (Fig. 2.24. 124).

Oxidised mortaria (MORTOX; MORTOX1-5). Classifying the mortaria has proved problematical as the sherds showed little similarity to the examples in the Chester fabric series. The main criteria for dividing the sherds has been on the basis of the trituration grits which appear to fall into four distinct groups (MORTOX1-4) although the group as a whole may be from one source. The same distinctions are made for the white-slipped examples which appear to replicate the same range of grits. Where the grits are absent or worn away the sherds have been classified as MORTOX (or where a white-slip is present MOWSOX), a dark orange ware of slightly variable texture ranging from fine to slightly sandy. One flange from [3031] has traces of white painted decoration. The slip varies in thickness from thin to a creamy white.

MO(WS) OX1: Fourteen sherds fall into this group. The sherds are very hard and well-fired, with an orange surface and occasionally a blue-grey inner core. The inner surface of the vessels have been scored roughly concentrically and a range of mixed coarse grits added which include rounded quartz, angular quartzite, and rare fragments of sandstone, flint, dark grey argillaceous grits, ?feldspar and occasional black cindery inclusions. One vessel from [3012] with a joining sherd from [3008] has a well-cut and impressed stamp across the rim E\/\I\/\Ico (Fig. 2.22. 37 and M5). A second vessel with a complete profile came from pit [5013] (Fig. 2.21. 24). A single white-slipped example shares the same type of grit (MOWSOX1).

MO(WS)OX2: A single oxidised sherd from pit [3012] with a sparse scatter coarse (up to 3 mm) grits comprising white quartz/quartzite and red-brown, rounded, fine grained ?sandstone inclusions. A further five white-slipped sherds were noted.

MO(WS)OX3: This is the most distinct of the group in that the trituration grits contain a scatter of sparkling gold (biotite) mica up to 1 mm in size mixed with rounded quartz, feldspar, brown argillaceous grains and black cindery inclusions. The maximum grit size is 2 mm with most falling below this. The fabric is well-fired, oxidised with a grey inner core. At x20 the paste is well-levigated with a sparse frequency of rounded quartz > 0.5 mm and rare angular fine black grains. Two oxidised sherds and 41 white-slipped sherds were noted in this fabric. One flange fragment from pit [2111] shows the edge of a potter's stamp with just the letter 'V' extant (Fig. 2.23. 98; M7).

MO(WS)OX4: A similar range of grits to MORTOX1 but with a higher percentage of fine, brown, argillaceous inclusions and a generally denser, finer grade of grit mainly 1-2 mm in size. Other inclusions include rounded to sub-angular quartz / quartzite, rounded quartz and rare ?flint. There are eight oxidised and twenty-two white-slipped examples in this group.

MORTOX5: A small group of six sherds have a slightly paler, brownish orange fabric with a slightly darker core. The fabric contains a moderate frequency of rounded, orange-brown ?ferruginous pellets up to 1 mm in size, occasional fine voids and rare grains of rounded quartz. The trituration grits comprise a mixture of sub-angular white and clear quartz/quartzite up to 3 mm across and rare rounded pale orange brown ?ferruginous grains. One sherd from pit [4041] has the edge of a stamp (Fig. 2.21. 6; M3).

Petrological analysis by Williams (1982) of a mortarium waster from the Chester ring-road identified the trituration grits as being composed of quartz quartzite, feldspar, sandstone and some fine-grained material which might broadly correlate with fabrics MO(WS)OX 1, 2 and 4 above.

LOCAL ?UNKNOWN WARES

Buff / pale wares (BUFF/PALE). A small group of buff or cream wares. Vessels includes examples of a flat-rim and hemispherical bowls, several jars and two lids.

Buff micaceous ware (BUFFMIC). A single, distinctly micaceous, buff ware sherd.

BB1 copies (BB1COPY). A small group of 19 sherds in a black sandy ware copying BB1 jars and flat rim bowls. This includes the wavy line around the neck of the jars.

Black-slipped grey ware (BSGREY). Six sherds only. Only one jar rim.

Fine black ware (BWF). A small group of 19 sherds in a fine black ware, some sherds with a dark reddish brown, fine fabric, with black polished surfaces including flared-rim bowls (Fig. 2.22.71), short squat rim beakers and a beaded rim jar.

Grey/oxidised ware (GYOX). This group of wares may represent underfired grey wares but collectively form a distinct group of quite sandy wares with grey exterior and oxidised interior surfaces. Forms include jars, flat rim bowls and lids. At least one sherd has rusticated decoration.

Grog-tempered wares (GROG). A slightly mixed group with examples of grey, black, brown sandy, pink with a black exterior wares all with a sandy fabric containing a sparse frequency of grog. The brown sandy types can be paralleled with Chester fabric 172. Most of the sherds are from jars and some have a finely rilled exterior. One grey grog vessel from pit [4012] is decorated with a burnished lattice.

Coarse-rock-tempered handmade wares (GYIGN; CWQZ). Two coarsely tempered body sherds are present, one with fragments of igneous rock from pit [4041]; the other with quartzite from [3038].

Limestone-tempered ware (LIMEOX/GY). A small group of three sherds, one oxidised, two reduced with fine sparse voids on the surfaces from leached limestone. Vessels include a lid-seated jar and a squat everted rim beaker.

Micaceous grey ware (GYMIC). A small group of three sherds including a beaded rim jar.

Unclassified mortaria (MORTWH; MORTBUFF). Five sherds of unclassified mortaria, two whitewares and three buff wares; the latter small fragments. One of the whitewares is a basesherd with coarse mixed grits.

Table 2.3: Summary of the main pottery forms present by EVE

	Form	EVE	EVE%
Finewares	dish	112	0.6
	bowl	106	0.5
	cup	65	0.3
Amphorae	amphorae	351	1.8
Tablewares/drinking	flagon	3425	17.4
	jug	47	0.2
	flask	35	0.2
	cups	99	0.5
	beakers	1633	8.3
Serving/cooking	jars	7566	38.5
	jar/beaker	193	1.0
	bowl/dishes	4250	21.6
	mortaria	657	3.3
	lids	1063	5.4
Specialised	tazzae	55	0.3
TOTAL		19657	100.0

Fine oxidised ware (OXIDF). Seven very fine oxidised wares, probably from a source on the Cheshire Plain. Featured sherds include two lids and a beaker.

Sandy oxidised ware (OXIDSY). A small group texturally distinct from the main group of oxidised ware, with example of a sharply everted rim jar, lid and the base of an unguent flask (Fig. 2.22. 36).

Shelly ware (SHELL). Six shelly ware sherds were recorded of which one was rilled. At least one sherd resembles Midlands late Roman shelly ware which, if correct, would suggest some late Roman intrusive material present.

Brown painted whiteware (WWBPT). A single very distinctive, hard fired, white pimply ware. The sherd is from a bowl copying samian form Dr 30 and is decorated with a reddish brown painted pattern (Fig. 2.23.82).

Forms

Table 2.3 gives a summary of the main forms present by EVE. There is quite a diverse range of forms with a spread of fine wares and tableware connected with the serving/ dispensing, or consumption, of food and liquids, and coarse wares linked with food preparation and cooking (jars, bowls, mortaria and lids). In addition there are tazzae possibly for more specialised purposes. Although jars dominate at 38.5% this is not an excessive amount. Significantly absent in the repertoire from the rims are storage jars. The main forms encountered are necked, everted rim forms with simple or expanded rims, beaded rim, triangular-rimmed, flat-rim and sharply everted neckless types. Amongst the latter are several with rusticated decoration. In terms of fabric 68.8% of the jars are in the grey wares, with almost equal amount of examples in oxidised fabrics and in DOR BB1 at 14% and 13% respectively. White-slipped examples contribute a further 1.8% with the remaining 2.4% distributed across other wares.

The second commonest group are the coarse ware bowls/ dishes at 21.6%. These are dominated by two main forms, the curved hemispherical bowl with a simple or beaded rim with either plain walls, or with cordons, which contribute around 31% of the bowl repertoire, and various permutations of the flat-rim bowl with plain or reeded rims and a straight or carinated body which effectively account for *c* 50% of the bowls. In terms of fabric, 39.9% occur in grey ware, 30% in oxidised ware, 15.7% in DOR BB1, 6% in micaceous oxidised ware and 2.7% in white-slipped ware.

The third commonest form is the flagon at 17.4%. This group is very much dominated by the ring-necked flagon, usually with an expanded upper ring which accounts for *c* 65% of the group. Most of the flagons are in white-slipped fabrics, 66.9% EVE with a further 23.8% in oxidised ware and 4.6% in grey ware. Verulamium examples contribute 2% EVE.

Beakers account for 8.3% EVE of the assemblage with the overwhelming majority, 63.8% EVE, made in grey ware. Oxidised examples make up 16%, mica-slipped ones 10.9% and white-slipped versions 1.9%. Continental imports contribute just 4.3%. Lids make up 5.4% of the total assemblage and again most of these are made in the local grey ware, 71.6%. A further 21% occur in oxidised ware and 4% in DOR BB1. The remaining 8.8% EVE comprise a range of types of which mortaria are the most prominent at 3.3%. On the basis of the fabrics 29% of the mortaria are non-local and this includes 16.6% from the Verulamium region.

The rest of the assemblage includes fine (samian) table wares, jugs, flask, cups and tazzae. Conspicuously absent are open or closed lamps, with only one small fragment of a closed moulded lamp noted. Other unusual vessels present which do not feature rims, include the wine strainer (Fig. 2.24. 102), an unguent flask and a feeding bottle or similar (Fig. 2.24. 109). The wine strainer is of a form similar to Marsh (1978) type 45.1 and is a form known to have been produced by the Holt potters (Grimes 1930, fig. 73, 215-6). The form has continental origins (Marsh 1978, 181) and is thus likely to be made by immigrant / military potters.

In conclusion, for most of the commoner vessel classes although a range of firing colours and surface finishes feature, the local grey ware was used more frequently than the other wares. The main exception is the flagons which most commonly occur in the white-slipped fabric. Given that it is non-local DOR BB1 makes a very significant contribution to the jar, bowl and lid classes.

Chronology

Most of the pottery assemblage came from the large quarry pits. The Roman pottery assemblage is remarkable for the homogeneity of the fabric range which appears to have quite a tight chronological range in from perhaps the later 1st century through to the early-middle 2nd century. Unfortunately it is notoriously difficult to sub-divide coarse ware pottery dating from within the later Flavian-Trajanic period (*c* AD 80-120), as forms and fabrics remain fairly uniform. The more refined dating therefore relies very much on the samian. It could be argued that this might have a slightly more extended period of use compared to coarse ware and therefore could be expected to have quite a few pieces of Flavian date still in circulation in the Trajanic and Hadrianic periods. It would be difficult to determine whether a discarded vessel of Flavian date is actually residual, or dating the context. A complete absence of other residual 1st-century fine ware imports of the type associated with the military fortresses at Kingsholm, Usk and Wroxeter is likely a reflection of the later date of foundation for the Chester fortress. Excavations at Goss Street within the fortress (Richmond and Webster 1950) did not appear to produce any pottery very different to the Delaware Street group. Unfortunately there are no published quantified assemblages from within the legionary fort for more direct comparison. Other wares are perhaps more likely to indicate the later rather than earlier end of the AD 80-120 date range, for example, the introduction of locally made fine wares such as mica-slipped and glazed wares and the appearance of DOR BB1. The following discussion is split into the allocated events in default of any clear ceramic divisions.

Phase 1: Event 1

The contexts allocated to Event 1 produced a very small assemblage of some 21 sherds weighing 1146 g. The group comprises some eight sherds of samian (South and Central Gaulish); six sherds of Baetican amphora including the grooved handle of a Haltern type 70; two mortaria (Holt and Cheshire Plain) and four sherds of coarse-ware. The only featured sherd is a rim from a ring-necked flagon in a white-slipped oxidised ware dating to the Flavio-Trajanic period.

Event 2 (Table 2.4)

The backfilling of the quarries produced some 4248 sherds of pottery weighing c 89 kg and with 67 EVE's. The individual groups from eleven of the larger pits are summarised in Table 2.4. These range from 106 sherds in pit [1019] to a maximum 2573 sherds from pit [4041].

In Area B pit [3030] produced 152 sherds (2164 g). The samian assemblage is quite high at 9.9% and suggests a Trajanic date. Oxidised wares account for 24.3% with 56.6% grey/ black wares. A single worn Verulamium mortarium also came from this feature but no DOR BB1. Pit [3030] was truncated by [3031] which yielded 226 sherds (2740 g). This group was slightly more fragmented compared to [3030], with an average sherd weight of 12.1 g, compared to 14.2 from pit [3030]. A later date might also be inferred by the presence of four sherds of DOR BB1 one of which is a plain-sided dish, a form which does not generally appear in southern assemblages until the mid 2nd century. Pit [3031] has one of the highest percentages of samian of any individual feature, 14.4% by count. It also contained several ring-necked flagons, a white-painted mortarium and a glazed sherd which was either a waster or had been burnt as the glazed had bubbled.

Further substantial groups were recovered from pits [3038/56] (Fig. 2.22. 38-9), [3058/37] and [3040] in this area (Table 2.4). These tend to show a slightly lower incidence of samian and other imports and a commensurate higher proportion of local grey, oxidised although DOR BB1 remains fairly high. The pottery from pit [3012] (Table 2.4) (Fig. 2.22. 36-7), although a moderately small group with just over 100 sherds, shows a particularly high percentage of samian (15.7%).

Pits [4002] and [4041] (Fig. 2.21. 1-23) in Area C produced 194 and 2573 sherds respectively. The assemblages are quite similar although [4041] contained a higher percentage of samian ware, 10.6% compared to 4.6% on sherd count. The fragmentation rates are also similar

TABLE 2.4: EVENT 2 FEATURES WITH LARGE POTTERY ASSEMBLAGES

Event 2 pits	1019		3012		3030		3031		3040		3038/56	3058/37
Wares	No%	Wt%	No%	Wt%	No%	Wt%	No%	Wt%	No%	Wt%	No%	Wt%
Samian	6.6	2.7	15.7	6.7	9.9	4.2	14.4	10.2	1.6	0.4	2.8	1.3
other finewares	0.0	0.0	0.0	0.0	0.0	0.0	0.5	0.1	0.0	0.0	0.0	0.0
North Gaulish mortaria	0.0	0.0	0.0	0.0	0.0	0.0	0.0	0.0	0.0	0.0	0.0	0.0
Baetican amphora (all)	10.4	34.5	7.5	27.8	2.0	25.2	1.9	27.6	0.0	0.0	4.8	16.5
Cadiz amphorae	0.0	0.0	0.0	0.0	0.0	0.0	0.5	2.4	0.0	0.0	0.0	0.0
Other amphorae	0.9	1.2	0.7	0.2	0.0	0.0	0.0	0.0	0.0	0.0	0.0	0.0
Gallic amphorae	0.0	0.0	0.0	0.0	0.7	1.2	0.0	0.0	1.6	0.7	0.8	3.0
Dorset BB1	3.8	1.5	5.2	2.9	0.0	0.0	1.9	0.9	8.0	4.7	16.1	9.4
Severn Valley ware	11.3	0.8	0.0	0.0	0.0	0.0	0.0	0.0	0.0	0.0	0.4	0.7
Verulamium white ware	0.0	0.0	0.0	0.0	40.0	0.0	0.0	0.0	0.0	0.0	1.2	6.3
?Wroxeter mortaria	0.0	0.0	0.0	0.0	0.0	0.0	0.0	0.0	0.0	0.0	0.0	0.0
British glazed ware	0.0	0.0	0.0	0.0	0.0	0.0	0.5	0.2	0.0	0.0	0.0	0.0
mica-slipped	2.8	2.3	3.0	0.6	0.7	0.1	0.5	0.4	3.2	3.2	0.0	0.0
grey /black wares	25.5	14.6	37.3	32.5	56.6	43.3	55.1	29.5	51.2	46.8	39.1	27.4
oxidised ware	20.8	14.9	14.9	10.0	24.3	22.8	13.9	20.1	17.6	23.5	26.6	25.5
white-slipped	15.1	16.7	11.9	12.5	5.3	3.1	8.3	5.8	14.4	11.4	4.4	4.5
whiteware	0.0	0.0	0.7	0.1	0.0	0.0	0.0	0.0	0.8	0.2	0.0	0.0
Holt mortaria	0.0	0.0	0.0	0.0	0.0	0.0	0.0	0.0	0.0	0.0	0.0	0.0
Cheshire Plain mortaria	0.0	0.0	2.2	5.6	0.0	0.0	0.5	0.4	0.0	0.0	0.4	1.7
Misc coarseware	0.0	0.0	0.0	0.0	0.0	0.0	2.3	2.4	0.8	0.9	1.6	1.1
Misc mortaria	2.8	10.6	0.7	1.0	0.0	0.0	0.0	0.0	0.8	8.1	0.0	0.0
Shelly ware	0.0	0.0	0.0	0.0	0.0	0.0	0.0	0.0	0.0	0.0	0.0	0.0
TOTAL	100.0	100.0	100.0	100.0	100.0	100.0	100.0	100.0	100.0	100.0	100.0	100.0
TOTAL SHERDS	106	2133	134	3436.5	152	2164	226	2740	125	1354	248	4476

and noticeably higher than the Area B pits with average sherd weights of 21.6 g and 20.3 g respectively. Sherds of note from [4002] include another waster or burnt sherd, an oxidised ware, one jar and two flat-rim bowls in DOR BB1, a roughcast decorated beaker in CNG CC and a sherd of eggshell white ware. Pit [4041] had a large quantity of amphorae including examples from Baetica (Dressel 20 and Haltern 70), Cadiz (Cam. 186), Italy (Dressel 2-4) and Gaul. In addition to the samian imported fine wares include a cornice rim roughcast beaker from Argonne and the mortaria a few sherds from North Gaul. DOR BB1 is well represented with several flat-rim bowls and short everted rim jars and a single lid. Other regional imports include Verulamium mortaria including one sherd with a stamp. Joining sherds (Fig. 2.22. 45) were recovered from pits [4002] and [4012]. The local grey wares include a waster with rusticated decoration. Also present were sherds from a mica-slipped bobble beaker, a pinch-mouthed flagon in oxidised ware and several white-slipped oxidised ring-necked flagons. Pit [1019] (Table 2.4) produced a small assemblage of 106 sherds again dominated by Flavio-Trajanic coarse wares although samian and CBM suggest a slightly later date of deposition.

In Area E pit [5013] (Fig. 2.21. 24-31) produced a moderately large assemblage of 395 sherds (19.5 kg) with several quite large sherds of Spanish amphorae and mortaria slightly skewing the overall sherd weight which is 49.3 g. The grey wares include a number of sherds with rusticated decoration. Apart from a small amount of Verulamium ware regional imports appear to be absent from this group. Pit [5020] with a small assemblage of just 69 sherds produced a Verulamium mortarium stamped by the potter Sollus who has a production period of around AD 70-100. Pit [5004] (Table 2.4) also shows an earlier phase of deposition with 119 sherds dominated by local grey wares (46.5%). The DOR BB1 includes jars with acute latticing, flat-rim bowls/ dishes and a single plain-walled dish. Samian levels are high at 11.7% with the latest sherds dating to the Hadrianic-early Antonine period.

Event 3 (Tables 2.5 and 2.6)

This event relating to waste discard not surprisingly produced the greatest amount of pottery, some 6212 sherds weighing c 129 kg and with 110.34 EVEs. Two pits on the south-eastern corner of Area A produced small assemblages of pottery [2097] and [2259] with just 22 and 11 sherds respectively. A larger assemblage was recovered from the upper fills of pit [2209/2241/2291] with a total 231 sherds (Fig. 2.22. 79; Fig 2.24. 106). The percentages of samian and Baetican amphorae are quite high but the assemblage is dominated by reduced grey or black local wares. Of particular note is the presence of an indented LNV CC beaker from [2209], the only example recorded from the site, which probably dates to the mid-later 2nd century and a tiny fragment of lamp from [2291]. Of the series of small refuse pits along the eastern edge of [2209] small groups of pottery were recovered from [2125], [2127], [2129], [2165] and [2208] (Fig. 2.24. 107). The coarse wares are not particularly distinguishable from previous groups but the samian hints at a mid 2nd century date.

Pit [2204], Area A (Table 2.5) singled out as a discrete feature yielded a modest assemblage of 186 sherds (3151 g). Although this contained samian dated to the Flavian-

4002		4041		5004		5013			
No%	Wt%	No%	Wt%	No%	Wt%	No%	Wt%	No%	Wt%
8.3	3.8	4.6	1.2	10.6	4.7	11.9	3.1	0.5	2.3
0.0	0.0	1.0	0.0	0.1	0.0	0.0	0.0	0.0	0.0
0.0	0.0	0.0	0.0	0.3	1.5	0.0	0.0	0.0	0.0
10.0	32.8	4.6	37.4	4.8	25.6	6.9	53.3	7.1	12.8
0.0	0.0	0.5	0.5	0.5	1.8	0.0	0.0	0.3	1.3
1.5	4.2	0.0	0.0	0.5	1.1	0.0	0.0	0.3	0.5
0.6	0.2	1.0	3.2	0.5	0.7	0.0	0.0	0.0	0.0
1.5	1.1	3.6	3.2	3.1	2.4	4.0	1.9	0.0	0.0
0.0	0.0	0.0	0.0	0.1	0.1	0.0	0.0	0.0	0.0
1.3	1.9	3.6	1.5	1.8	3.4	0.0	0.0	0.3	0.4
0.0	0.0	0.0	0.0	0.1	0.6	0.0	0.0	0.0	0.0
0.2	0.1	0.0	0.0	0.0	0.0	0.0	0.0	0.0	0.0
0.6	0.5	1.5	1.0	2.6	1.4	2.0	1.3	0.0	0.0
47.5	30.3	52.6	32.3	46.5	30.0	46.5	27.2	62.8	27.4
16.7	13.7	15.5	12.2	18.9	14.2	18.8	9.7	19.2	15.2
7.3	4.3	7.7	3.1	7.2	8.5	8.9	3.3	3.0	27.3
0.0	0.0	1.5	2.0	0.6	0.6	0.0	0.0	1.3	0.3
0.2	0.5	0.0	0.0	0.4	2.0	0.0	0.0	0.0	0.0
1.7	4.4	1.0	0.8	0.4	0.9	0.0	0.0	1.5	11.3
2.7	2.1	0.5	0.6	0.7	0.3	0.0	0.0	3.8	1.1
0.0	0.0	0.5	0.7	0.2	0.1	0.0	0.0	0.0	0.0
0.0	0.0	0.0	0.0	0.1	0.1	1.0	0.1	0.0	0.0
100.0	100.0	100.0	100.0	100.0	100.0	100.0	100.0	100.0	100.0
483	8603	194	4194	2573	52284	101	4055	395	19487.5

Trajanic period, it also contained a DOR BB1 flat-rim bowl, which if not intrusive, cannot be earlier than Trajanic placing it at the very start of the import period. Interestingly the group produced largely grey wares, 72.6% by count, but very little oxidised ware (4.8%). White-slipped wares contributed 5.9% count mainly comprising ring-necked flagons. Samian is also quite prominent accounting for 5.9% by count although only 2.2% by weight. Amphorae make up 8.4% by weight.

The series of wide shallow pits truncating the north edge of [2209] all produced moderately small groups of pottery apart from [2292] which produced some 419 sherds (Table 2.5) (Fig. 2.24. 112-14). The sherds are well fragmented with an average weight of just 10.3 g and are dominated by local oxidised, reduced and white-slipped wares. DOR BB1 is well represented at 15.5% count, mainly flat-rim bowls. Samian at 10.3% is also well-represented. A sherd of an imported Italian white-ware jar came from pit [2225] (Fig. 2.24. 111). Sub-circular pit [2110] (Table 2.5) also produced a large assemblage of 732 sherds (Fig. 2.23. 85-104). The range of fabrics and forms is very similar to [2292] with samian at 10.7%, DOR BB1 at 11% and a dominance of local wares. The latter, in addition to the usual jars and flat-rim bowls and dishes witnessed from other deposits, has a single example of a plain-walled dish perhaps indicative of a slightly later date in the 2nd century. There are several sherds of white eggshell ware including a cup (Fig. 2.23. 88) and imported cornice rim beakers from Central Gaul and the Argonne (Fig. 2.23. 87).

Pottery was also recovered from pit [3012] (Table 2.5) (Fig. 2.24.115) with 506 sherds with some links observed between layers. Again the samian betrays a high level of residuality in its chronological range with the latest sherds dated to the Hadrianic- early Antonine period. Local wares again dominate and jars contribute 46% of the vessel range by EVE followed by bowls/ dishes at 19% EVE. The DOR BB1 includes a beaded rim, plain-walled dish.

In Area C a further 836 sherds belonging to this event was recovered from pit [4041] (Table 2.6) with nearly half the sherds being grey wares. These are accompanied by several sherds of Dressel 20 and Cam. 186 amphorae. Regional imports are dominated by DOR BB1 which mainly features jars and a single lid. Another large group came from [4002] (Table 2.5) (Fig. 2.22. 44-8) with 579 sherds. Of note from this group are four sherds of Rhodian amphorae and several pieces of Baetican amphora some with burnt interior surfaces suggesting re-use. Reduced wares account for just over 46% of the group. It is suggested from the large amount of food waste (animal bone) and other artefacts recovered from pits [4002] and [4041] that some of the waste may have come from the kitchen area of the fort. An examination of the range of forms present from each pit on the basis of recorded rims shows some differences from the overall site figures (Table 2.1). Pit [4002] has the greatest diversity of form with most of the defined types present. Jars account for 41% (EVE) which is broadly comparable to the site percentage of 38% (Table 2.1). Beakers, tazzae and mortaria are all slightly higher whilst flagons or less well represented. The biggest difference lies in the percentage of bowls/ dishes which account for 34% of the [4002] assemblage compared to the overall site figure of 21.6%. The assemblage from [4041] (Fig. 2.22. 50-60) is less diverse with no examples

TABLE 2.5: EVENT 3 PITS WITH LARGE POTTERY ASSEMBLAGES 1

Event 3 pits	1019		2110		2204		2209/2241/2291	2292	
Wares	No%	Wt%	No%	Wt%	No %	Wt %	No%	Wt%	No %
Samian	9.3	4.2	10.7	5.9	5.9	2.2	4.7	3.5	10.3
other finewares	0.4	0.1	0.5	0.1	0.0	0.0	0.0	0.0	0.0
Imported mortaria	1.0	2.8	0.1	0.2	0.0	0.0	0.9	0.0	0.0
Baetican amphora (all)	7.2	35.5	2.1	9.8	0.5	8.4	21.8	34.2	0.7
Cadiz amphorae	0.0	0.0	0.4	0.2	0.0	0.0	0.0	0.0	0.0
Other amphorae	0.4	0.7	0.3	2.1	0.0	0.0	0.8	0.3	0.5
Gallic amphorae	2.9	1.8	0.1	0.5	0.0	0.0	0.1	2.3	0.2
Dorset BB1	8.5	4.0	11.0	9.9	2.2	1.2	4.3	7.8	15.5
Severn Valley ware	0.0	0.0	0.1	0.1	0.0	0.0	0.0	0.0	0.0
Verulamium white ware	0.2	1.2	0.1	0.2	0.0	0.0	0.0	0.0	0.5
?Wroxeter mortaria	0.0	0.0	0.0	0.0	0.0	0.0	2.4	0.0	0.0
British glazed ware	0.0	0.0	0.0	0.0	0.0	0.0	0.0	0.0	0.0
mica-slipped	2.3	1.0	1.8	2.2	0.5	0.7	3.0	3.0	0.0
grey /black wares	36.0	23.6	25.2	20.4	72.6	66.5	33.5	11.5	5.0
oxidised ware	18.0	11.1	30.7	28.1	4.8	0.9	14.6	21.1	22.4
white-slipped	10.7	8.0	10.7	15.2	5.9	7.4	6.7	7.3	19.3
whiteware	1.2	0.9	2.1	1.0	0.0	0.0	0.2	0.0	16.7
Holt mortaria	0.4	1.9	0.0	0.0	0.0	0.0	0.0	0.8	0.0
Cheshire Plain mortaria	1.0	2.9	0.8	2.1	2.2	3.3	5.2	5.6	0.0
Misc coarseware	0.2	0.2	3.2	2.0	5.4	9.5	1.8	2.4	0.5
Misc mortaria	0.0	0.0	0.0	0.0	0.0	0.0	0.0	0.0	8.4
Shelly ware	0.2	0.2	0.0	0.0	0.0	0.0	0.0	0.0	0.0
TOTAL	100.0	100.0	100.0	100.0	100.0	100.0	100.0	100.0	100.0
TOTAL SHERDS	484	10638	732	13994.5	186	3151	231	4744	419

TABLE 2.6: EVENT 3 PITS WITH LARGE POTTERY ASSEMBLAGES 2

Event 3 pits	4041		5004		5007	
Wares	No%	Wt%	No%	Wt%	No%	Wt%
Samian	10.8	4.9	8.7	4.6	3.5	4.1
other finewares	0.0	0.0	0.0	0.0	0.0	0.0
Imported mortaria	0.1	0.7	0.0	0.0	0.0	0.4
Baetican amphora (all)	4.7	35.3	2.4	12.9	34.2	40.5
Cadiz amphorae	0.6	1.2	0.0	0.0	0.0	1.7
Other amphorae	0.1	0.1	0.3	1.1	0.3	0.0
Gallic amphorae	0.5	0.8	0.5	0.6	2.3	0.6
Dorset BB1	3.1	1.7	10.6	6.9	7.8	8.9
Lower Nene Valley cc	0.0	0.0	0.0	0.0	0.0	0.0
Severn Valley ware	0.0	0.0	0.0	0.0	0.0	0.0
Verulamium white ware	1.9	1.6	0.7	5.4	0.0	0.1
?Wroxeter mortaria	0.0	0.0	0.0	0.0	0.0	0.1
British glazed ware	0.0	0.0	0.0	0.0	0.0	0.0
mica-slipped	1.7	1.0	0.3	0.1	3.0	1.5
grey /black wares	49.3	31.8	46.2	28.3	11.5	11.7
oxidised ware	17.8	12.6	21.0	24.8	21.1	15.1
white-slipped	7.9	5.9	8.0	5.6	7.3	3.2
whiteware	0.7	0.4	0.2	0.0	0.0	0.1
Holt mortaria	0.1	0.5	0.3	1.2	0.8	0.0
Cheshire Plain mortaria	0.5	1.3	0.9	8.5	5.6	10.1
Misc coarseware	0.2	0.3	0.0	0.0	2.4	0.7
Misc mortaria	0.0	0.0	0.0	0.0	0.0	1.2
Shelly ware	0.0	0.0	0.0	0.0	0.0	0.0
TOTAL	100.0	100.0	100.0	100.0	100.0	100.0
TOTAL SHERDS	836	16185	587	16227	579	18002

3012		4002		4012		
Wt %	No%	Wt%	No%	Wt%	No%	Wt%
8.6	9.3	8.1	9.5	3.9	8.7	4.7
0.0	0.0	0.0	0.2	0.0	0.0	0.0
0.0	0.0	0.0	0.0	0.0	0.2	0.9
2.5	5.9	23.2	7.1	37.0	2.1	21.8
0.0	0.0	0.0	0.3	0.4	0.0	0.0
3.9	0.4	1.4	1.2	1.2	0.4	0.8
0.1	0.4	0.5	0.2	0.5	0.2	0.1
10.6	8.1	7.2	3.3	2.4	3.1	4.3
0.0	0.0	0.0	0.0	0.0	0.0	0.0
0.9	0.4	2.3	1.0	0.9	0.2	0.0
0.0	0.0	0.0	0.2	1.6	0.2	2.4
0.0	0.2	0.0	0.0	0.0	0.0	0.0
0.0	2.0	1.7	3.6	3.8	3.7	3.0
4.5	36.6	22.8	46.1	24.1	49.4	33.5
19.5	22.9	16.9	16.6	13.9	17.0	14.6
23.1	12.1	10.4	8.6	4.4	8.9	6.7
11.0	0.4	0.4	0.3	0.2	0.4	0.2
0.0	0.2	0.6	0.2	0.6	0.0	0.0
0.0	1.0	3.8	0.9	4.6	1.5	5.2
0.9	0.0	0.0	0.5	0.2	3.9	1.8
14.4	0.2	0.7	0.2	0.2	0.0	0.0
0.0	0.0	0.0	0.0	0.0	0.0	0.0
100.0	100.0	100.0	100.0	100.0	100.0	100.0
4294	506	8388	579	12807	482	8921

of mortaria or tazzae rims. Jars account for 40.2% similar to pit [4002] and again bowls are particularly prominent at 51.9%. Although samian is well represented on the fabric totals the number of vessel rims present contributes less then 1% by EVE.

Area C smaller refuse pits [4012, 4045, 1018 (Fig. 2.22. 40-1), 1019 (Fig. 2.23. 80-4)] all produced pottery but only [4012] (Fig. 2.24. 108) and [1019] produced sizeable groups (Table 2.4) with 482 and 484 sherds respectively. A mortarium spout from [4012] joins fragments from pits [4004] and [4046] may suggest a common source for the rubbish perhaps a midden or disturbance of already deposited material during the pit digging. Both groups conformed to the pattern seen elsewhere with moderately high percentages of samian around 9%, several amphora including a Rhodian handle from [4012] and high percentages of local grey, oxidised and white-slipped local ware. The samian again reflects a potentially high level of redeposition but with a small number of Hadrianic-early Antonine pieces from [1019] and Hadrianic sherds from [4012] which provide the dating evidence. The latest coarse ware from [1019] is a single sherd of Midlands pink-grog-tempered ware usually seen as appearing in the later 2nd century. Jars dominate the forms at 46.4% EVE followed by bowls at 20% and flagons at 13.6%.

Pit [5004] in Area E continued to be backfilled with waste in Event 3 with a further 587 sherds (Table 2.6) (Fig. 2.22 and 2.24. 61-78). If pit [5004] is analysed in more detail it shows subtle differences compared to, for example, pit [4002]. As with [4002] and [4041] jars dominate at 38.7% and bowls/ dishes come second at 22.9% significantly less than [4002] and [4041]. In [4002] the proportion of hemispherical bowls is much higher in [4002] accounting for around 47.3% of the bowl group with 31% flat/reeded-rim bowl. In [5004] the proportion of flat or reeded-rim bowls is 58% of the bowls with just 20% hemispherical bowl. Beakers are also more prolific in [5004] accounting for 19% of the assemblage by EVE compared to 4.9% from [4002]. In the later the forms are nearly all simple everted or cornice-rim, whilst the former are predominately globular forms with short everted rims.

Other larger groups from Area E include those from pit [5007] with 579 sherds (Table 2.6) (Fig. 2.24. 116-38) and [5004] with 587 sherds (Table 2.6). Pit [5007] has a lower incidence of samian (3.5%) but a very marked number of amphora sherds mainly from globular Dressel 20's and Haltern 70's both used for transporting olive-oil. DOR BB1 is well represented in both these pits at 7.8% and 10.6%. One marked difference lies in the amount of grey wares in [5007] which are particularly low at 11.5% with oxidised wares accounting for 21.1%. By contrast grey wares account for 46.2% of the assemblage from [5004].

Group 113 other features

Linear ditch [2130] produced a small assemblage of 37 sherds weighing 405 g. The sherds are thus quite fragmentary with an average sherd weight of 10.9 g and contain few featured pieces. Six pieces of South Gaulish samian of Flavian date are present but there is also one sherd of DOR BB1, which if not intrusive, suggests an early 2nd century date at the earliest. There are almost equal numbers of grey and oxidised ware, the only other pieces being two possible scraps of Gallic amphora. Featured sherds are limited to two jars and one bodysherd with vertical, rusticated decoration. The fragmentary nature of the sherds might indicate a level of residuality present.

Gully [2244] contained a single grey ware bodysherd and perpendicular ditch [2200/2252] 46 sherds mainly unfeatured coarse wares including DOR BB1. Of the four refuse pits cutting this ditch only [2197] and [2301] produced pottery and this only amounted to nine very small sherds including further 2nd-century DOR BB1 suggesting that most or all this material is redeposited.

Summary of pit assemblages

The site has produced a very large rich assemblage of pottery with a moderately short chronological span. As a consequence the analysis of the pottery has produced a large amount of data which can be manipulated in a number of different ways. It is possible to detect some variations between the different pit assemblages in terms of proportions of wares or forms but it is difficult to assess what these differences might mean; they could be chronological; indicate different sources of waste within the fort; redistribution of midden material or other

TABLE 2.7: WARES BY EVENT

	Event 1		Event 2		Event 3	
Wares	No%	Wt%	No%	Wt%	No%	Wt%
Samian	38.1	4.5	8.9	4.0	9.3	4.6
other imported finewares	0.0	0.0	0.1	0.0	0.1	0.0
Imported mortaria	0.0	0.0	0.1	0.6	0.1	0.4
Imported coarseware	0.0	0.0	0.0	0.0	*	0.0
Baetican amphora (all)	33.3	77.2	5.2	25.5	4.7	26.5
Cadiz amphorae	0.0	0.0	0.3	1.2	0.2	0.5
Other amphorae	0.0	0.0	0.4	1.0	0.3	0.8
Gallic amphorae	0.0	0.0	0.6	0.8	0.6	0.6
Dorset BB1	0.0	0.0	3.5	2.3	8.6	6.1
Lower Nene Valley cc	0.0	0.0	0.0	0.0	*	0.0
Midlands pink grog-tempered	0.0	0.0	*	0.0	*	0.0
Severn Valley ware	0.0	0.0	0.3	0.1	*	0.0
Verulamium white ware	0.0	0.0	1.1	2.3	0.6	1.3
?Wroxeter mortaria	0.0	0.0	0.1	0.4	*	0.3
British glazed ware	0.0	0.0	*	0.0	*	0.0
lamp	0.0	0.0	0.0	0.0	*	0.0
mica-slipped	0.0	0.0	1.5	1.0	2.7	2.0
grey /black wares	9.5	1.2	46.8	31.7	37.9	23.9
oxidised ware	0.0	0.0	18.1	15.9	21.2	17.7
white-slipped	9.5	3.8	10.0	7.0	9.6	7.5
whiteware	0.0	0.0	0.5	0.4	0.7	0.3
Holt mortaria	4.8	6.5	0.2	0.9	0.2	1.1
Cheshire Plain mortaria	4.8	6.6	0.9	4.3	1.2	4.7
Misc coarseware	0.0	0.0	0.9	0.5	1.7	1.3
Misc mortaria	0.0	0.0	0.1	0.0	*	0.0
Shelly ware	0.0	0.0	0.1	0.0	*	0.0
TOTAL	100.0	100.0	100.0	100.0	100.0	100.0

*= present

factors. There is clearly a high level of residuality which may be masking other patterns. Table 2.7 summarises the main wares from the three defined events. The groups from Events 2 and 3 are generally very close. The higher incidence of DOR BB1 in Event 3 may be a chronological feature as is the only sherd of LNV CC from the site.

DISCUSSION

The assemblage from Delamere Street is remarkable for a number of reasons: it is a very well preserved assemblage of some quality; chronologically it is very limited with negligible Roman pottery later than the later 2nd-early 3rd century; it has a very significant quantity of samian present and it intimates local kiln production which is likely to be of legionary origin. This raises some interesting questions in terms of how this material relates to the XX Legionary pottery production site at Holt located some 12 km to the south, and the civilian kiln complex at Wilderspool both active at the same time.

The assemblage is very much dominated by a group of wares which must have been produced locally either from Holt, or more likely, from Chester itself. The group is mixed in that whilst there are a small number of clear waster pieces present along with several over-fired sherds indicating pottery production nearby, many of the discarded vessels show traces of use, wear and repair indicating domestic refuse. It would seem therefore, that the stone quarries formed a useful dumping ground for domestic and possible industrial waste.

Swan (1984, fiche 1.239-41) documents two instances where quantities of wasters and under-fired and over-fired potsherds have been found in Chester, one at City Road and one at St Anne/George St both recovered from salvage work during road construction in 1970-1. The pottery is described as white-slipped oxidised wares including ring-necked flagons, mortaria and bowls dating to the later 1st-early 2nd century. It was suggested that the finds came from kilns on the fringe of the adjacent *vicus* but no definite structural evidence has yet been identified. In an unpublished paper, Carrington and Ward (1982) detail the evidence from these salvage finds. The pottery is described as being made from iron-rich clay containing quartz sand typical of that found in the Cheshire Plain and in all cases was fired grey and in several cases with a white slip.

In addition to the Holt kilns a number of civilian kilns have been investigated at Wilderspool, Warrington (Hartley and Webster 1973; Swan 1984). Subsequent reassessment of the dating of the site suggests a production period spanning *c* AD 90-160 (Webster 1991, 13). Both sites were producing an identical range of forms to one another and to the range of vessels documented from Delaware Street. A petrological study of some of the Chester kiln wasters and local wares from Chester, along with samples of allied forms from Holt and Wilderspool was undertaken by David Williams (Williams 1982). The samples from all three sources were petrologically very similar with the same range of inclusions, highlighting potential problems in trying to attribute material in this area.

Although few assemblages of pottery have been published in detail from Chester the Delaware assemblage appears to be broadly similar to a smaller group of 408 sherds from a site in City Road (Garner 2005, table IV.3) also located outside the fortress walls. Samian accounted for 18% of this group, Dorset black burnished ware contributed 6% of the assemblage and Dressel 20 amphorae 2.2%. In the local range of wares oxidised wares are far more common than grey wares and make up 65% of the Roman assemblage. Further groups of pottery from the site of the mansion the south of the fortress (Ward and Carrington 1981) have been analysed by minimum vessel estimates to track changes in supply through time. The authors saw three potential customers; the military garrison, the staff of official buildings such as this and the civilians at the *canabae*. Superficially the range of material for the period 100-30 looks quite similar to the Delaware assemblage and distinguishing between these three customers may not be very clear cut from a pottery perspective.

Trade and supply

Table 2.8 summarises some comparative data from three other towns which developed out of military fortresses, Wroxeter, Gloucester and Carlisle. One link between these sites is *Legio* XX, perhaps initially based at Kingsholm, Gloucester in the Neronian period and subsequently moved to Wroxeter, probably via Usk and thence to Chester and eventually Carlisle. At Gloucester *Legio* XX was replaced by *Legio* II to construct the city centre fortress from where the data on Table 2.8 is derived. Legionary pottery production is attested at both Kingsholm and Gloucester city (Timby 1991), at Wroxeter and Holt. In each case, as far as is possible, the comparative data is drawn from later 1st-early to mid 2nd-century phases. The letter 'p' indicates a ware is present but in too small a quantity to register as a percentage.

At Chester continental fine ware imports are very much dominated by samian table wares which effectively account for 9.3% of the assemblage by sherd count. 4.4% by weight. This is quite a high percentage but completely in keeping with that to be expected from a military town. At Wroxeter samian formed 16% count, 9% weight of the Period 2 (later 1st-mid 2nd century) assemblage (Faiers 2000, table 4.20) (Table 2.8). Gloucester at around the same time has around 6.2% (based on unpublished figures from Berkeley Street), whilst at Carlisle, for an amalgamation of periods 5 and 6 (Trajanic-mid Hadrianic) (Taylor 1985, tables 41-4), it formed 17.5% on sherd count, 6.1% by weight.

Surprisingly perhaps there is only a very small quantity of other continental fine ware present in the Chester assemblage and this is limited to a few beaker sherds from the Argonne, Central Gaul and Cologne and two white ware jug/ flagon sherds probably from Italy. This is in great contrast to some of the more southern-based urban centres which were receiving much larger numbers of beakers from these sources in the early 2nd century, for example Silchester and London, but is not that dissimilar to Wroxeter, Gloucester and Carlisle at this time which all have a low count in terms of colour-coated wares. At Delaware Street the amphorae are very much dominated by Baetican vessels transporting olive-oil, particularly the Dressel 20 globular form. There are however, a number of other types present from Southern Spain, Italy, France, North Africa and the Aegean transporting products such as olive oil, garum and wine. Imported continental mortaria are not common with eleven sherds of North Gaulish white ware and two sherds of Rhone Valley type representing the continental imports.

In terms of other military towns (Table 2.8) all the sites were receiving small amounts of North Gaulish mortaria and significant quantities of Baetican amphorae. Other amphora types are difficult to assess as these have not always been identified. Wroxeter has a marked presence of Fishbourne type 148.3 not noted at Chester whilst both Gloucester and Wroxeter have sherds of the *Camulodunum* 189 carrot amphora. Both Rhodian and Gallic wine amphorae are also noted as present at Wroxeter and Gloucester. Small quantities of Rhone Valley and Aoste mortaria are recorded from Wroxeter and Rhenish mortaria from Carlisle and Wroxeter.

In terms of the regional mortaria there are a few vessels from the Verulamium region and possibly a small number of pieces from Wroxeter at Chester. Carrington (1981, 465) notes that almost all the mortaria reaching Chester in the 1st century are from North Gaul or the Verulamium region. In the early 2nd century most of the vessels in use come from the Cheshire Plain kilns and by the 3rd century the main supplier was Mancetter-Hartshill. Significantly no examples of the latter are present in the Delaware Road assemblage. After locally made types Verulamium-region mortaria were the second commonest type present in the Period 2 assemblage at Wroxeter (Faiers 2000, 266) with

TABLE 2.8: COMPARATIVE DATA FOR WARES FOUND AT ROMAN MILITARY SITES

			Chester	Wroxeter	Gloucester	Carlisle				
	Code	Description	No %	Wt %	No%	Wt%	No%	Wt%	No%	Wt%
IMPORTS	SAMIAN	South Gaulish samian	9.4	4.4	16.0	9.0	6.2	3.7	17.5	6.1
	ARG CC	Argonne colour-coat	0.1	0.0	0.0	0.0	0.2	0.4	0.0	0.0
	CNG CC	Central Gaulish colour-coat	0.1	0.0	0.9	0.3	0.0	0.0	0.0	0.0
	KOL CC	Cologne colour-coat	0.0	0.0	p	p	p	p	0.6	0.3
	FWIMP	other fineware imports	0.0	0.0	0.6	2.0	0.2	0.1	0.6	0.2
mortaria	NOG WHM	North Gaulish mortaria	0.1	0.6	p	p	0.4	6.3	0.3	2.7
	GLG OX	Gallia Lugdunensis (Rhone Valley)	0.0	0.1	p	p	0.0	0.0	0.0	0.0
	Rhineland	Rhineland mortaria	0.0	0.0	p	p	0.0	0.0	0.2	1.5
amphorae	BAT AM	Baetican amphora (all)	5.2	29.6	1.2	6.6	0.4	4.6	3.5	32.4
	CAD AM	Cadiz amphorae	0.2	0.8	0.0	0.3	0.0	0.0	0.0	0.0
	CAM AM	Campanian fabric ?Dr 2-4	0.0	0.1	0.1	0.1	0.2	0.6	0.0	0.0
	Cam 189	Camulodunum type 189 (carrot)	0.0	0.0	0.2	0.2	p	p	0.0	0.0
	Dressel 7	Dressel 7? South Spain	0.0	0.1	p	p	0.0	0.0	0.0	0.0
	Fishbourne 148.3	Amphora	0.0	0.0	1.7	3.2	0.0	0.0	0.0	0.0
	GAL AM	Gallic amphorae	0.7	0.9	0.1	0.2	p	p	0.0	0.0
	RHO AM1	Rhodian	0.1	0.2	p	p	p	p	0.0	0.0
	NAF AM	North African amphora	p	p	0.0	0.0	0.0	0.0	0.0	0.0
	AMP	unassigned amphorae	0.2	0.5	2.6	7.5	0.2	1.1	1.0	3.0
REGIONAL	DOR BB1	Dorset black burnished ware	6.6	5.0	11.0	10.0	7.1	7.1	3.7	2.9
	LNV CC	Lower Nene Valley colour-coat	p	0.0	p	p	p	p	0.0	0.0
	PNK GT	Midlands pink grog-tempered	p	p	0.0	0.0	0.0	0.0	0.0	0.0
	MAL RT	Malvernian	0.0	0.0	1.0	1.4	0.2	0.2	0.0	0.0
	SVWOX?	Severn Valley ware	0.1	0.1	6.0	7.0	11.3	18.6	0.0	0.0
	VER WH	Verulamium white ware	0.8	0.6	p	p	p	p	0.0	0.0
Local fine	GLAZE	British glazed ware	0.0	0.0	0.0	0.0	0.0	0.0	0.0	0.0
	MICOX	mica-slipped oxidised ware	2.3	1.7	p	p	1.5	1.4	0.0	0.0
	OXIDRC	roughcast oxidised/colour-c ware	0.1	0.0	0.0	0.0	p	p	0.0	0.0
Local cw	GREY	grey /black wares	40.3	28.9	31.0	27.0	45.4	32.5	71.8	45.3
	OXID	oxidised ware incl finer fabrics	19.1	18.3	19.0	17.0	0.0	0.0	0.0	0.0
	WSOXID	white-slipped oxidised ware	9.7	8.0	0.0	0.0	18.6	12.4	0.0	0.0
	WW	whiteware	0.5	0.4	5.0	5.0	0.0	0.0	0.0	0.0
mortaria	Local mortaria	Local mortaria	1.2	0.4	0.0	0.0	0.0	0.0	0.7	4.2
	OTHER	other coarsewares and mortaria	2.9	0.0	3.4	3.0	7.8	10.8	0.2	1.3
TOTAL			**99.9**	**100.0**	**99.7**	**99.6**	**99.8**	**99.9**	**100.0**	**100.0**

very small quantities noted from Gloucester. Examples of mortaria from the West Midlands and Gloucester were present in the Carlisle assemblages (Taylor 1991).

Of the regional imports Dorset black burnished ware is far the most significant at Chester accounting for 6.6% by sherd count of the total assemblage. Vessels mainly comprise jars, bowls and lids. It is quite likely that small quantities of DOR BB1 were being distributed in the Trajanic period. At Castle Street, Carlisle, the ware appears for the first time in Period 5 dated to *c* AD 105 on dendro-chronological and other grounds, where it accounts for 2.8% by count, 4.2% EVE of the pottery in use (Taylor 1991, 384 and table 41). Webster suggested a pre-Hadrianic date for the earliest occurrence of the ware at Caersws, Powys (Webster 1989). It seems likely that there was a small amount of material being distributed from the Trajanic period with the bulk of the wares coming in from the Hadrianic period. At Wroxeter it formed 11% by count of the Period 2 assemblage which dates broadly from the later 1st century – mid 2nd century (Faiers 2000, table 4.20). At Gloucester it contributed around 7.1% in the early colonia period at the Berkeley Street site and collectively for periods 5-6 at Carlisle it accounted for 3.7%

Other regional wares are very sparse. Odd sherds of Lower Nene Valley colour-coated ware and Verulamium white wares were reaching Chester, Wroxeter and Gloucester. The latter two sites, as might be expected were receiving higher numbers of Severn Valley wares, 6% at Wroxeter and 11.3% at Gloucester compared to less than 1% at Chester.

In all cases the assemblages are dominated by local coarsewares with a mixture of oxidised (orange) and reduced (grey) wares. In terms of form composition the range of vessels is very similar as they are at many other sites at this time with flanged, carinated and hemispherical bowls, everted rim rusticated jars, ring-necked flagons and flanged mortaria. At Wroxeter, like Chester, the grey wares dominate at 31% compared to 40% at Chester. Oxidised wares account for around 19% at both sites. At both Gloucester and Carlisle the both firing colours have been subsumed into one or more fabric codes, although at the former white-slipped wares were distinguished and appear slightly more prolific at 18.6%. At Gloucester, and probably Wroxeter, kilns were established during the military occupation in the Flavian period, if not earlier, and continued to flourish into the Trajanic-Hadrianic periods. The same range of wares, oxidised, reduced and white-slipped made in the Flavian period were joined in the early 2nd century by mica-slipped wares and other experimental types such as the glazed wares. A similar repertoire is seen at Holt.

It would appear, therefore, that these sites on the western side of Britain were receiving a remarkably similar range of continental and regional traded wares. All appear to have been provided with an almost identical complement of coarse wares, both in firing colours, surface finishes and forms and most of this seems to have been produced within or peripheral to the towns for local consumption. The uniformity of the form repertoire and the typology of the vessels suggests potters with some military connections who probably originally came from the continent in the employ of the army and perhaps set up civilian enterprise to supply the expanding settlements.

FOUR GRAFFITI SHERDS *by Roger Tomlin*

A number of sherds displayed signs of graffiti which indicate different characteristics. Graffiti G1, G2 and G3, were made after firing, and thus relate to ownership. G4 was made *before* firing, and thus relates to manufacture (Fig 2.25).

G1, 4015, Pit [4041], Area C. Body sherd of a south-Spanish oil amphora (Dressel 20), from the shoulder like most ownership-inscriptions on Dressel 20.

(i) Two vertical incisions, that to the right perhaps being an uncompleted first attempt at (ii).

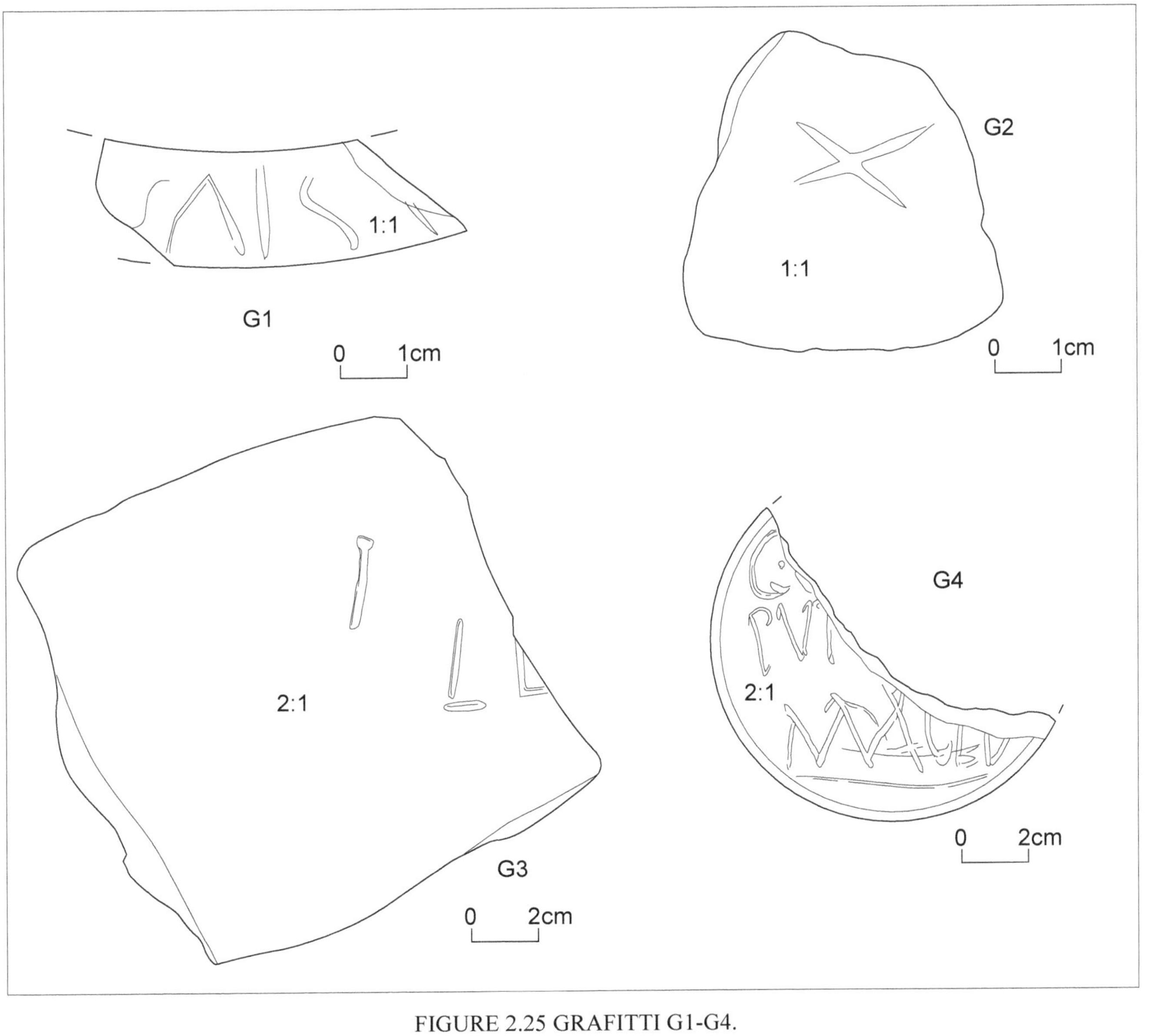

FIGURE 2.25 GRAFITTI G1-G4.

(ii) L ° L[...]

The second letter is broken, and is either L or E. The first letter, which is certainly L, is followed by a more or less triangular incision which is deeper and different in character than the casual damage elsewhere. It is not certainly deliberate, but if so, it would mark the initial L as an abbreviated praenomen, *L(ucius)*. Next would then be a nomen, with nomina beginning with L being much commoner than those with E, and finally a cognomen. All three names may even have been reduced to their initials. The graffito is therefore quite likely to be the name of a Roman citizen, and thus probably a legionary's. But if the 'triangle' is only casual damage, the graffito would be a single name beginning with LE.

G2, 3011, Pit [3012], Area B. Base sherd in grey fabric with oxidised surface. Scratched underneath, two intersecting lines:

X

This 'cross' identified the vessel, as an illiterate mark of ownership. Such marks are common.

G3, 4006, Area C, U/S. Rim sherd of a black-burnished (1) flat-rim dish. Scratched on the rim, so as to be read from outside:

[...]..A...[...]

The central letter is clearly A with a vertical mid-stroke, but the others are ambiguous because they were awkward to make or have been damaged subsequently. The upper curve of the first letter suggests S or (if the mark below represents a return) C or G. After it the second letter may be S or even L. Then A, and after it, either S or, if the scratches are deliberate, F. Finally an incomplete S, or L, or the diagonal of V completed by the vertical stroke at the very end.

G4, 5016, Area E, U/S. Crescent-shaped sherd 114 mm in diameter, amounting to almost half the base of a jar in a reddish tile-like fabric. Inscribed before firing, lines 1 and 2 in neat and stylish capitals; line 3 in larger less well-formed capitals:

G ° V[...]

PVD[...]

MACEIA

Probably *G(aius) V[al(erius) G(ai) f(ilius)]* | *Pud[ens]* | *Maceia (tribu)*

'Gaius Valerius Pudens, son of Gaius, of the *Maecia* voting-tribe.'

The name of the legionary who made this jar in a legionary tile-works.

In line 1, the well-formed G is followed by a medial point. G, not C, for the praenomen *G(aius)* is unusual, but for another example from Chester, see *RIB* 539. It is followed by part of the diagonal of V. The nomen is not necessarily *Valerius*, but this is frequent in northern Italy and Gallia Narbonensis, and is typical of legionaries in the early Principate: there are many British examples, including *RIB* 478, 479, 480, 539, 540, 541 and 542 from Chester. Written in full, it would be difficult to accommodate, but it is often abbreviated to VAL. This would allow space for the patronymic G F, *G(ai) f(ilius)*, as might be expected in the formal nomenclature of a legionary in the late first century.

In line 2, the third letter consists of a down-stroke curving at the ends like that of P, but with part of a second, arc-like stroke above it and below; this must be D, since the only other possibility, B, would have left some trace mid-way. The cognomen *Pudens* is frequent, examples including the legionary L(ucius) Valerius Pud[ens] at Chester (*RIB* 542)

In line 3, the first three letters are certain, with the lower loop of C perhaps written twice. The fourth letter is probably E: L is possible, but a continuing diagonal descender might have been expected, not a separate short horizontal stroke. It is followed by two converging strokes, probably IA: the angle is too steep for D, and the only other possibilities, V and PA, make no sense either. MACEIA is apparently a slip for MAECIA, *Maecia (tribu)*. The voting-tribe should properly follow the patronymic, as in *RIB* 257 (Lincoln), *G(aius) Valerius G(ai) f(ilius) Maec(ia tribu)*, but it is sometimes found in delayed position: British examples include *RIB* 539 already cited from Chester. In this graffito, however, it seems to have been added as an afterthought or correction.

By the end of the 1st century AD, the voting-tribe had long been a formality, but it was regularly included in the epitaphs of legionaries at Chester and elsewhere. It emphasised that they were Roman voters. As late as the 3rd century, Danubian legionaries and Praetorian guardsmen still fabricated a 'pseudo-tribe' from their city of origin so as to appear authentically Roman in nomenclature.

Catalogue of illustrated sherds

Event 2 (1-33 Fig 2.21, 34-71 Fig 2.22, 72-105 Fig. 2.23, 106-142 Fig. 2.24)

1. Reeded flat-rim bowl. Fabric: OXID. Pit [1018] (1015).
2. Pulley-wheel flagon. Fabric: WSOXID. Pit [4041] (4030).
3. Flat-topped jar decorated with incised wavy lines on the rim top and below the shoulder. Fabric: OXID. Pit [4041] (4030).

4. Hemispherical bowl decorated with random, slightly irregular blobs of white paint. Fabric: OXIDWPT. Pit [4041] (4030).
5. Small lid. Fabric: GREY. Pit [4041] (4030).
6. Ring-necked flagon. Fabric: WSOXID. Pit [4041] (4030).
7. Fragment of mortarium rim with the edge of a stamp. Fabric: MORTOX5. Pit [4041] (4030).
8. Short everted rim beaker with a wide grooved shoulder. The upper zone is polished smooth; the body is lightly roughened. Fabric: GREY. Pit [4041] (4030).
9. Collared-rim flagon. Fabric: WSOXID. Pit [4041] (4030).
10. Jar with an angular rim and a zone of ridging in the upper body. Fabric: OXID. Pit [4041] (4030).
11. Beaker bodysherd with orange-red painted decoration. Fabric: WW. Pit [4041] (4030).
12. Neckless jar with short everted rim. Sooted exterior with a sandy white interior. Fabric: VERWH. Pit [4041] (4030).
13. Handle from a Dressel 20 amphora. The piece has been ground flat on the longitudinal axis for use as a grinder or similar. Originally stamped (See also Fig. 00.A1). Fabric: BAT AM. Pit [4041] (4030).
14. Pinch-mouthed flagon. Fabric: OXID. Pit [4041] (4033).
15. Narrow-necked jar. Fabric: OXID. Pit [4041] (4033).
16. Flagon. Fabric: VER WH. Burnt. Pit [4041] (4033).
17. Lid. Fabric: DOR BB1. Burnished exterior. Pit [4041] (4033).
18. Hemispherical bowl with a lid-seated rim. Fabric: OXID. Pit [4041] 4046).
19. Conical bowl with a squat reeded rim. Rouletted on the upper rim surface. Fabric: MICOX. Pit [4041] (4046).
20. Large beaded, collared flagon. Fabric: WSOXID. Pit [4041] (4046).
21. Hemispherical bowl with white painted decoration. Fabric: OXIDWPT. Pit [4041] (4046).
22. Beaded rim flagon. Fabric: WW with pink core. Pit [4041] (4046).
23. Handled, reeded rim jug. Fabric: OXID. Pit [4041] (4046).
24. Spout or tapered base. Fabric: MICOX. Pit [4041] (4046).
25. Mortarium. Fabric: MORTOX1. Rounded quartzite and dark grey, argillaceous trituration grits. Worn towards the base. Pit [5013] (5014).
26. Large shallow dish. Fabric: OXID. Pit [5013] (5014).
27. Short, everted rim, globular beaker with rouletted decoration. Fabric: GREY/BW. Pit [5013] (5014).
28. Flared rim jar. Fabric: OXID. Pit [5013] (5014).
29. Bowl. Fabric: GREY. Pit [5013] (5014).
30. Carinated cup. Fabric: GREY. Pit [5013] (5014).
31. Short, everted rim, globular beaker. Fabric: WWF. Pit [5013] (5014).
32. Deep bowl with a flanged rim and a shallow applied spout. Fabric OXID. Pit [5013], (5012).
33. Hemispherical bowl with vertical combed decoration. Fabric: GREY. Pit [3037] (3035).
34. Globular jar/beaker . The exterior is decorated with grey barbotine rings and dots. Fabric: GYF. Pit 3058 (3049).
35. Wide-mouthed jar. Fabric: OXID. Pit [3058] (3049).
36. Mortarium with the potter's stamp SOLLVSF (made by Sollus) impressed across the flange. Fabric: VER WH. Pit [5020] (5011). (See also Fig. 00. M1).
37. Base of an unguent flask. Fabric: OXIDSY. Pit [3012] (3009).
38. Mortarium. Very hard fired, orange ware with grey core and mixed trituration grits. Stamped across the flange. Fabric: MORTOX1. Pit [3012] (3011) with joining sherd from (3006).
39. Squared rim jar. Fabric: OXID. Pit [3038] (3039).
40. S hallow dish. Fabric: OXID. Pit [3038] (3039).

Event 3

41. Hemispherical bowl with beaded lip. Fabric: MICOX. Pit [1018] (1015).
42. Globular jar with a short everted rim. Fabric: GREY. Pit [3030] (3029).
43. Bowl with an internal ledge/ lid seating. Fabric: GREY. Pit [3030] (3029).
44. Large, cordoned, bowl. Fabric: MICOX. Quarry [4002] (4007).
45. Mortarium stamped across the flange [IITA\]. Fabric: WRX WH. Quarry [4002] (4004). Second identical stamp from the same vessel (not illustrated) from pit [4041] (4046) and a joining spout from [4012] (4011).
46. Carinated, flared rim bowl. Fabric: OXID. Quarry [4002] (4007).
47. Tazza. Fabric: MICOX. Quarry [4002] (4004).
48. Tazza. Fabric: OXID. Internally blackened from use. Quarry [4002] (4004).
49. Bowl. Fabric: OXID. Cleaning (4006).
50. Hemispherical bowl with white painted decoration. Fabric: OXIDWPT. Pit [4041] (4016).
51. Carinated, flat-rim bowl. Fabric: GREY. Pit [4041] (4016).
52. Beaker/jar bodysherd with white painted ring-and-dot decoration. Fabric: OXIDWPT. Pit [4041] (4016).
53. Beaker with short everted rim. Fabric: MICOX. Pit [4041] (4016).
54. Large ring-necked flagon. Fabric: GREY. Pit [4041] (4016).
55. Deep hemispherical bowl. Fabric: OXID. Pit [4041] (4016).
56. Everted rim jar. Fabric: GREY. Pit [4041] (4016).
57. Flat reeded-rim bowl. Fabric: OXID. Pit [4041] (4016).

58. Hemispherical, facetted, bowl. Fabric: GREY. Pit [4041] (4028).
59. Rolled-rim jar. Fabric: GREY. Pit [4041] (4028).
60. Moulded-rim flagon. Fabric: WSOXID. Pit [4041] (4029).
61. Mortarium with a stamp to the left side of the spout. Fabric: VER WH. Retrograde stamp BRVCIF. Pit [5004] (5019).
62. Moulded rim beaker decorated with vertical white painted stripes. Fabric: GREY. Pit [5004] (5002).
63. Jar with rusticated decoration. Fabric: GREY. Pit [5004] (5002).
64. Deep hemispherical bowl with a shallow spout. Fabric: OXID. Pit [5004] (5002).
65. Grooved rim dish. Fabric: DOR BB1. Pit [5004] (5002).
66. Slightly dropped flat-rim dish with burnished-line decoration. Fabric: DOR BB1. Pit [5004] (5002).
67. Shallow dish. Fabric: OXID. Pit [5004] (5002).
68. Expanded rim jar with a pushed out ridge below the rim. Fabric: OXID. Pit [5004] (5002).
69. Triangular-rimmed narrow-necked jar. Fabric: OXID. Pit [5004] (5002).
70. Flared rim ?bowl/beaker. Fabric: BWF. Polished black exterior with a dark brown core. Pit [5004] (5002).
71. Reeded-rim carinated bowl. Fabric: GREY. Pit [5004] (5002).
72. Short, everted rim globular beaker. Fabric: GREY. Pit [5004] (5002).
73. Lid-seated beaker. Fabric: GREY. Pit [5004] (5002).
74. Flanged cup. Fabric: OXID. Pit [5004] (5002).
75. Everted rim jar. Fabric: DOR BB1. Pit [5004] (5002).
76. Beaded rim jar. Fabric: DOR BB1. Pit [5004] (5002).
77. Short, everted rim, jar. Fabric: DOR BB1. Pit [5004] (5002).
78. Small beaker. Fabric: MICOX. Ditch [2241] (2272).
79. Wide-mouthed jar with short everted rim and much smeared rusticated decoration. Fabric: GREY. Pit [1019] (1008).
80. Dolium-type flat-topped jar. Fabric: OXID. Pit [1019] (1008).
81. Bowl copying a Drag 30 form. Decorated with horizontal bands of brown paint on the upper zone and a latticed design on the body. Fabric: WW. Pit [1019] (1008).
82. Beaker bodysherd with depressed dimples and a pearlescent bluish-green glaze. Fabric: GLAZE. Pit [1019] (1008).
83. Verulamium mortarium. Originally stamped across the flange with only the corner of the stamp surviving showing a herring-bone border. Fabric: VER WH. Pit [1019] (1008).
84. Deep bowl decorated with an incised lattice. Fabric GREY. Pit [2110] (2111).
85. Single-handled flagon with a conical mouth and cordoned neck. Fabric: WSOXID. Pit [2110] (2111).
86. Cornice rim beaker with roughcast decoration. Fabric: ARG CC. Pit [2110] (2111).
87. Small hemispherical bowl or cup. Fabric: WWEGG. Pit [2110] (2111).
88. Indented beaker. Fabric: OXID. Pit [2110] (2111).
89. Flat-rim dish. Fabric: DOR BB1. Pit [2110] (2111).
90. Flat-rim bowl with a carinated base. Fabric: DOR BB1. Pit [2110] (2111).
91. Carinated flat-rim bowl. Fabric: OXID. Pit [2110] (2111).
92. Bowl with a reeded rim and rouletted decoration. Fabric: OXID. Pit [2110] (2111).
93. Jar with short, everted rim and lattice decoration. Fabric: OXID. Pit [2110] (2111).
94. Large double-handled flagon. Fabric: WSOXID. Pit [2110] (2111).
95. Short, everted rim, globular beaker. Fabric: GREY. Pit [2110] (2111).
96. Flared-rim flagon. Fabric: OXID (granular). Pit [2110] (2111).
97. Mortarium with a broken flange and part of a potter's stamp. Fabric: MOWSOX3. Pit [2110] (2111).
98. Carinated, flat reeded-rim bowl. Fabric: GREY. Pit [2110] (2112).
99. Small beaker with short everted rim. Fabric: GREY. Pit [2110] (2112).
100. Everted rim jar. Fabric: OXID. Pit [2110] (2112).
101. Carinated bowl with a strainer. The vessel has an omphalos base and a circular hole at the centre of the strainer. The upper part of the bowl has been broken off. Similar to Marsh (1978) type 45. The form has continental origins and is thought to have been a wine strainer. Fabric: MICOX. Pit [2110] (2112).
102. Flagon. Fabric: WW. Pit [2110] (2113).
103. Everted rim jar. Fabric: OXID. Pit [2110] (2113).
104. Cup-mouthed flagon. Fabric: WSOXID. Layer (2156).
105. Cordoned bowl. Fabric: MICOX. Layer (2156).
106. Large double-handled flagon. Finger depression at base of handle. Fabric: WSOXID. Pit [2208] (2207).
107. Necked bowl. Fabric: WWF. Pit [4012] (4003).
108. Nozzle. Uncertain whether vertical or horizontal. Possibly from a feeding bottle. Fabric: OXID. Pit [2191] (2212).
109. Single-handled flagon. Fabric: WSOXID. Pit [2221] (2285).
110. Jar with a heavy, triangular, rim. Fabric: CAM WH. Pit [2255] (2254).
111. Beaker with short, everted, rim and ridged neck. Fabric: OXID. Pit [2292] (2285), Group112.
112. Beaded-rim flagon. Fabric: WSOXID. Pit [2292] (2285).

113. Ring-necked flagon, warped and cracked waster. Fabric: WSOXID. Pit [2292] (2283).
114. Lid-seated jar. Fabric: WW. Pit [3012] (3008).
115. Narrow-mouthed necked jar with a slight internal lid-seating. Decorated with single burnished, horizontal, lines. Fabric: GREY. Pit [5007] (5005).
116. Double-handled flagon with a slightly bifid rim. Fabric: OXID. Pit [5007] (5005).
117. Hemispherical bowl with a slightly dropped, flat rim. Fabric: GREY. Pit [5007] (5005).
118. Ring-necked flagon with an expanded upper ring. Fabric: WSOXID. Pit [5007] (5005).
119. Ring-necked flagon. Fabric: WSOXID. Pit [5007] (5005).
120. Flat-rim bowl decorated with spaced vertical burnished lies. Fabric: DOR BB1. Pit [5007] (5005).
121. Ridged wall bowl. Fabric: MICOX. Partly burnt upper zone internally. Pit [5007] (5005).
122. Barrel-shaped jar/large beaker with a cornice rim. Fabric: OXID. Pit [5007] (5005).
123. Mortarium. Fabric: HOL OX. Well-fired. Sparse quartzite trituration grits. Pit [5007] (5005).
124. Large jar decorated with an acute burnished lattice. Fabric: DOR BB1. Pit [5007] (5005).
125. Short-necked jar. Fabric: DOR BB1. Pit [5007] (5005).
126. Small jar/ mug. Fabric: DOR BB1. Pit [5007] (5005).
127. Wide-mouthed jar. Fabric: GREY. Pit [5007] (5005).
128. Jar with a short, everted rim, and rusticated decoration. Fabric: GREY. Pit [5007] (5005).
129. Jar with burnished line decoration. Fabric: DOR BB1. Pit [5007] (5005).
130. Flat-rim bowl with burnished line decoration. Fabric: DOR BB1. Pit [5007] (5005).
131 132. 133. Deep plain-rimmed bowl with intersecting chevron decoration. Fabric: DOR BB1. Pit [5007] (5005).
132. Lid. Sooted on the exterior. Fabric: GREY. Pit [5007] (5005).
133. Lid with a sooted lip. Fabric: OXID. Pit [5007] (5005).
134. Lid with burnished line squiggles on the interior, burnished exterior. Fabric: DOR BB1. Pit [5007] (5005).
135. Bowl decorated with diagonal red painted, burnished lines on the outer collar. Fabric: WIL OX. Pit [5007] (5005).
136. Short everted rim beaker/jar with rusticated decoration. Fabric: OXID. Pit [5007] (5005).
137. Beaker with short everted rim and rouletted decoration. Fabric: WSOXID. Pit [5007] (5005).

Unstratified

138. Amphora rim, *Camulodunum* type 186. Fabric: CAD AM. Layer (2246).
139. Beaded rim globular jar. Fabric: GREY. Context (3015). Void.
140. Large, hemispherical bowl. Fabric: OXID. Cleaning (4006).
141. Carinated bowl with vertical, grooved rim. Decorated with stamp-impressed decoration in the form of leaves joined by comb-impressed tendrils. Fabric: GREY.

Context (5016).

Graffiti (Fig 2.25)

G1. Basesherd with a post-firing cross on the underside. Burnt. Fabric: ?GREY. Pit [3012] (3011). Event 2.

G2. Bodysherd from a Dressel 20 amphora with part of a graffito inscribed after firing. Fabric: BAT AM. Pit [4041] (4015). Event 3.

G3. Flat-rim dish. Post-firing graffiti scratched into rim surface. Fabric: DOR BB1. Cleaning (4006). Event 3.

G4. Base from a closed form with a pre-firing graffiti on the underside (see Tomlin below). Fabric: WSOXID. Context (5016).

Stamps

Amphora

A1. Slightly worn stamp on a re-used handle from a Dressel 20 amphora. The name is probably PORTAN[V] (Rodriguez 1997). Pit [4041] (4030). Event 2.

Mortaria

M1. Mortarium with the potter's stamp SOLLVSF (made by Sollus) impressed across the flange. Fabric: VER WH. Pit [5020] (5011). Event 2. Probably the same die as one from Verulamium (Hartley1972, fig 146. no 38). At least four dies are known belonging to this potter with products widely distributed over Britain. Typologically the production is dated to the period AD 70-100/110.

M2. Mortarium. Very hard fired orange ware with grey core and mixed trituration grits. Stamped across the flange. Fabric: MORTOX1. Pit [3012] (3011).Event 2.

M3. Fragment of mortarium rim with the edge of a stamp. Fabric: MORTOX5. Pit [4041] (4030).Event 2.

M4. Mortarium with a stamp to the left side of the spout. Fabric: VER WH. Retrograde stamp BRVCIF. Brucius or Bruccius. Thought to have worked in the period AD 80-100 (Hartley 1972, 374; fig. 145.13). Pit [5004] (5019). Event 3.

M5. Verulamium mortarium. Originally stamped across the flange with only the corner of the stamp surviving showing a herring-bone border. Fabric: VER WH. Pit [1019] (1008). Event 3.

M6. Mortarium stamped across the flange [IITA\]. Fabric: WRX WH. Quarry [4002] (4004). Second identical stamp (not illustrated) from pit [4041] (4046) and a joining spout from [4012] (4011). Event 3.

M7. Mortarium with a broken flange and part of a potter's stamp. Fabric: MOWSOX3. Pit [2110] (2111). Event 3.

ROMAN CERAMIC BUILDING MATERIALS

by Alison Heke

The Roman ceramic building material was analysed in terms of quantity, range, condition, source and date-range and was recorded to the level set out in 2.6 of the Archaeological Ceramic Building Materials Group (ACBMG) minimum standards draft document for the recovery, curation, analysis and publication of ceramic building material (Hunter-Mann 2001), with the addition of 'part' (e.g. middle, edge, end, corner, etc) as a sub-term of 'form'. Fragments were identified by eye, except for unusual fabrics and surface features, which were examined microscopically at x 20 magnification. The assemblage was quantified by fragment count and weight and measurements (in mm) were taken of any complete dimensions. Rubbings were made of all stamps, signatures and other markings. Small find numbers were assigned to stamps, signatures and atypical or unusual forms. Where possible, *tegula* lower cutaways were identified and assigned to dated groups after Warry (2006). The archive is available for consultation in hard copy (paper records) and also as an MS Access database. The assemblage from the evaluation (site code DEL 02) is discussed alongside that from the excavation (site code DEL 06).

The analysis comprises a description of the assemblage, organised by Event, followed by a discussion of the nature and character of the assemblage. The catalogues comprise only those objects chosen for illustration. They are ordered by Event number and are numbered consecutively.

Quantity

The site produced a relatively large assemblage of ceramic building material; comprising 4,591 fragments weighing 276,206 g, with an average fragment weight of 60.2 g. Hand-collection produced 3,185 fragments/ 273,937 g (69.4% by fragment count and 99.2% by weight of the total assemblage). The remaining 1,406 fragments/2,269 g (30.6% by fragment count and 0.8% by weight of the total assemblage) were retrieved from wet sieving. 67 fragments weighing 14,069 g (1.5% by fragment count and 5% by weight of the total assemblage), with an average fragment weight of 209.9 g, came from the evaluation; 4,524 fragments weighing 262,137 g (98.5% by fragment count and 95% by weight of the total assemblage), with an average fragment weight of 57.9 g, were recovered from the excavation. This lower average fragment weight for the excavation partially reflects the amount of material recovered from samples (in comparison, the fragmentation index for the hand-collected assemblage alone from the excavation (DEL 06) amounts to 83.5g).

Table 2.9 lists the total site assemblage and the average weight per fragment. There is a much lesser degree of fragmentation amongst the material recovered from the evaluation than from the excavation, even accounting for the bias introduced by the numbers recovered from samples. This is probably because the greater numbers involved from the excavation statistically give a more accurate indication of the level of brokenness of the assemblage.

The bulk of the assemblage (4,267 fragments/257,717 g) was recovered from Roman contexts. The greatest proportion, by both fragment count and weight, came from Event 3 (55.2% by count and 54.3% by weight of the total). The second largest group came from Event 2 (31.6% by count and 35.4% by weight of the total). Less than 1% by fragment count and weight was recovered from contexts belonging to Event 1. The post-Roman, unphased and unstratified contexts produced 7.1% by fragment count and 6.7% by weight of the total (*see* Table 2.9).

Range

A wide range of forms were recovered. These comprise roof tiles (*tegulae* and *imbrices*/ridge* tiles), bricks, box tiles, facing tiles, herringbone floor (*opus spicatum*) bricks, other small floor bricks, fragments of portable oven and indeterminate forms (see Table 2.10). [*curved roof tiles have been classified as *imbrices* only when the fragments exhibit an obvious taper; otherwise they have been recorded as *imbrices*/ ridge tiles. However, it is probable that the vast majority of these fragments are in fact *imbrices*].

Tegulae form the largest component, by weight (53.3%), of the entire site assemblage, closely followed by indeterminate forms (17%), *imbrices*/ridge tiles (17.4%) and bricks (9.4%). The remaining forms together comprise just 3.2% by weight of the total assemblage: herringbone-floor (*opus spicatum*) and other small floor bricks 0.6%,

Table 2.9: Total site assemblage (by Event)

	Evaluation			*Excavation*			Total		
	No frags	Wt (g)	Av frag wt (g)	No frags	Wt (g)	Av frag wt (g)	No frags	Wt (g)	Av frag wt (g)
Phase 1 (Roman): Event 1	0	0	0	21	103	4.9	21	103	4.9
Phase 1 (Roman): Event 2	15	2093	139.5	1435	95740	66.7	1450	97833	67.5
Phase 1 (Roman): Event 3	41	10515	256.5	2491	139454	55.9	2532	149969	59.2
Phase 1 (Roman): Other Group 112	0	0	0	171	8173	47.7	171	8173	47.7
Phase 1 (Roman): Other Group 113	0	0	0	93	1639	17.6	93	1639	17.6
PR, unphased and unstrat	11	1461	132.8	313	17028	54.4	324	18489	57.1
Total	**67**	**14069**	**209.9**	**4524**	**262137**	**57.9**	**4591**	**276206**	**60.2**

TABLE 2.10: TOTAL SITE ASSEMBLAGE (RANGE OF FORMS)

	Box tile		Brick		Facing tile		Facing tile or Box tile	Floor brick		Half-box		Imbrex/ ridge tile		
	No frags	Wt (g)	No frags	Wt (g)	No frags	Wt (g)	No frags	Wt (g)	No frags	Wt (g)	No frags	Wt (g)	No frags	Wt (g)
DEL 02	2	151	3	1005	1	209	1	675	0	0	0	0	11	1660
DEL 06	11	1327	82	24894	2	492	4	468	4	1681	2	1211	464	45208
Total	13	1478	85	25899	3	701	5	1143	4	1681	2	1211	475	46868

Imbrex		Indet		Op spic brick	Portable oven	Tegula		Total			
No frags	Wt (g)	No frags	Wt (g)	No frags	Wt (g)	No frags	Wt (g)	No frags	Wt (g)	No frags	Wt (g)
0	0	13	720	1	277	0	0	35	9372	67	14069
2	706	3135	46429	7	1306	5	676	806	137739	4524	262137
2	706	3148	47149	8	1583	5	676	841	147111	4591	276206

TABLE 2.11: ROMAN CERAMIC BUILDING MATERIAL QUANTIFIED BY EVENT, FORM, NO OF FRAGMENTS AND WEIGHT

	Roman: Event 1	Roman: Event 2		Roman: Event 3		Roman: Other Group 112 contexts		Roman: Other Group 113 contexts		PR, unphased and unstrat		Total		
Form	No frags	Wt (g)	No frags	Wt (g)	No frags	Wt (g)	No frags	Wt (g)	No frags	Wt (g)	No frags	Wt (g)	No frags	Wt (g)
Tegula	0	0	299	50472	447	83015	32	4734	1	13	62	8877	841	147111
Imbrex	0	0	0	0	1	502	0	0	0	0	1	204	2	706
Imbrex/ridge tile	1	17	203	19812	218	20689	19	2587	5	788	29	2975	475	46868
Box tile	0	0	5	870	7	532	0	0	0	0	1	76	13	1478
Facing tile	0	0	0	0	3	701	0	0	0	0	0	0	3	701
Facing or box tile	0	0	0	0	4	1090	0	0	0	0	1	53	5	1143
Half-box tile	0	0	2	1211	0	0	0	0	0	0	0	0	2	1211
Brick	0	0	32	9713	45	14422	0	0	2	430	6	1334	85	25899
Herringbone-Floor brick	0	0	2	232	4	1004	0	0	0	0	2	347	8	1583
Small floor brick	0	0	0	0	2	1152	0	0	0	0	2	529	4	1681
Indet	20	86	905	15335	1800	26647	119	622	85	408	219	4051	3148	47149
Portable oven	0	0	2	188	1	215	1	230	0	0	1	43	5	676
Total	**21**	**103**	**1450**	**97833**	**2532**	**149969**	**171**	**8173**	**93**	**1639**	**324**	**18489**	**4591**	**276206**

box tiles 0.5%, half-box, facing tile/ box tile 0.4% each, and facing tiles and portable oven fragments 0.3% each by weight of the total assemblage.

Condition

The assemblage is of mixed condition, ranging from fresh to abraded and weathered. This mix occurs across the site and also within contexts. There is no obvious difference in condition between the Events or between the assemblages recovered from the evaluation and the excavation. The material retrieved from post-Roman contexts and found unstratified is of similar condition. A small proportion of the assemblage is burnt (6.1%), some possibly through use (eg some of the oven and box tile/ facing tile fragments) but most following use or re-use. The burning may have happened as part of the demolition process or during rubbish disposal. There is some evidence of re-use – 51 fragments (1.1%) are mortared on their broken edges/ surfaces. Some small fragments are also coated in *opus-signinum* type mortar and are probably pieces of cement mix or concrete.

Source

It is probable that most of the assemblage derives from the legionary kilns at Holt, situated 12 km south of Chester. The kilns were established by the Twentieth Legion, which was based at Chester from the early AD 90s, and are known to have been producing ceramic building material and pottery for the legionary fortress from *c* AD 90/ 100. Although it is not certain when production of ceramic building material ended at Holt, it is clear that the Twentieth Legion or its contractors continued to produce ceramic building material into the third century (Warry 2006, 80).

Date range

Datable lower cutaways from the *tegula* assemblage comprise mainly (59.6%) group As (which were produced between *c* AD 90/100-120), with lesser amounts of group Bs (7.7%), Cs (3.8%) and Ds (15.4%). Group B *tegulae* broadly date between *c* AD 100-180 but are thought to date at Chester to between *c* AD 120-140 (Warry 2006, 156).

Table 2.12: Residual Roman tile from the quarry watching brief

Tegulae		Imbrex/Ridge		Box tile		Half-box tile		Indet		Total	
No	Wt (g)	No	Wt (g)	No	Wt (g)	No	Wt (g)	No	Wt(g)	**No**	**Wt (g)**
3	1468	1	391	1	178	2	1680	4	103	**11**	**3820**

Group C *tegulae* date between *c* AD 160-260 and group D *tegulae* between *c* AD 240-380 (Warry 2006, 63).

The presence of thin-walled, knife-scored box tiles alongside facing tiles and half-box tiles (*tegulae hamatae*) is notable as these forms were generally in use in the late-1st century and were largely replaced by combed and roller-stamped box tiles in the early 2nd century (Black 1996, 62; Pringle 2006, 128; Ward 1999, 48).

Description

Phase 1: Roman (dated late-1st to mid-2nd century AD)

This phase is represented by a series of large quarry pits, smaller refuse pits and two linear ditches..

The bulk of the assemblage came from Phase 1 and comprises 92.9% by fragment count and 93.3% by weight of the total assemblage), with a relatively low average fragment weight of 60.4 g, although it should be noted that this includes material retrieved from samples. 56 fragments weighing 12,608 g were recovered from the evaluation and 4,211 fragments weighing 245,109 g came from the excavation. The forms identified are listed in Table 2.12.

It should be noted that a large number of tiny fragments of indeterminate form were recovered from samples (ie 1,312 fragments/2,194 g or 30.8% by count and 0.9% by weight from Phase 1). This figure has clearly distorted the fragment count percentages for all the Phase 1 forms. The weight percentages have therefore been used instead as a more reliable indicator of the relative amounts of each form.

Tegulae are the most common form (53.6% by weight of the Phase 1 assemblage), followed by almost equal amounts of *imbrices*/ridge tiles (17.0% by weight) and indeterminate forms (16.7% by weight), with bricks at 9.5%. Small but significant numbers of box tiles, half-box tiles and herringbone-floor (*opus spicatum*) bricks each comprise 0.5% each by weight of the Phase 1 assemblage; facing or box tiles and small floor bricks 0.4%; facing tiles 0.3% and *imbrices* and portable oven fragments 0.2% each by weight of the Phase 1 assemblage.

Event 1

The quarrying activity represented by Event 1 produced just 21 fragments weighing 103 g, less than 1% each by count and weight of the Phase 1 assemblage. All were retrieved from (2282) and the only identifiable form comprises a single fragment of *imbrex*/ridge tile. The remainder are indeterminate. The *imbrex*/ridge tile is fairly fresh in condition; the indeterminate fragments are all abraded and weathered. None are burnt and there is no evidence of re-use. None of the fragments were recovered from samples.

Event 2

The activity represented by Event 2 (the initial backfilling of the quarry pits with rubble and quarry waste) produced 1,450 fragments weighing 97,833 g (33.6% by count and 37.2% by weight of the Phase 1 assemblage). Roof tiles outnumber all other forms from Event 2 and *tegulae* are the most common form, comprising 51.6% by weight of the Event 2 assemblage. *Imbrices*/ridge tiles comprise 20.3% by weight of the total; indeterminate forms 15.7% and bricks 9.9%. Small, but significant, amounts of box tile, half-box tile, herringbone-floor bricks and portable oven fragments were also recovered. 296 fragments/301 g (20% by fragment count and just 0.3% by weight of the Event 2 assemblage) were recovered from samples.

Tegulae

The evaluation produced eight fragments of *tegula* weighing 1,510 g, from layer (52) and fill (53), with the largest group by weight (79.4%) coming from (52). The *tegulae* from the evaluation are of mixed condition, ranging from fairly fresh to slightly weathered and abraded. There is no evidence of re-use. Three fragments show signs of burning.

A middle fragment with hobnail imprints in the upper surface, SF 201, was recovered from (52). These impressions are rare on civilian tiles but occur on approximately 10% of military *tegulae*. It is possible that the marks were made by the officer in charge of the tile-works testing the hardness of the tile with his foot (Warry 2006, 16).

All the *tegulae* from the excavation were recovered from pits: [1019], [2209], [3030], [3031, [3038], [3040], [3058], [4002], [4041] and [5004], with the largest group coming from [4041] (44.8% by weight of the total number of *tegulae* from Event 2). Pit [3038] produced 19.4% by weight of the total, Pit [3058] 8.9%, Pit [1019] 5.4%, Pit [4002] 3.1% and Pit [2209] 2.6%. The remaining 15.8% of the total came from the other pits. A single (flange) fragment, weighing 9 g, was recovered from samples.

The assemblage from the excavation is also of mixed condition, ranging from fresh or fairly fresh to weathered, battered and abraded. 40 fragments show signs of burning.

One fragment has been re-used, as patchy white mortar is present on the broken edges and surfaces.

The only datable form comprises an almost complete group A 26 lower cutaway from fill (3039) of pit [3038], dated to *c* AD 90/100-120.

Pit [1019]

A partial group A lower cutaway came from fill (1014). A corner fragment with curved signature (made using four fingers) arising from the end, SF 207, came from fill (1013). The same context also produced a corner with an upper cutaway, sanded on the same alignment as the sanded stepdown band, which implies that the stepdown was formed after the cutaway.

Pit [2209]

Fill, (2196) produced a probable (partial) group C or D lower cutaway, dated *c* AD 160-380. The latter form may be intrusive as it post-dates the end of the phase.

Pit [3038]

46 fragments weighing 9,517 g, came from the single fill (3039) of pit [3038]. The group includes a middle fragment with two wide, parallel finger marks, SF 238, and a very worn middle fragment with overlapping finger or paw prints in the ?upper surface, SF 239. A flange fragment has faded cream slip on the top and inner edge of the flange. There is also a corner fragment with a sanded stepdown band and an end piece with a faint, lightly sanded stepdown.

Pit [3058]

Notable forms from fill (3049) comprise a corner with sanded stepdown band overlain by a knife-trimmed upper cutaway, indicating that the cutaway was formed after the stepdown (which is the opposite sequence of events to the example from (3039). Warry has pointed out that the stepdown is not part of the process for forming the upper cutaway, as has sometimes been suggested (Warry 2006, 14). A flange fragment also has a sanded stepdown band. The only datable form comprises a probable (partial) group D lower cutaway (*c* AD 240-380). This may also be intrusive as it post-dates the end of the phase.

Pit [4002]

Tegulae of note comprise an end piece painted/decorated with cream slip on the upper surface from fill (4038), and two partial group A or B lower cutaways, one from (4038), the other from fill (4051).

Pit [4041]

Fill (4030) produced 42 fragments weighing 7,801 g. The group includes a middle piece decorated with faded red slip on the upper surface a partial sanded upper cutaway, and a middle fragment with sanded stepdown band; the latter is a feature of military production, occurring on approximately 57 % of Chester/Holt/Tarbock *tegulae* (Warry 2006, 14). Datable forms comprise two Warry group A or B *tegula* with partial lower cutaways, dated between *c* AD 90/100-140.

Fill (4033) produced 42 fragments weighing 7,474 g. The group includes a corner fragment with a triple, curved signature arising from the end, SF 264, with a partial group A lower cutaway. An end piece is decorated with faded cream slip on the upper surface. Other lower cutaways comprise two partial group A 28s and two partial group A 26s. Two upper cutaways are partially sanded.

Fill (4046) produced 47 fragments weighing 6,668 g. The group includes a middle fragment with faded cream slip on the upper surface and a group A 28 lower cutaway.

Pit [5004]

Notable forms comprise a middle piece with a double, curved signature, SF 267 from (5003). There is also a sanded, partial upper cutaway from this context. The only datable form is a partial group A or B lower cutaway (*c* AD 90/100-140), also from (5003).

Notable forms from the remaining pits comprise three group A lower cutaways (including one form A 26) from fill (3041) of pit [3040], as well as two sanded upper cutaways from the same context, a complete group A 26 lower cutaway from fill (3029) of pit [3030] and two adjoining fragments with sanded stepdown band from fill (3033) of pit [3031].

Imbrices/ridge tiles

The evaluation produced four fragments of *imbrex*/ridge tile weighing 398 g, from layer (52) and fill (53). The tiles are generally fresh or fairly fresh in condition; only one fragment (from (52) is slightly abraded. There is no evidence of re-use and there are no signs of burning. The only form of note from the evaluation comprises an edge fragment with signature parallel to the edge, SF 202, from (52).

Almost all the *imbrices*/ridge tiles from the excavation came from pits [1019], [1022], [2165], [2209], [2241], [3030], [3038], [3040], [3058], [4002], [4041] and [5004], with the largest group coming from [4041]. Layer (3048) produced a single fragment weighing 108 g. A single (middle) fragment, weighing 72 g, was recovered from samples. The *imbrices*/ridge tiles from the excavation are of mixed condition, ranging from fresh to abraded and weathered. 16 fragments show signs of burning. One (middle) fragment has been re-used, as grey/cream sandy mortar is attached to the underside and one broken edge.

Pit [3038]

Fill (3039) produced the largest group of *imbrices*/ridge tiles, with 38 fragments weighing 3,831 g. The group includes two edge pieces with signatures parallel to the edge, SFs 240 and 241, and a middle fragment with a ?signature across the top of the gable, SF 242. Five fragments have traces of white lime-wash and mortar attached to the edges and upper surfaces, indicating that these tiles were originally mortared into position. Four fragments bear patchy traces of cream slip on the upper surface.

Pit [4002]

Notable forms comprise an end piece with a signature across the top of the gable, SF 253, from (4013), a middle fragment with faded cream slip on the upper surface, also from (4013), and an end fragment with white mortar attached to the upper surface, from (4038). The last indicates that the tile was probably originally mortared into position.

Pit [4041]

Fill (4030) produced 33 fragments weighing 3,784 g. Notable forms comprise signatures across the top of the gable, SF 261, and parallel to the edge, SFs 262 and 265. Three tiles are painted on the upper surface, two with cream slip, the other with red slip – all faded.

Fill (4033) produced 24 fragments weighing 3,162 g. Notable forms comprise two fragments with faded cream slip on the upper surface and a middle piece with faded red slip on the upper surface.

Fill (4046) produced 34 fragments weighing 2,190 g. The group includes 11 fragments with patchy cream slip painted onto the upper surface.

Imbrices/ridge tiles of note from fill (4038) comprise an end piece with white lime-wash and mortar attached to the upper surface, suggesting that the tile was originally mortared into position.

Notable forms from the remaining pits comprise a middle fragment with patchy cream slip on the upper surface, from fill (2164) of pit [2165], an edge piece with a signature parallel to the edge, SF 245, from fill (3049) of pit [3058], a middle fragment with faded cream slip on the upper surface, also from (3049), and two middle pieces with faded red slip on the upper surface, from fill (5003) of pit [5004].

Box tile (*Tubuli*)

The evaluation produced a corner fragment of box tile (*tubulus*), weighing 128 g, from layer (52). It is relatively thin-walled (with a thickness of 14 mm), fairly fresh in condition and with a lightly sooted interior, presumably through use. The exterior is sanded but there is no evidence of the type of keying used. The edge only of a knife-cut air vent is present and the shape therefore cannot be determined.

The excavation produced four probable fragments of box tile weighing 742 g. Two are fairly fresh in condition; two others are slightly abraded and partially burnt. There is no evidence of re-use.

A middle fragment was recovered from fill (3009) of pit [3012]. Both the interior and exterior surfaces are sanded but there is no evidence of the type of keying used. With a thickness of 23 mm, it is possible that the fragment is actually a piece of facing tile. A middle fragment from fill (3041) of pit [3040] is thin-walled, with a thickness of 13 mm, and has a sanded, knife-scored exterior. There is also the edge of a knife-cut air vent (shape uncertain). A corner fragment, SF 244, was recovered from fill (3049) of pit [3058] (Cat 1). The form is unusual, with thick walls (32 mm) and two air vents. One of these is oval in shape, the other may also be oval in shape but the edge only survives. The angled corner suggests that the form may be multi-sided. Both surfaces are smooth and the fabric is atypical for Holt. The form presumably had a specialised function, but what exactly this may have been is uncertain. A corner fragment, with sanded exterior surfaces and a thickness of 21 mm, came from fill (4046) of pit [4041]. No evidence survives of the type of keying used.

Half-box tile (*Tegulae hamatae*)

Two adjoining fragments of half-box tile (*tegula hamata*) weighing 1,211 g were recovered from the excavation, from fill (4033) of pit [4041]. Together they comprise a slightly abraded edge fragment with sanded exterior keyed with large diamond-shaped knife-scoring, as is usual for these tiles (Brodribb 1987, 67). The flange is broken and there is a partial central cut-away, which is lightly sanded (*cf* Brodribb 1987, 66 fig 27 b).

Brick

Bricks in a range of thicknesses (and therefore of forms) were recovered from the excavation, comprising 32 fragments weighing 9,713 g. All were recovered from pits: [1019], [2209], [3012], [3031], [3038], [3040], [3058], [4002], [4041] and [5004].

The assemblage is of mixed condition, ranging from fresh to abraded and battered. Seven fragments show signs of burning and two have been re-used, as cream/white Roman mortar is attached to broken edges and surfaces.

Pit [4041]

Fill (4046) produced the largest group of bricks, with 5 fragments weighing 1,082 g. The thickest form (63 mm) may be from either a *tegula bipedalis* (average thickness: 60 mm), which is the largest and thickest of the bricks, or from a Holt *pedalis*. The Holt forms range in thickness from 50 to 88 mm, although they generally average at a thickness of 43 mm on most British sites. Two other fragments have complete thicknesses of 52 mm and another has a thickness of 44 mm. The thicker pieces may be from

Holt *bessales*, Holt *pedales* or from the *sesquipedales* (average thickness: 52 mm); the thinner form may be from a *lydion* (average thickness: 41 mm).

Fill (4030) produced four fragments weighing 937 g. A notable form comprises a possible *tegula bipedalis* or Holt *pedalis* (with a thickness of 63 mm) with a looping signature, SF 263. Signatures are relatively rare on bricks, although they are more common on the three larger brick forms (the *lydion*, *sesquipedalis* and *tegula bipedalis*) (Brodribb 1987, 102).

There are few forms of note from the remaining pits. Fill (2196) of pit [2209] produced a possible *tegula bipedalis* or Holt *pedalis* (with a thickness of 65 mm) and an indeterminate brick with a thickness of 49 mm. Other complete thicknesses comprise 45 mm and 49 mm, from fill (3033) of pit [3031] and a corner (two adjoining fragments) from fill (5003) of pit [5004], with a thickness of 39 mm.

Herringbone-floor (*opus spicatum*) bricks

These small rectangular bricks were laid with the long, thin edge uppermost and arranged in a herringbone pattern to produce a hardwearing floor surface (Brodribb 1987, 50-3). The bricks made at Holt conform to a playing-card shape and average in size at 113mm in length; 75 mm in width and 32mm in thickness (Grimes 1930, 135). The excavation produced two adjoining fragments weighing 232 g, from fill (4033) of pit [4041]. At width: 72 mm and thickness: 36 mm, the brick is close to the average for Holt. It has a much abraded (worn and rounded) underside.

Indeterminate forms

The evaluation produced two indeterminate fragments weighing 57 g, from fill (53). One is fresh in condition; the other is slightly abraded and lightly burnt or reduced. There is no evidence of re-use. The only form of note comprises SF 203, a middle fragment, possibly from a *tegula*, with a single curved and two straight signatures.

The excavation produced 903 indeterminate fragments weighing 15,278S g. Apart from ten fragments weighing just 3 g from layer (3052) (all from samples), the remaining indeterminate forms were recovered from pits [1018], [1019],1022], [2110], [2165], [2209], [2241], [3012], [3027], [3030], [3031], [3037], [3038], [3040], [3056], [3058], [4002], [4041], [5004] and [5013]. The largest group (by weight) came from pit [4041] (42.6% by weight of the total number of indeterminate forms from the excavation in Event 2). 294 fragments weighing 220 g were recovered from samples.

The assemblage is of mixed condition, ranging from fresh to abraded, battered and weathered. 45 fragments show signs of burning. Seven fragments have been re-used, with mortar attached to broken edges. Three of these are coated in mortar and may have been used as aggregate for *opus signinum*.

Fragments of note comprise a fragment from fill (3009) of pit [3012], with faded cream slip on the upper and lower surfaces; a fragment with overlapping looped and straight finger marks, SF 243, from fill (3039) of pit [3038]; and a middle fragment (of ?*tegula*) pierced by a nail hole, which is not apparently pre-formed and is partial only, SF 208 from fill (1014) of pit [2019]. Approximately one in every four *tegulae* has a partial pre-formed nail hole and most are at the top of the tile. A few are blind, apparently only to be broken through when required (Warry 2006, 17), which may have been the case with this example.

Portable ovens (*Clibani*)

Two fragments of portable oven (*clibanus*) were recovered from pit [3012]. They comprise a wall fragment, SF 235, from fill (3010), which is slightly abraded and weathered and has a smooth interior (Cat 2). The wall thickness is 26 mm and the exterior surface is decorated with diagonal finger-wiping. It may be part of the same oven as wall fragment, SF 236, from fill (3011), although the pieces do not adjoin. SF 236 has a thickness of 25 mm and is abraded, weathered and battered. It has a smooth interior and the exterior (part only of which survives) is decorated with diagonal finger-wiping (Cat 3).

Portable ovens made at Holt are coil-built and have a smoothed/wiped interior. The exterior surface is either left smooth or sometimes decorated with distinctive diagonal finger-wiping, as shown on the example from Prestatyn, published in Blockley 1989, 164 fig 87. The only published example of a portable oven from Chester is a thumb-impressed rim fragment from 25 Bridge Street (Garner *et al* 2008,170, Ill 5.1.5). It may have been used for the production of hot snacks for visitors to the fortress baths.

Catalogue (Fig 2.26)

1 Corner of ?box tile (?*tubulus*) with two ?oval air vents and an angled corner (?multi-sided form). Th: 32 mm; Event 2, (3049), SF 244.

2 Wall of portable oven (*clibanus*) with diagonal finer-wiping on exterior surface. Coil-built. Possibly the same oven as (3011), SF 236. Th: 26 mm; Event 2, (3010), SF 235.

3 Wall of portable oven (*clibanus*) with diagonal finer-wiping on exterior surface. Coil-built. Possibly the same oven as (3010), SF 235. Th: 25 mm; Event 2, (3011), SF 236.

Event 3

The deposition of rubbish into the partially infilled quarry pits (Event 3) produced 2,532 fragments weighing 149,969 g (59.3% by fragment count and 54.1% by weight of the Phase 1 assemblage). As with Event 2, roof tiles predominate and *tegulae* are the most common form,

FIGURE 2.26 TILE.

comprising 55.4% by weight of the Event 3 assemblage. Indeterminate forms comprise 17.8% by weight of the total, *imbrices*/ridge tiles 13.8% and bricks 9.6%. Small, but significant, amounts of small floor bricks, facing or box tiles, herringbone-floor bricks, facing tiles, box tiles, *imbrices* and portable oven fragments were also recovered (*see* Table 2.11). 896 fragments/1,767 g (35.4% by count but only 1.2% by weight of the Event 3 total) were recovered from samples.

Tegulae

The evaluation produced 22 fragments of *tegulae* weighing 7,219 g, from fills (4) and (31), and layers (15), (28), (49), (70) and (82), with the largest group coming from (70) (43.2% by weight of the total number of *tegulae* from the evaluation in Event 3). Layer (28) produced 27.5% by weight of the total and fill (31) 18.6% by weight of the total. The remaining 10.7% came from fill (4) and layers (15), (49), and (82). The *tegulae* from the evaluation are in mixed but relatively good condition, ranging from fresh to slightly abraded and weathered. Only two fragments are soft and powdery due to weathering. There is no evidence of re-use and only one fragment is burnt.

The group from (70) includes a middle fragment with a fairly complex signature of overlapping straight and curved finger marks, SF 204. A more typical signature, SF 200, comprises an end fragment from (31) with a single, curved signature arising from the end of the tile. Two fragments are decorated with a coloured slip wash. These comprise a fragment from (4) with traces of red slip on the upper surface and a corner from (82) with a wash of pale cream slip over the flange and upper surface.

The only datable form from the evaluation comprises a group A *tegula* from (28), dated *c* AD 90/100-120 (Warry 2006, 4 and 63).

Almost all the *tegulae* from the excavation (425 fragments weighing 75,796 g), were recovered from pits: [1018] [1019], [1022], [2110], [2125], [2191], [2216], [2221], [2234], [2243], [2255], [2259], [2291], [2292], [2293], [2295], [2352], [3012], [4002], [4012], [4041], [5004], [5007 [5013] and [5020], with the largest group coming from [4041] (22.3% by weight of the total number of *tegulae* from the excavation in Event 3). The remainder came from layer (2326) and scoop [4045]. 15 fragments weighing 836 g were recovered from samples.

As with the *tegulae* from the evaluation, the assemblage is of mixed condition, ranging from fresh or fairly fresh to abraded, battered and weathered. 35 fragments show signs of burning. Nine fragments have been re-used, as white lime-wash or mortar is present on broken edges.

Pit [4041]

Fill (4015) produced the largest group of *tegulae*, with 60 fragments weighing 13,680 g. The group includes signatures SF 255, a middle fragment with a double curved (probably circular) signature, SF 256 a middle fragment with a single straight finger mark overlapped by double straight finger marks and SF 254, an end piece with a triple, looping signature. Two *tegulae* have partially-sanded upper cutaways, although it is very rare for upper cutaways to be covered in moulding sand (Warry 2006, 20). Datable forms comprise an A 28 *tegula* and a (partial) probable group A form (both dated *c* AD 90/100-120), a group B 6 (dated *c* AD 120-140) and a partial probable group D 1 (dated *c* AD 240-380). The last is probably intrusive.

Tegulae of note from the remaining fills (4016), (4028), (4029) and (4049) comprise SF 260, a middle fragment with double, curving signature; a corner piece with lightly sanded stepdown band; and a partial, sanded upper cutaway, all from (4016).

Pit [1019]

Fill (1008) produced 46 fragments weighing 8,846 g. The group includes a middle fragment with faint, overlapping curved finger marks, SF 209. Datable forms comprise two group D 16 lower cutaways (dated *c* AD 240-380), a partial group C 56 lower cutaway (dated *c* AD 160-260) and a group A lower cutaway (dated *c* AD 90/100-20). The group C and D *tegulae* are presumably intrusive.

One of the group D *tegulae* has cream slip wash painted onto the top of the flange only. There are also two partial, lightly sanded upper cutaways. One end fragment has patchy cream mortar attached to the end and underside, probably from where the tile was originally mortared into position.

Pit [2110]

Fill (2111) produced two group A lower cutaways and a partial group ?B lower cutaway.

Fill (2112) produced an end piece with a faint triple, curved signature arising from the end, SF 213, and an end fragment with a partial curved signature arising from the end, SF 214. A probable (partial) group A lower cutaway was also recovered.

One of the three upper cutaways recovered from fill (2113) is sanded.

Pit [2191]

All of the *tegulae* from pit [2191] was retrieved from fill (2192) and comprises 15 fragments weighing 2,582 g. The group includes an abraded end fragment with a very worn/rounded edge, SF 221, which may have been re-used as a rubber/grinder/smoother.

Pit [2209]

32 fragments weighing 5,780 g came from fill (2175)/(2272). Notable forms comprise a fragment with faded

cream slip on the upper surface and a partially sanded upper cutaway.

Pit [2292]

The largest group was retrieved from fill (2285), with 10 fragments weighing 1,664 g. The group includes a corner fragment with a complete group A lower cutaway, which has been re-used and subsequently burnt. Forms of note from the remaining fills comprise a flange fragment with faded cream slip on the upper surface from (2284), a fragment with faded cream slip on the upper surface from (2283), and a possible (partial) group C or D lower cutaway (dated *c* AD 160-380), also from (2283). The last is presumably intrusive as it post-dates the end of the phase.

Pit [3012]

The largest group was retrieved from fill (3005), comprising 15 fragments weighing 2,694 g. These include a middle fragment with a faded single straight ?signature, SF 234, a flange fragment with a partial sanded upper cutaway and a group A 28 lower cutaway. A middle fragment with sandy mortar attached only to the upper surface may originally have been mortared into place.

Fill (3006) produced a probable (partial) group D 16 lower cutaway (dated *c* AD 240-380), which may also be intrusive. There is also a sanded upper cutaway. The group from fill (3007) includes a complete group A 26 lower cutaway (dated *c* AD 90/100-120), and a sanded, partial upper cutaway.

Pit [4002]

Fill (4009) produced 22 fragments weighing 3,594 g. The only datable form is a group A 26 lower cutaway (*c* AD 90/100-120).

Fill (4004) produced 18 fragments weighing 2,854 g. Notable forms comprise a middle fragment with single, curved signature, SF 249, and a partial group A 26 lower cutaway.

Fill (4007) produced 10 fragments weighing 1,224 g. One fragment has partially vitrified mortar attached to the broken edges, indicating that it was burnt after re-use. The only datable form comprises a partial group B 6 lower cutaway.

Pit [5004]

Pit [5004] produced 16 fragments weighing 1,957 g from fills (5002) and (5019), including a single fragment/39 g from samples. There are three sanded, partial upper cutaways, one from (5002) and two from (5019). A middle piece from (5002) has patchy white mortar attached to the sanded underside but not on the broken edges, indicating that it was originally mortared into position.

Notable forms from the remaining pits comprise SF 206, an abraded and weathered fragment with a double curved signature from fill (1010) of pit [1022]; two fragments with white mortar attached to the underside and side of the flange from fill (1015) of pit [1018], indicating that these tiles were originally mortared into position; a partially sanded upper cutaway, also from (1015); a complete group A 26 lower cutaway from fill (2290) of pit [2291]; and a complete group C 56 lower cutaway from fill (5014) of pit [5013], dated *c* AD 160-260. The last is probably intrusive.

Pit [5007]

All the *tegulae* from pit [5007] (29 fragments weighing 7,344 g) came from fill (5005). The group includes a flange fragment with a curved signature, SF 268, and a middle piece with a single curved signature, SF 269. SF 268 also has faded cream slip painted onto the upper surface and the side of the flange. One of the two upper cutaways from (5005) is sanded. Two other fragments are much worn at the junction of the base and the side of the flange and may have been re-used as rubbers/grinders/smoothers. Datable forms comprise two group A lower cutaways, an A 26 and an A 28 (both dated *c* AD 90/100-120). The A 26 tile has patchy white *opus-signinum* type mortar attached to the end, upper and lower surfaces, but not to the broken edges, indicating that it may originally have been mortared into position.

Notable forms from the remaining contexts comprise a sanded stepdown band from fill (2199) of pit [2216]; a flange fragment, SF 224, with a very worn and abraded underside, which may have been re-used as a rubber/grinder/smoother, from fill (2242) of pit [2243]; a partial group A lower cutaway, also from (2242); a middle fragment, SF 232, with a partial, ?single curved signature from fill (2311) of pit [2293]; and a sanded upper cutaway from fill (2294) of pit [2295]. An end fragment with faded cream slip on both the upper surface and underside was recovered from layer (2326).

Imbrices

The excavation produced one definite *imbrex* weighing 502 g, comprising a corner fragment, fairly fresh in condition, from fill (4015) of pit [4041].

Imbrices/ridge tiles

The evaluation produced seven fragments of *imbrex*/ridge tile weighing 1,262 g, from fill (31) and layer (70). The tiles are generally fresh or fairly fresh in condition. There is no evidence of re-use and there are no signs of burning.

Almost all the *imbrices*/ridge tiles from the excavation (211 fragments weighing 19,427 g) came from pits: [1018], [1019], [1022], [2110], [2191], [2209], [2234], [2243], [2255], [2259], [2292], [2293], [2295], [2299], [3012], [4002], [4012], [4041], [4045], [5004]and [5007], with the largest group coming from [4041] (23.7% by fragment count and 28.2% by weight of the total number of imbrices/ridge tiles from Event 3). A single fragment, weighing 42 g was retrieved from layer (2326).

Five fragments weighing 55 g were recovered from samples. The *imbrices*/ridge tiles from the excavation are of mixed condition, ranging from fresh to abraded, battered and weathered. 17 fragments show signs of burning. One fragment has been re-used, as pink *opus signinum*-type mortar is attached to broken edges.

Pit [1019]

Fill (1008) produced the largest group, with 25 fragments weighing 1,891 g. Three fragments have patchy cream slip on the upper surface. An end piece has patches of cream mortar attached to the underside, indicating that the tile was originally mortared into position on a roof.

Pit [2110]

Fill (2113) produced an unusual, roughly finished tile, SF 215, with a raised ridge at the upper edge of the smoothed end; the latter is thumbed over at one edge. Another notable form comprises an unusual ?signature of overlapping single lines, SF 212 from fill (2111). Signatures on *imbrices* are normally on the side of one corner of the wider end or across the top of the gable (Brodribb 1987, 101). The same context also produced a middle piece with faded cream slip on the upper surface.

Pit [2209]

15 fragments/1,959 g came from fill (2175)/(2272). These include a middle piece with signature across the top of the gable, SF 218, and a corner fragment with faded cream slip on the upper surface, both from (2175).

Notable forms from the remaining pits comprise a middle fragment with faded cream slip on the upper surface from fill (1010) of pit [1022] and an edge fragment with signature parallel to the edge, SF 266, from fill (5002) of pit [5004].

Pit [2292]

Notable tiles comprise a corner fragment with three finger marks across the top of the gable, SF 229, a middle fragment with a signature parallel to the edge, SF 230, and an edge with faded cream slip on the upper surface, all from fill (2283). A middle piece with faded cream slip on the upper surface came from fill (2285).

Pit [2293]

Fill (2286) produced a middle fragment with cream sandy mortar attached to the upper surface, which indicates that the tile was originally mortared into position. A middle fragment with faded cream slip on the upper surface was recovered from fill (2311).

Pit [3012]

The only notable piece comprises an edge fragment with faded cream slip on the upper surface from fill (3003).

Pit [4002]

Notable forms comprise an edge fragment with patchy cream slip on the upper surface from (4004) and a middle fragment with patchy cream slip on the upper surface from (4007).

Pit [4041]

Fill (4015) produced the largest group of *imbrices*/ridge tiles, with 39 fragments weighing 4,486 g. The group includes a corner fragment with a faint ?signature across the top of the gable, SF 257, a corner fragment with signature parallel to the edge, SF 258, and an end piece with a ?signature across the top of the gable, SF 259. One middle fragment has faded cream slip on the upper surface. Another fragment has been re-used, with pink *opus-signinum* type mortar attached to broken edges.

Imbrices/ridge tiles of note from the remaining fills comprise an edge fragment with faded cream slip on the upper surface from (4016).

Pit [5007]

All the *imbrices*/ridge tiles, comprising 11 fragments weighing 1,622 g, came from fill (5005). The group includes a corner fragment with a signature parallel to the edge, SF 270.

Notable forms from the remaining contexts comprise a corner fragment with a ?signature parallel to the edge, SF 226 from fill (2254) of pit [2255], two fragments with cream slip on the upper surface from fill (2192) of pit [2191], and an edge piece from fill (4011) of pit [4012], with faded cream slip painted onto the upper surface. Faded cream slip appears on the upper surface of a middle fragment from layer (2326).

Box tile (*Tubuli*)

The evaluation produced a corner fragment of box tile (*tubulus*) weighing 23 g from layer (81). It is relatively thin-walled (with a thickness of 15 mm), slightly abraded in condition, and with sanded interior and exterior surfaces (Chester type 12).

The excavation produced six probable fragments of box tile, weighing 509 g, with four fragments from pit [4041]. They are generally fresh or fairly fresh in condition. One fragment, from fill (4015) of pit [4041], is slightly abraded and partially burnt; another, from fill (5002) of pit [5004], is misfired/burnt and partially vitrified.

Pit [3012]

Fill (3006) produced a corner fragment with the edge of a knife-cut air vent (shape uncertain). There is no evidence for keying on the sanded exterior but the thinness of the walls (12 mm) indicates that it is probably an early, thin-walled form.

Pit [4041]

Box-tile fragments were recovered from fills (4015) and (4028). Three end fragments from (4015) comprise two with sanded exterior surfaces and thicknesses of 19 mm and 20 mm; the former has the edge of a knife-cut air vent (shape uncertain). Neither fragment is keyed. The third end piece has an unsanded exterior with knife-scored keying. It is fairly thick (24 mm) for a knife-scored box tile and therefore may be a facing tile or half-box tile instead.

A corner fragment with sanded exterior surfaces came from (4028) (thickness 18 mm). There is no surviving evidence for the type of keying used.

Pit [5004]

Fill (5002) produced a corner fragment of box tile weighing 50 g. It is relatively thin-walled, with a thickness of 14 mm. The sanded exterior has no surviving evidence of the type of keying used.

Facing tile (*Parietales*)

The evaluation produced a corner fragment of facing tile (*parietalis*) weighing 209 g from fill (31). It is slightly abraded and weathered and is sooted on the sanded, exterior face. There is a notch in the side of the tile, for attachment to a wall using a nail or cramp. There is no evidence of knife-scored keying on the sanded face, which, along with the nail- or notch-hole in the side, is normally also a distinctive feature of this form (Brodribb 1987, 58).

An end fragment weighing 675 g, probably also from a facing tile (thickness 25 mm), was retrieved from layer (70). The exterior face is sanded and keyed with thick, roughly made, lattice or diamond knife-scoring. The tile is slightly abraded.

Two fragments of facing tile weighing 492 g were recovered from the excavation. Both are slightly abraded and weathered. A middle fragment from fill (2111) of pit [2110] has a thickness of 30 mm. The sanded, exterior face is keyed with lattice or diamond-shaped knife-scoring. Another middle fragment, SF 252, was recovered from fill (4011) of pit [4012]. It has a thickness of 26 mm and the sanded face is keyed with deep diamond-shaped knife-scoring. The smooth interior face, unusually, bears a double circular signature.

The excavation produced three fragments of probable facing tile weighing 415 g, from pits [2291], [4002] and [4041]. They are of mixed condition. An edge fragment (thickness 21 mm) from fill (2290) of pit [2291] has a sanded, knife-scored exterior which is also sooted, possibly through use. A middle piece (thickness 24 mm) with sanded knife-scored exterior (partially burnt) came from fill (4004) of pit [4002]. A middle fragment (thickness 22 mm) was recovered from fill (4016) of pit [4041]; it also has a sanded, knife-scored exterior.

Brick

The evaluation produced two fragments of brick weighing 709 g, from fill (4) and layer (70). Both are slightly abraded edge fragments. The piece from (4) has a thickness of 50 mm and is therefore more likely be part of either a Holt *bessalis*, a Holt *pedalis* or a *sesquipedalis* (*see* above for the relative average thicknesses of these forms). The piece from (70) has been re-used, with pink/white *opus-signinum* type mortar attached to broken edges. With a thickness of 40 mm, it is perhaps more likely to be a *lydion*.

Bricks in a range of thicknesses/forms were recovered from the excavation, comprising 43 fragments weighing 1,371 g. All were recovered from pits [1018], [1019], [2110], [2191], [2209], [2255], [2291], [2292], [2293], [4002], [4012], [4041] and [5007], with the largest group by weight coming from [4041] (11.0% by weight of the total number of bricks from Event 3). Five fragments have thicknesses ranging between 36 and 42 mm and are more likely to be either *lydion* bricks or thin (for Holt) *bessales*. Nine fragments have thicknesses ranging between 45 and 56 mm, and are likely to be from *bessales*, *pedales* or *sesquipedales*. Seven fragments range in thickness between 58 and 62 mm and are more likely to be either Holt *pedales* or tegulae *bipedales*.

The assemblage is of mixed condition, ranging from fresh to abraded, battered and weathered. 11 fragments show signs of burning and one, from fill (4015) of pit [4041], is also heat-cracked. Two fragments have been re-used, as they are coated in a grey/brown sandy ?Roman mortar.

Pit [2209]

Fill (2175) produced three fragments, probably from the same brick (although they do not adjoin), with a thickness of 61 mm. It may be either a *tegula bipedalis* or a Holt *pedalis*.

Pit [4002]

Forms include three with thicknesses of >55 mm from fills (4004), (4005) and (4009); they may be from either *tegulae bipedales* or Holt *pedales*.

Pit [4041]

Fill (4015) produced six fragments weighing 960 g. These include three possible *tegulae bipedales* or Holt *pedales*, with thicknesses of 59 mm, 62 mm and >57 mm.

Notable forms from the remaining fill, (4029), comprise a fragment with white *opus-signinum* type mortar attached to the sanded underside, indicating that it may originally have been mortared into position, and a possible *tegula bipedalis* or Holt *pedalis*, with a thickness: of 62 mm.

Forms of note from the other pits comprise a fragment with a double, curved signature, SF 210, from fill (2111) of pit [2110]. One fragment, from fill (1008) of pit [1019], has an abraded upper surface, probably from use-wear. An edge fragment from fill (2192) of pit [2191] has patches of pink *opus-signinum* type mortar attached to the edge and underside, but not to broken edges, indicating that it may originally have been mortared into position. A corner fragment (thickness: 60 mm) of either a *tegula bipedalis* or a Holt *pedalis*, with a partial single straight ?signature, SF 231, was recovered from fill (2285) of pit [2292].

Herringbone-floor (opus spicatum) bricks

The excavation produced four fragments weighing 1004 g, from pits [1019], [2209] and [2291].

A complete herringbone-floor brick weighing 306 g was recovered from fill (1008) of pit [1019] (Cat 5). It is slightly abraded and lightly sooted and may have been laid flat rather than on edge, as one wide face is abraded in comparison to the other face and edges. At length: 104 mm, width: 65 mm and thickness: 25 mm, it is smaller than average for Holt but closer to the average generally for width (62 mm) and thickness (26 mm). Another complete brick, SF 273, weighing 385 g, was recovered from fill (2290) of pit [2291] (Cat 4). One long edge is slightly worn, presumably through use-wear. At length: 109 mm, width: 67 mm and thickness 32 mm, it is smaller than the average length and width for Holt but the same average thickness.

Two pieces, probably from the same brick, although they do not adjoin, came from fill (2175) of pit [2209]. With a width of 67 mm and a thickness of 29-30 mm, the pieces are smaller than average for Holt but wider than the average generally (width: 62 mm). However, they are close to the average thickness for Holt but thicker than the average generally (26 mm).

Small floor bricks

Two complete small floor bricks, weighing 1,152 g, were recovered from the excavation. Both came from fill (3008) of pit [3012]. Both are fairly fresh in condition. One has a slightly abraded upper surface, probably through use-wear. They are closely similar in size: one is length: 110 mm, width: 88 mm and thickness: 32 mm (Cat 6); the other is length: 110 mm, width: 96 mm and thickness: 30 mm. These dimensions are comparable to the small floor bricks produced at Holt (length: 113 mm, width: 88 mm and thickness: 32 mm). They were probably laid flat (i.e. wide face down) on their sanded undersides.

Portable ovens (Clibani)

The excavation produced a single wall fragment of portable oven (*clibanus*), SF 219, weighing 215 g, from fill (2175) of pit [2209] (Cat 7). It is fairly fresh in condition, although the interior is sooted, probably though use. The oven is coil-built and has a smoothed/wiped interior. The exterior surface is decorated with distinctive diagonal finger-wiping. It has a thickness of 23 mm.

Indeterminate forms

The evaluation produced seven indeterminate fragments weighing 418 g, from fill (31) and layers (49), (64) and (70). They are of mixed condition, ranging from fairly fresh to abraded. One fragment is lightly burnt. There is no evidence of re-use.

The excavation produced 1,793 indeterminate fragments weighing 26,229 g. Apart from a single fragment, weighing 102 g, from ditch [2200] and two fragments, weighing 333 g, from scoop [4045], the remaining indeterminate forms were recovered from pits: [1018], [1019], [1022], [2110], [2125], [2129], [2167], [2191], [2204], [2208], [2209], [2211], [2216], [2221], [2234], [2243], [2255], [2259], [2291], [2292], [2293], [2295], [2299], [2301], 2317], [2325], [2352], [3012], [], [4002], [4012], [4041], [5004], [5007] and [5013]. The largest groups came from pits [4041] (15.5% by weight of the total number of indeterminate forms from Event 3), [1019] (11.1% by weight), and [4002] and [3012] (both 9.9%). 876 fragments weighing 876 g were recovered from samples.

The assemblage is of mixed condition, ranging from fresh to abraded, battered and weathered. 48 fragments show signs of burning. 21 fragments have been re-used, with mortar attached to broken edges. 12 of these, including two cemented together, are coated in mortar and may have been re-used as aggregate for *opus signinum*.

Fragments of note comprise an edge fragment from fill (1008) of pit [1019], SF 220, which appears to have been deliberately smoothed to form a curved shape (of uncertain function); two other pieces from (1008) are decorated, one with cream slip on the underside, the other with red slip on the upper surface. Other fragments of note comprise a middle piece with single curved signature, SF 211, from fill (2111) of pit [2110]; a fragment roughly shaped into a disc, SF 223, from fill (2212) of pit [2191]; a middle fragment with a partial, deeply-impressed ?animal print in the upper surface, SF 216, from fill (2113) of pit [2110]; a middle piece also from (2113) with patchy cream slip on the upper surface; a middle piece with a partial curved signature, SF 217, from fill (2128) of pit [2129]; a fragment with very worn/abraded upper surface which may have been re-used as a rubber/grinder/smoother, SF 222, from fill (2201) of pit [2200]; a middle piece (of ?*tegula*), with faded red slip on the upper surface, from fill (3006) of pit [3012]; a middle piece with a fragmentary (indeterminate) legionary stamp, SF 250, from fill (4005) of pit [4002]; and a chip with a single curved signature, SF 271, from fill (5005) of pit [5007].

Catalogue (Fig. 2.26)

4 Complete herringbone-floor (*opus spicatum*) brick. L: 109 mm; W: 67 mm; Th: 32 mm; Event 3, (2290), SF 273.

5 Complete herringbone-floor (*opus spicatum*) brick. L: 104 mm; W: 65 mm; Th: 25 mm; Event 3, (1008) [*Not illustrated*].
6 Complete small floor brick. L: 110 mm; W: 88 mm; Th: 32 mm; Event 3, (3008), SF 274.
7 Wall of portable oven (*clibanus*) with diagonal finger-wiping on the exterior surface. Coil-built. Th: 23 mm; Event 3, (2175), SF 219.

Other Group 112 contexts

Group 112 contexts, for which no Event numbers have been assigned, comprise 171 fragments with a weight of 8,173 g (4% by fragment count and 3.2% by weight of the Phase 1 assemblage) (*see* Table 2.11). A limited range of forms was recovered, comprising *tegulae*, *imbrices*/ridge tiles and indeterminate forms. A wall fragment of portable oven was also retrieved. As with the assemblages from Events 2 and 3, roof tiles predominate. *Tegulae* comprise 57.9% by weight of the assemblage from this group, *imbrices*/ridge tiles 31.7% and indeterminate forms just 7.6% by weight of the total. The assemblage is of mixed condition, ranging from fresh to very abraded. 14 fragments are burnt and two have been re-used, as mortar is attached to broken edges and surfaces. 37 fragments (21.6%) weighing 29 g (0.4%) were recovered from samples.

Most of the group (56.1% by fragment count and 69.3% by weight) came from fill (4003) of pit [4012]. The remainder was recovered from fills (2098) and (2106) of pit [2097] and layers (2156), (2185), (2276), (3002) and (4001). The only datable form comprises a probable group B *tegula*, dated *c* AD 120-140, from (4003).

Tegulae

Layer (4001) produced just 3 fragments weighing 514 g, all fairly fresh in condition. Fill (4003) of pit [4012] produced 21 fragments weighing 2,204 g. The group includes an end fragment with a single curved signature arising from the end, SF 246, a middle fragment with a double, curving signature, SF 247, and, a middle fragment with the very faint impression of a dog's paw print in the upper surface, SF 249. There is also a corner piece with a probable (partial) group B6 lower cutaway.

Notable forms from the remaining contexts comprise a middle fragment with a single curved signature, SF 233, from layer (3002).

Imbrices/ridge tiles

Layer (4001) produced just four fragments weighing 830 g, either fresh or fairly fresh in condition, including an edge fragment with patchy cream slip painted on the upper surface.

Notable forms from the remaining contexts comprise a middle fragment with faded cream slip on the upper surface, from fill (2098) of pit [2097], and a middle fragment from layer (2156) with a line of three roughly circular spots (drips) of red slip on the upper surface.

Portable ovens (Clibani)

A single wall fragment of portable oven, SF 248, weighing 230 g was recovered from fill (4003) of pit [4012]. It has a slightly abraded exterior and a fairly fresh interior. Both surfaces are smooth and lightly sooted, possibly through use. The wall thickness is 33 mm (Cat no 8).

Indeterminate forms

There are no fragments of note from this group.

Catalogue (Fig. 2.26)

8 Wall of portable oven (*clibanus*). Coil-built. Th: 33 mm; Phase 1: Other Group 112 contexts, (4003), SF 248.

Group 113: Other features

Other features in Phase 1 (Group 113) produced 93 fragments with a weight of 1,639 g (2.2% by fragment count and 0.6% by weight of the Phase 1 assemblage) (*see* Table 2.11). A limited range of forms was recovered, comprising *imbrices*/ridge tiles, bricks and indeterminate forms. A single fragment of ?*tegula* was also retrieved. As with the assemblages from Events 2 and 3 and other Group 112 contexts, roof tiles predominate. *Imbrices*/ ridge tiles comprise 48.0% by weight of the assemblage from Group 113, bricks 26.2% and indeterminate forms 24.9% by weight of the total. The assemblage is of mixed condition, ranging from fresh to abraded and weathered. Six fragments are burnt and two have been re-used, as mortar is attached to broken edges and surfaces. 69 fragments (74.2%) weighing 29 g (4.9%) were recovered from samples.

Most of the group came from fills (2248), (2249), (2250) and (2277) of ditch [2252]. The remainder came from fill (2131) of pit [2130]. Little of note was recovered.

The post-Roman and unstratified assemblage

Post-Roman and unstratified contexts from the evaluation and excavation together produced 324 fragments weighing 18,489 g (7.1% by fragment count and 6.7% by weight of the total assemblage). Eleven fragments weighing 1,461 g came from the evaluation and 313 fragments weighing 17,028 g were recovered from the excavation. The assemblage is of mixed condition, ranging from fresh to very abraded. 18 fragments are burnt and three have been re-used, as mortar is attached to broken edges and surfaces. 87 fragments (26.9%) weighing 74 g (0.4%) were recovered from samples.

A wide range of forms was retrieved, although roof tiles predominate. *Tegulae* comprise 48.0% by weight of the post-Roman and unstratified assemblage, with *imbrices*/ridge tiles at 16.1%. Indeterminate fragments account for 21.8% by weight of the total, with bricks at 7.2%. Small, but significant, amounts of herringbone-floor bricks, and box tiles were also recovered, as well as single examples each of *imbrex*, small floor brick (two adjoining fragments) and portable oven.

Tegulae

The evaluation produced five fragments of *tegulae* weighing 643 g (44.0% by weight of the total post-Roman and unstratified assemblage from the evaluation). Notable forms comprise a probable (partial) group A lower cutaway from (17) and a complete group D lower cutaway from (35).

The excavation produced 57 fragments of *tegulae* weighing 8,234 g (48.4% by weight of the total post-Roman and unstratified assemblage from the excavation). Datable forms comprise three group A lower cutaways (all partial) from (3048), (4024) and (4036) and a complete group D lower cutaway from (2137). The former are all dated to *c* AD 90/100-120, the latter is dated to *c* AD 240-380.

Other notable forms comprise a middle fragment with a double, curved signature, SF 228, from (2280), and a middle piece with faded red slip on the upper surface, from (4047). Of the six upper cutaways recovered, four are sanded and one, from (2205) also has a sanded stepdown band overlying the cutaway. The only item of note found unstratified comprises an end fragment with a faint double curved signature, SF 272.

Imbrices/ridge tiles

These were all recovered from the excavation and comprise 29 fragments weighing 2,975 g (17.5% by weight of the total post-Roman and unstratified assemblage from the excavation). The group includes SF 251, a corner piece from (3048) with a signature parallel to the edge. Faded or patchy cream slip appears on the upper surfaces of two middle fragments, from (2005) and (2319) and on a corner fragment from (4006). The only item of note found unstratified comprises a middle fragment with faded cream slip on the upper surface.

Bricks

The evaluation produced a single fragment of brick weighing 296 g (20.3% by weight of the total post-Roman and unstratified assemblage from the evaluation). The excavation produced five fragments weighing 1,038 g (6.1% by weight of the total post-Roman and unstratified assemblage from the excavation). The only form of note comprises an edge fragment with faded cream slip on the upper surface from (2205). It has a thickness of 62 mm and is probably from either a *tegula bipedalis* or a Holt *pedalis*.

Small floor bricks

Two adjoining pieces, forming a complete (except for a missing corner) small floor brick weighing 529 g were recovered from the excavation. Both pieces came from (2319). The brick is comparable in both form and size to the examples recovered from Event 3. It has a sanded underside, with smooth edges and upper surface. It has a length: 109 mm, width: 93 mm and thickness: 33 mm. This also compares well with the small floor bricks from Holt (length: 113 mm, width: 88 mm and thickness: 32 mm).

Herringbone floor (Opus spicatum) bricks

The evaluation produced a complete (except for a missing corner) herringbone-floor brick, weighing 277 g, from (78). At length: 103 mm, width: 68 mm and thickness: 29 mm, it is slightly smaller than average for the examples from Holt (length: 113 mm; width: 75 mm; thickness: 32 mm). One long edge (upper surface) is abraded, presumably through use-wear.

A single piece, with two surviving corners, weighing 70 g, was recovered from the excavation, context (4047). It is slightly abraded and weathered. At width: 62 mm and thickness: 24 mm, it is smaller than average for Holt (*see* dimensions above) but close to average for the rest of Britain (width: 62 mm; thickness: 26 mm).

Portable ovens (Clibani)

The excavation produced a single wall fragment of portable oven weighing 43 g, SF 237, from (3018). It is fairly fresh but is partially burnt/reduced on the exterior surface and one broken edge. The interior surface is missing and the exterior surface is also decorated with diagonal finger-wiping.

Indeterminate forms

Three indeterminate fragments weighing 223 g were recovered from the evaluation (15.3% by weight of the post-Roman and unstratified assemblage from the evaluation). The only piece of note is a middle fragment with a deeply impressed, single curved signature, SF 205 from (78). The assemblage from the excavation comprises 215 fragments weighing 3,806 g (22.4% by weight of the post-Roman and unstratified assemblage from the excavation), but nothing of note was recovered. The only item of note found unstratified comprises an edge fragment with faded cream slip surviving on the edge and underside.

Catalogue (Fig.2.26)

9 Wall of portable oven (*clibanus*) with finger-wiping on exterior surface; interior surface missing. Coil-built. Post-Roman (3018), SF 237.

Discussion

Phase 1 produced 4,267 fragments of Roman ceramic building material weighing 257,717 g (92.9% by fragment count and 93.3% by weight of the total assemblage). Although roof tiles, particularly *tegulae*, dominate the assemblage, a wide range of forms was recovered (*see* Table 2). The Phase 1 assemblage is of mixed condition, ranging from fresh to badly abraded and weathered. 253 fragments show signs of burning (representing 5.9% of the Phase 1 assemblage) and 49 fragments have been re-used (just 1.1% of the Phase 1 assemblage).

The presence of bricks in a range of thicknesses implies a range of forms. The range of thicknesses indicates that most are probably fragments of either *bessales*, *pedales* or *tegulae bipedales*, the three forms most commonly used in hypocaust construction. The small square *bessalis* was used to form brick-stack *pilae*, the larger square *pedalis* (which measured exactly one Roman foot square) was generally used as a capping brick for such *pilae*, and the *tegula bipedalis*, the largest of the square bricks, was generally used to bridge the gaps between *pilae*. Some may have come from the *sesquipedalis*, a large square brick measuring one and a half Roman feet square, which was also used for flooring and often found bedded into mortar, or from the *pedalis*, a large rectangular brick, similar in size to the *tegula*, which was commonly used for levelling or binding courses in walls (Brodribb 1987, 34-43).

Although small in number, the presence of box tiles (*tubuli*), facing tiles (*parietales*) and half-box tiles (*tegulae hamatae*) is also notable, as these forms will have originated from centrally-heated rooms, which were therefore presumably occupied by individuals of some wealth or status. Three of the box tiles are thin-walled forms, although only one fragment has knife-scored keying present on its wide face. In London and the South East, these thin-walled, knife-scored box tiles pre-date the early second century (Black 1996, 60; Ward 1999, 48; Pringle 2006, 128). They are often found in association with facing tiles and half-box tiles. Black also notes that hypocausts used to heat domestic rooms are extremely rare before *c* AD 150 and that the vast majority of box tiles made before the mid-second century (especially the thin-walled forms) are more likely to have come from bath-buildings (Black 1996, 64). It is therefore possible that this small group of early box tiles and facing tiles, together with the single half-box tile recovered from Phase 1, indicates the presence of a bath-house, possibly an extra-mural bath-building, somewhere in the vicinity of the site, although it is not possible to be more definitive about such a small group.

The *tegulae* assemblage from Phase 1 includes 48 datable lower cutaways, of which 30 (62.5%) are group As, five (10.4%) are group Bs, four (8.3%) are group As or Bs, three (6.25%) are group Cs, five (10.4%) are group Ds and one (2.1%) is a group C or D. This suggests that the bulk of the *tegula* assemblage (81.2%) dates to between *c* AD 90/100-140 and was presumably deposited sometime in the early to mid second century. The high percentage of group As indicates that they were probably deposited close to the date of transition from group A to group B (ie *c* AD 120). Of the remaining 18.8%, the majority (10.4%) are group D *tegulae*, which suggests that there was another phase of deposition post-dating the mid-third century AD. As Phase 1 is thought to come to an end in the mid second century, the presence of these later forms is slightly surprising. It is possible that some may be later intrusions but it is unlikely that all can be accounted for in this way.

A small group of *tegulae* (1.7% of the Phase 1 *tegula* assemblage) bear traces of coloured slip, generally on their upper surfaces, sometimes including the tops and inner edges of flanges, and more rarely on their undersides. This is sometimes red but is usually cream in colour (on 11 out of a total of 13 examples so decorated). A higher proportion of *imbrices*/ridge tiles (11.7% of the Phase 1 *imbrex*/ridge tile assemblage) are similarly decorated. Cream slip occurs on 47 of the 52 fragments, red on the remaining five. Warry notes that some *tegulae* appear to have been coloured by the application of slip and that the colours are limited to red, white/cream and dark blue. In most cases, the slip appears on both sides (indicating the tile was dipped in slip); on others, the slip appears to have been painted onto the upper surface only. He also writes that coloured tiles seem to be mainly associated with later sites (Warry 2006, 17). The only datable form decorated in this way is a group D *tegula* with cream slip on the upper surface, from (1008) in Event 3.

Warry has noted that the presence of moulding sand on upper cutaways is rare (Warry 2006, 20) and therefore the proportion of sanded upper cutaways is surprising, as these account for 36.0% of the total number of upper cutaways present in the *tegula* assemblage. The proportion for Phase 1 is almost identical; at 35.7% (25 are sanded out of a total of 70).

Other notable forms comprise a small group of herringbone-floor and other small floor bricks. Herringbone-floor bricks were commonly used for the construction of hard-wearing floors and are known to have been used extensively in both the fortress baths and the western extra-mural bath-house in Chester. The bricks were generally laid with the narrow edge down in a thick bed of concrete and arranged in a herringbone pattern. The resulting floor surface was both hard wearing and long lasting. Herringbone floors are not particularly common in Britain - plain brick floors paved with large square or rectangular bricks are more usual. Herringbone flooring is known to have been used in bathhouses at other British sites, as well as in a ?kitchen, a ?stable or cart shed and in other rooms of unknown function (Brodribb 1987, 52).

Perhaps the most surprising of the forms recovered are the fragments of portable oven. The only published example of this form from Chester is a thumb-impressed rim fragment from 25 Bridge Street (Garner *et al* 2008, 170, Ill 5.1.5).

The portable ovens made at Holt are barrel-shaped, have a large opening at the top with a thick, thumb-impressed rim and a semi-circular opening near the base. The outer surface of some ovens is also decorated with diagonal finger-wiping. Only two have been found near-complete, one from Holt itself (Grimes 1930, 184 and 212 fig 60, no 9); the other from Prestatyn, a Roman industrial settlement near Chester, which was also supplied by the Holt kilns (Blockley 1989, 164 fig 87).

The assemblage is distributed fairly evenly across the site and there do not appear to be any significant patterns of dispersal in terms either of date range or the range of forms. The assemblage presumably represents rubbish disposal from both the fortress and the nearby extra-mural settlement. It is not possible to link specific forms with specific buildings or structures anywhere in the vicinity of the site. However, it is possible to speculate that the oven fragments may have originated from the cook-houses of the second-century stone rampart buildings discovered at the nearby site of Abbey Green, which is located to the south of the site, within the eastern sector of the north wall of the fortress. (McPeake, Bulmer & Rutter 1980, 19).

The post-Roman and unstratified assemblage adds little to the picture outlined above. It generally reinforces both the date-range and range of forms already discussed. Most notably, the post-Roman assemblage includes an additional portable oven fragment, SF 237, from (3018).

THE ROMAN VESSEL GLASS *by Hilary Cool*

The Roman vessel glass from these excavations is extremely limited both with regard to date and to types represented. As can be seen from Table 2.13 approximately two thirds by fragment count consists of fragments of blue--green bottles with tablewares being much less common. In what follows the types represented will be briefly considered from the point of view of typology and dating. Then how this assemblage compares to other glass assemblages from Chester will be considered.

TABLE 2.13: THE ROMAN VESSEL GLASS BY COLOUR AND EVENT (FRAGMENT COUNT)

Colour	**2**	**3**	**0**	**Total**
Cast	1	1		2
Light colours		2	1	3
Colourless	1		1	2
Blue/Green	20	34	8	62
Bottle	35	67	17	119
Total	**57**	**104**	**27**	**188**

The earliest piece is a small fragment of a cast vessel (No. 1). This would have come from a shallow bowl made from canes with an emerald green ground in which there was a red centre surrounded by rings of opaque yellow dots. This type of vessel was in use at the time of the conquest, going out of use during the third quarter of the first century (Price and Cottam 1998, 47-8). The other example of cast glass (No. 2) comes from a blue--green pillar moulded bowl, a very common type generally out of use by the end of the first century (Price and Cottam 1998, 44-6).

There were several fragments from jugs, mostly from the very common types in use during the later first and into the second century (Isings Forms 52 and 55 - Price and Cottam 1998, 150-57). No. 3 came from the globular form, the handle fragments Nos. 4 and 5 could have come from either type and the body fragment No. 6 was probably from a conical jug. The globular form went out of use in the early second century but the conical form continued in popular use into the middle of the second century. The body fragments Nos. 10 and 11 have the typical colours and ribbed decoration seen on jugs of this type and on the contemporary collared jars.

No. 8 is a slightly distorted fragment that seems most likely to have come from a spouted jug with a pulled out rather than pinched-in rim (Price and Cottam 1998, 159-60). If this tentative identification is correct, then a second or third century type is appropriate.

No. 12 has the typical rim formation of a collared jar (Fig 2.27). The thickness of the glass strongly suggests that it does not belong to tableware form contemporary with globular jugs such as No. 3 (Price and Cottam 1998, 137-8), but to a robust container and so is one of the far less commonly identified square jars with this rim form (Price and Cottam 1998, 135-6) dateable to the later first and second centuries.

The rest of the assemblage consists of fragments from blue--green bottles (Price and Cottam 1998, 191-200). These came into common use late in the first century. The cylindrical form went out of use early in the second century. The prismatic forms continued in use throughout the second century and in some cases into the third century. Table 2.14 summarises the distribution of the bottle glass by weight in Events 2 and 3. The cylindrical bottles make up 7% and 18% respectively of the Event 2 and 3 bottle glass. For the prismatic forms the equivalent figures are 47% and 43%. The relatively strong presence of cylindrical bottles in the Event 3 contexts suggests that there was appreciable quantities of late first to early second century material in the rubbish that made up those fill. If they had been derived from rubbish with an appreciable mid to late second century element, then the prismatic figures should have been much higher

TABLE 2.14: THE BLUE/GREEN BOTTLES BY TYPE AND EVENT (WEIGHT IN G.)

Bottle type	**2**	**3**	**Total**
Rims etc	119	220	339
Cylindrical	17	104	121
Hexagonal	44	15	59
Prismatic	63	181	244
Square	13	49	62
Total	**256**	**569**	**825**

It is possible to place the vessel glass assemblage more in context than it has been possible to do for the small finds. The large body of vessel glass from the city remains unpublished in any detail, but in what follows I have drawn on my records from the 1980s when I wrote reports on many assemblages from excavations for the Chester Archaeological Service.

Table 2.15 presents a simple presence/absence listing for the forms which were common at Chester in the later first to mid second centuries for a number of sites where the type of occupation within the fortress at the time is known and which have assemblages of more than 100 fragments.

With the exception of the probable square collared jar, all of the forms represented at Delamere Street are ones that have been found at Chester before. This even extends to the cast polychrome piece No. 1 (not included in Table 2.15). Precisely similar fragments have been found at Abbey Green (1975-78 excavations unpublished) and from the most recent excavations at the amphitheatre. This presumably means that these vessels were still in circulation when the fortress was founded which would be late in the generally accepted date range for the type.

The first six columns of the table relate to tableware forms, the last three to containers. As can be seen in comparison

TABLE 2.15: PRESENCE/ABSENCE OF COMMON LATE FIRST TO EARLY SECOND CENTURY FORMS AT CHESTER

	EVE	Type	1	2	3	4	5	6	7	8
Delamere Street	5.84		X	X	-	-	-	X	X	X
Abbey Green 75-8	65.07	Barracks	X	X	X	X	X	X	X	X
Crook St. 1963-4	11.01	Barracks 1st	X	(?)	X	-	X	X	X+	-
Crook St. 1973-4	20.66	Barracks 1st	X	X	X	X	-	X	X	-
Hunter St School	16.44	Barracks	X	(?)	-	-	-	_	X	X
Goss St 1973	10.88	Tabernae	-	X	X	-	X	X	X+	
Old Market Hall	29.90	Praetorium/ workshops	X	X	-	-	X	X	X	X

Key to Columns
1 – blue/green pillar moulded bowl; 2 – Isings Forms 52/55 jugs; 3 - externally ground colourless beakers; 4 - lightly tinted beakers; 5 - tubular-rimmed bowls; 6 – cylindrical bottles, 7 – square bottles; 8 – hexagonal bottles.
X indicates presence; - indicates absence
Note (+) indicates the presence of prismatic bottles which are probably square bottles. Barracks 1st indicates first cohort barracks

TABLE 2.16: A FUNCTIONAL COMPARISON OF THE DELAMERE STREET EVENT 2 AND 3 ASSEMBLAGES WITH CONTEMPORARY ONES FROM THE INTERIOR OF THE FORTRESSES AT CHESTER AND YORK (DATA FOR THE CHESTER AND YORK ASSEMBLAGES TAKEN FROM COOL AND BAXTER 1999, TABLE 2)

Site	Cup	Bowl	Jar	Flask	Jug	Bottle	Total EVE
Event 2	-	0.20	0.18	-	0.14	1.00	1.42
Event 3	-	-	-	-	0.42	2.36	2.78
Chester	4.00	3.52	0.69	0.94	1.54	3.78	14.47
York	6.40	6.00	1.44	2.20	0.84	3.08	19.96

to the other sites the common tableware forms are scarce at Delamere Road. The scarcity of tablewares is thrown into even starker relief if the assemblage is compared using EVE measurements of the different functional forms. This is done in Table 2.16 using aggregated data from contexts from the interiors of the Chester and York fortresses that are precisely contemporary with the Delamere Street assemblage

As can be seen, this is an overwhelmingly utilitarian assemblage, and that is true for both events. The bottles were all purpose containers for items suitable for the kitchen for not only liquids but solids as well given the presence of wide-mouthed examples (see no. 17). This is at some odds with the samian pottery where the incidence of decorated bowls led Wild to suggest that it might represent the detritus from officers' dining tables (*see Chapter 3*). If this was the case then one could have expected colourless or lightly tinted glass beakers to be represented but they are absent and there is only a single sherd of colourless glass in the assemblage (see Table 2.13). As can be seen from Table 2.15 column 3, facet-cut and other externally ground colourless beakers are a regular occurrence within the fortress (for type see Price and Cottam 1998, 80-82). It is these that one might have expected to grace officers' tables.

With glass it is always possible that an assemblage might be biased because parts of it might have been being collected for re-cycling as cullet to be made into new vessels and articles. That might be thought to be a particular problem in the case of primary rubbish which this clearly is. This does not appear to be the case here though as there are many large and substantial pieces of precisely the types that would have been useful for re-melting. The production of colourless glass is much more specialised than that of blue/ green glass and it would be highly unlikely that colourless glass would have been selected for recycling and the blue/ green ignored. It seems reasonable to conclude, therefore, that the Delamere Street assemblage does represent an unbiased sample of the rubbish that was being generated. If that was the case then it points to it being derived from kitchen premises and in the light of this is it interesting to note that a fragments of a *clibani*, portable ovens, were also present (*see Chapter 4*). The mis-match between the vessel glass and the samian is fascinating. Could it be perhaps because though we as modern archaeologists are fascinated by bright shiny orange pottery and endow it with importance, it was less highly regarded by its users? This is perhaps a heretical view but probably one which should be considered more often.

As with the small finds the vessel glass shows no chronological or functional differences between the two events, suggesting the rubbish is derived from similar specialised sources for both. Equally the distinctive and frequently very common vessel forms that become common during the latter third of the second century are entirely absent here. If the Event 3 assemblages derive from the clearance of middens in the later second century, these were old middens that had not been added to for some decades.

Catalogue (Fig 2.27)

1 Hemispherical bowl; body fragment. Cast polychrome. Cane with translucent emerald green, opaque red central dot surrounded with ring of small opaque yellow dots around it. Dimensions 16 x 14mm, wall thickness 3mm. 4007, Pit 4002, Group 112, Area C, SF41, Event 3 (ID G216). Not illustrated.

2 Pillar moulded bowl; lower body fragment. Blue/green. Part of two ribs dying out. Surfaces abraded. Dimensions 27 x 18mm. EVE 0.2. Weight 3g. 5003, Pit 5004, Group 112, Area E, Event 2. (ID G192). Not illustrated.

3 Globular jug, handle fragment. Blue/green. Lower handle attachment; one side of straight handle with three prominent ribs; attachment retains convex-curved body. Present length 35mm, width of attachment 52mm. EVE 0.14. Weight 23g. 1012, Pit 1016, Group 112, Trench 1, Area C, Event 2. (ID G9). Not illustrated.

4 Jug; handle fragment. Blue/green. Straight handle with pronounced central rib. Present length 27mm, section (excluding rib) 31 x 5.5mm. EVE 0.14. Weight 8g.: 4003, Pit 4012, Group 112, Area C, Event 3, SF36. (ID G232). Not illustrated.

5 Jug; handle fragment. Light yellow/green. Small section of straight ribbon handle with single rib. Present length 8mm, section (without rib) 31 x 4mm. EVE 0.14. Weight 1g. 4007, Pit 4002, Group 112, Area C, Event 3 (ID G153). Not illustrated.

6 Conical jug (?); body fragment. Blue/green. Straight side sloping out; diagonal ribs. Dimensions 60 x 40mm, wall thickness 2.5mm. Weight 7g. 4001, Group 112, Area C, unphased. (ID G148). Not illustrated.

7 Jug; handle fragment. Blue/green. Straight handle broken at angle. Length 46mm, section 25 x 8mm. EVE 0.14. Weight 14g. 4006, Group 112, Area C, unphased (ID G161). Not illustrated.

8 Spouted jug; rim fragment. Blue/green. Rim edge fire-rounded; part of pulled-up edge; tooling mark in side. Dimensions 32 x 24mm, wall thickness 4.5mm. Weight 5g. 4004, Pit 4002, Group 112, Area C, Event 3 (ID G130). Not illustrated.

9 Jug or flask; cylindrical neck fragment. Blue/green. Weight 2g. 4030, Pit 4041, Group 112, Area C, Event 2 (ID G172). Not illustrated.

10 Body fragment. Light green. Slightly convex-curved; shallow optic ribs. Dimensions 26 x 11mm, wall thickness 1.5mm. 4001, Group 112, Area C, unphased. (ID G142). Not illustrated.

11 Body fragment. Amber. Part of one very shallow rib. Dimensions 8 x 7mm, wall thickness 1.5mm. 2285, Pit 2292, Group 112, Area A, Event 3. (ID G 206). Not illustrated.

12 Tubular-rimmed jar; rim fragment. Blue/green. Rim edge first bent in, then out and down; side broken as it bends out. Rim diameter 130mm, present height 28mm. EVE 0.18. Weight 34g. 1016, Pit 1022, Group 112, Trench 1, Area C, Event 2. (ID G3). (Fig. 2.27.12)

13 Bottle; rim, neck and handle fragment. Blue/green. Rim edge bent out, up, in and flattened; cylindrical neck; upper part of angular ribbon handle applied to neck with upper attachment trailed back and then back to neck and up to the underside of the rim. Rim diameter 52mm, handle section 52 x 8mm, present height 34mm. Weight 72g. EVE 0.42. 4001, Group 112, Area C, unphased (ID G106). Not illustrated.

14 Bottle; rim fragment. Blue/green. Rim bent out, up, in and flattened; tiny fragment of trail from handle attachment on outer edge. Rim diameter 45mm. EVE 0.14. Weight 4g. 5005, Pit 5007, Group 112, Event 3. (ID G198). Not illustrated.

15 Bottle; rim and neck in two joining fragments. Blue/green. Rim out and down, up and in; upper part of cylindrical neck. Rim diameter 45mm. Weight 18g. EVE 0.14. E603 SF1 (ID G239). Not illustrated.

16 Bottle; rim and neck fragment. Blue/green. Rim bent out, up and in with edge sloping in; cylindrical neck; scar from upper attachment of handle on neck; horizontal scratch marks between rim and handle attachment; some vertical scratches on neck. Present height 50mm. Weight 17g. 4009, Pit 4002, Group 112, Area C, Event 3. (ID G140). Not illustrated.

17 Bottle; rim fragment. Blue/green. Rim bent out, up, in and flattened. Rim diameter 85mm. EVE 0.14. Weight 14g. 4046, Pit 4041, Group 112, Area C, Event 2. (ID G133). Not illustrated.

18 Bottle; neck fragment. Blue/green. Cylindrical neck broken at base with junction of shoulder; with fragment of upper handle attachment on side. Neck diameter c. 60mm; length 43mm. Weight 9g. 5019, Pit 5004, Group 112, Area E, Event 3. (ID G181). Not illustrated.

19 Bottle; neck fragment. Blue/green. Small fragment of cylindrical neck retaining part of folded upper handle attachment. Weight 3g. 4046, Pit 4041, Group 112, Area C, Event 2(ID G171). Not illustrated.

20 Bottle; neck fragment. Blue/green. Cylindrical neck broken at junction with shoulder. Neck diameter c. 80mm, present height 40mm. EVE 0.14. Weight 14g. 4028, Pit 4041, Group 112, Area C, Event 3. (ID G144). Not illustrated.

21 Bottle; neck fragment. Blue/green. Neck diameter c. 50mm. Weight 12g. 4029, Pit 4041, Group 112, Area C, Event 2. (ID G120). Not illustrated.

22 Bottle; neck fragment. Blue/green. Cylindrical curving out to base. Neck diameter c. 50mm, present height 26mm. Weight 16g. 4046, Pit 4041, Group

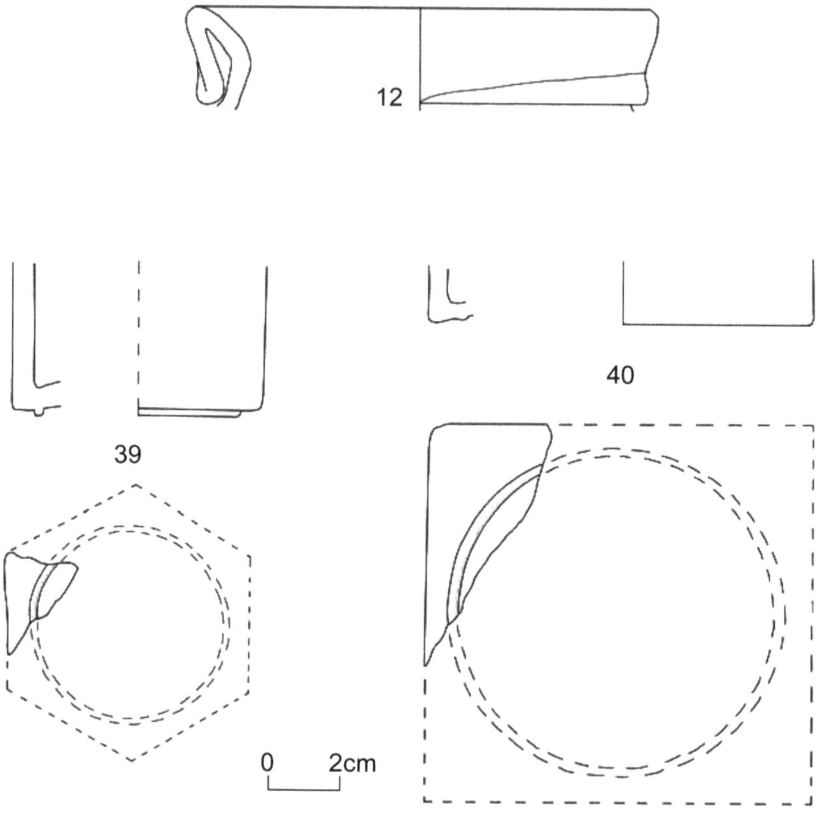

FIGURE 2.27 ROMAN GLASS 12,39 AND 40.

112, Area C, Event 2. (ID G134). Not illustrated.

23 Bottle; neck fragment. Blue/green. Cylindrical neck curving out to shoulder; some tooling marks at base of neck. Neck diameter c. 40mm, present height 26mm. Weight 16g. 5019, Pit 5004, Group 119, Area E, Event 3. (ID G180). Not illustrated.

24 Bottle; neck fragments (2). Blue/green. Cylindrical neck beginning to curve to shoulder; some tooling marks at base. Neck diameter 30-35mm, present height 40mm. Weight 9g. 5002, Pit 5004, Group 112, Area E, Event 3. (ID G189). Not illustrated.

25 Prismatic bottle; complete handle. Blue/green with green impurities. Angular reeded handle; retaining outer part of upper part originally attached to neck trailed back and then back to neck. Height of handle 47mm, maximum section 33 x 6mm. EVE 0.28. Weight 31g. 4044, Pit 4045, Group 112, Area C, Event 3. (ID G113). Not illustrated.

26 Bottle; handle fragment. Blue/green. Upper part of angular handle retaining outer part of upper attachment. Handle width 42 x 6mm. Weight 27g. 4009, Pit 4002, Group 112, Area C, Event 3. (ID G108). Not illustrated.

27 Bottle; handle fragment. Blue/green. Upper part angular handle, attached to neck, folded back and then trailed back to neck. Handle section c. 50 x 7mm. Wieght 22g. 4033, Pit 4041, Group 112, Area C, Event 2. (ID G111). Not illustrated.

28 Bottle; handle fragment. Blue/green. Upper part of reeded handle. EVE 0.14. Weight 9g. 4015, Pit 4041, Group 112, Area C, Event 3 . (ID G138). Not illustrated.

29 Bottle; handle fragment. Blue/green. Edge of angular ribbon handle; upper part retains part of folded attachment to (missing) neck lower part retains one rib and handle may have been reeded. Weight 26g. 3049, Pit 3058, Group 112, Area B, Event 2. (ID G90). Not illustrated.

30 Bottle; handle fragment. Blue/green. Edge of upper part of angular reeded handle. Weight 4g. 2063, Group 103, Area A, Phase 4. (ID G35). Not illustrated.

31 Bottle; handle fragment. Blue/green. Edge of upper part of angular handle with some scarring from folded upper handle attachment. Weight 10g. 2206, Pit 2204, Group 112, Area A, Event 3. (ID G234). Not illustrated.

32 Bottle; handle fragment. Blue/green; black impurities. Edge of upper part of angular handle retaining small part of fold from upper attachment. Weight 4g. 3048, Group 100, Area B, Phase 5. (ID 84). Not illustrated.

33 Bottle; shoulder fragment. Blue/green. Retaining small part of lower edge of reeded handle attachment. Also 1 body fragment. 2111, Pit 2110, Group 112, Area A, Event 3. (ID G61). Not illustrated.

34 Bottle; handle fragment. Blue/green. Lower part of reeded handle, attachment retains part of shoulder. Handle section 42 x 6.5mm, present height 42mm.

EVE 0.28. Weight 23g. 5003, Pit 5004, Group 112, Event 2. (ID G193). Not illustrated.

35 Bottle; chip from folded upper handle attachment. Weight 3g. 4029, Pit 4041, Group 112, Area C, Event 3. (ID 174). Not illustrated.

36 Cylindrical bottle; handle and shoulder fragment. Blue/green. One side of lower part of reeded handle; attachment retaining part of shoulder with convexity on upper body; vertical scratch marks. Present height 76mm, width of handle 78mm. EVE 0.28. Weight 85 g. 2206, Pit 2204, Group 112, Area A, Event 3. (ID G45). Not illustrated.

37 Cylindrical bottle; base fragment. Blue/green. Side curving into edge of base. Some vertical scratch marks on side and wear on base. Present height 35mm. EVE 0.28. Weight 13g. 2096, Drain Cut 2095, Group 106, Area A, Phase 4. (ID G43). Not illustrated.

38 Cylindrical bottle; base fragment. Blue/green. Fragment from edge of concave base. Base diameter c. 140-150mm. EVE 0.14. Weight 5g. 4030, Pit 4041, Group 112, Area C, Event 2. (ID G168). Not illustrated.

39 Hexagonal bottle; lower body and base fragment. Blue/green. Base design - part of one concentric circle extant. Base worn. Bottle width c. 80mm (width of extant side 40mm), present height 43mm, EVE 0.28. Weight 23g. 4039, Pit 4040, Group 100, Area C, Phase 5. (ID G128). (Fig. 2.27.39)

40 Square bottle; base fragment. Blue/green. Corner fragment, base design - circular moulding. Present height 16mm, diameter of circle c. 95mm, base width c. 105mm. EVE 0.14. Weight 17g. Sf 36 : 4003, Pit 4012, Group 112, Area C, Event 3. (ID G233). (Fig. 2.27.40)

41 Prismatic bottle; lower body and base fragment. Blue/green. Base design - part of one concentric circle extant. Base worn. Bottle width c. 100mm, diameter of extant moulding 80mm, present height 15mm, EVE 0.14. Weight 15g. 2283, Pit 2292, Group 112, Area A, Event 3. (ID G52). Not illustrated.

THE ROMAN SMALL FINDS *by Hilary Cool*

Introduction

The excavations at Delamere Street have produced a group of small finds that are of great value both for our understanding of the occupation at Chester and for the material culture of Roman Britain more generally. The deposition of what appears to be primary rubbish over a relatively short period in the later 1st and earlier 2nd centuries has provided valuable information about the introduction of certain types. This is also one of the larger groups of Roman finds to have been published from Chester in the recent past. It is a matter of great regret that the majority of the excavations in the city carried out during the 1960s and 1970s such as those at Abbey Green (McPeake *et al* 1980) remain unpublished, so our knowledge of the material culture of the fortress is limited. This makes it more difficult to place the Delamere Street finds in their correct context against the background of finds being used elsewhere in the vicinity. In what follows information has been drawn from other excavations at Chester, but the publications of many of these belong to the earlier part of the 20th century and even, in one case, to the late-19th century and so the documentation does not always provide the details we would normally expect from modern publications. That said, however, it is a tribute to those early excavators that the sites were published and do provide this useful information.

This report deals with all the metalwork other than the nails, and all the stone, frit and glass small finds. The soil conditions at Delamere Street where such that the copper alloy was extremely corroded. Even after extensive X-radiography and conservation it has not always been possible to establish fine details of some of the pieces. This probably means that the assemblage is missing some elements. As will be noted in the discussion of the military equipment, only the more substantial *lorica segmentata* has been identified as being present. Other forms of armour were unlikely to have survived in a recognisable state. Some features of the composition of the small finds assemblage though, cannot be entirely explained by the corrosive environment. Table 2.17 sets out the catalogued Roman finds by material and Event. The material that has been catalogued includes everything from the Roman contexts that can be identified. The Roman contexts have been taken to include cleaning layers attributed to Group 112 but not otherwise phased. Items that are typologically Roman but which come from post-Roman or unstratified contexts have also been included. As can be seen the small finds can contribute nothing useful to any assessment of the activities and timing of Event 1. The only stratified item associated with it was a piece of relatively undiagnostic perforated sheet. For Events 2 and 3, by contrast there are relatively large assemblages with more potential to contribute to our understanding of the activities.

Table 2.18 shows the small find assemblage according to function. As can be seen the amounts of personal ornaments and military equipment are almost the same. Given the former normally dominates Roman assemblages this is very good evidence that the rubbish being deposited in these pits was being derived from within the fortress.

TABLE 2.17: THE ROMAN FINDS BY MATERIAL AND EVENT

Material	EVENT 1	EVENT 2	Event 3	0	Total
Copper alloy	1	27	41	10	79
Lead alloy	-	1	3	1	5
Iron	-	1	4	1	6
Glass	-	1	1	-	2
Frit	-	1	4	1	6
Stone	-	2	1	-	3
Composite	-	2	3	-	5
Total	**1**	**35**	**57**	**13**	**106**

TABLE 2.18: THE ROMAN FINDS BY FUNCTION AND EVENT

Function	1	2	3	0	Total
Personal	-	8	11	3	22
Toilet	-	1	4	-	5
Textiles	-	-	1	-	1
House	-	1	-	-	1
Tools	-	3	3	-	6
Fasten	-	11	19	8	38
Military	-	8	12	1	21
Misc	1	3	7	1	12
Total	**1**	**35**	**57**	**13**	**106**

In what follows the finds are discussed first by functional category and then according to what they tell us about the people who originally used them. Material of post-Roman date was generally confined to modern items, but one medieval object of note is discussed at the end.

Personal ornaments

Six bow brooches, one penannular brooch and several fragments of brooch spring and chain were found. As is to be expected at Chester where the beginning of occupation is conventionally placed in the early to mid 70s, the brooch types typical of the Claudio-Neronian army are rare and only represented by a fragmentary Nauheim Derivative one-piece brooch (No. 1). This has a rod bow and is thus an example of the variant that did continue in use into the Flavian period (Bayley and Butcher 2004, 147 Hull Type 10/11). Chester provides good evidence for the continued use of Nauheim Derivative brooches into the later first century as in addition to this example, there is also a similar one from Hunter Street (Newstead 1928, 72 no. 7) and one with a wider strip bow from Linenhall Street (Thompson 1969, 15 no. 8) in a late first or early second century context.

There are two examples of Polden Hill brooches where a separate spring is held by a bar that runs through the spring and is lodged in the end plates of the spring cover. This is the typical western method of fixing the spring in place in the two-piece Colchester Derivative brooches typical of the later part of the first century and into the second century (Bayley and Butcher 2004, 159-60). One of these (No. 2) is a conventional example of a Hull Type 98 with the characteristic mouldings either side of the bow at the head. The perforated catch plate would suggest a date in the first century for its manufacture. The second (No. 3) is a far less common variant. It has the typical features of a Polden Hill brooch, again with a perforated catch plate, but centrally on the bow there is a moulded button typical of the trumpet series of brooch (see No. 5 below). The combination of features seen on this strange hybrid is most interesting as Bayley and Butcher's work on the metal compositions of trumpet brooches has shown that the trumpet variants with flat-backed central mouldings as here share their compositions with brooch types typical of the western and southern parts of the country. This led them to suggest that such trumpet brooches might be being made by the same people who appeared to be manufacturing the Polden Hill brooches in the West Midlands (Bayley and Butcher 2004, 163). This hybrid brooch would support this suggestion and it may well come from an early phase in the production of both types as the catch plate clearly had a complex perforation pattern, very probably two sections divided by a dog-leg bar. This is a pattern that belongs to early members of the Colchester Derivative family, and Mackreth has argued that it is not to be expected late in the first century (Mackreth 2000, 146 no. 2). There is what appears to be a small triangular cell below the button and so this brooch was probably enamelled as well.

The fragmentary hinged 'T'-shaped brooch no 4 would also fit this late first century date as it belongs to Bayley and Butcher's initial series of this South-Western type, lacking the head loop and enamel decoration typical of the developed series second century (Bayley and Butcher 2004, 159).

Appropriately the one trumpet brooch in the group (No. 5) belongs to the variant Hull named the Chester Type (Hull Type 154) with lentoid mouldings above and below the half round central button. These fall into Bayley and Butchers's (2004) Group C trumpet brooches. Despite being called the Chester type the distribution has long suggested that they might well be being made in the Gloucestershire area (Mackreth 1998, 118 nos. 17-20), and this was confirmed by the recovery of a mould and a finished brooch of the type that fitted into it from Dymock in Gloucestershire, close to the Herefordshire border (Cool 2007, 173). They were in use during the later first and earlier second centuries. A similar example appears to have been found at the Amphitheatre though the quality of the illustration and description leaves much to be desired (Thompson 1976, 189 no. 3)

The final bow brooch to be considered (No. 6) is enamelled and has a fantail. A range of fantail brooches were being manufactured in Britain in the later first and second centuries (Bayley and Butcher 2004, 169, fig. 143; for a typical range see Cool and Philo 1998, 31 fig. 10), but this piece has various features that sets it apart from those. In the first case the upper bow is formed by an asymmetrical diamond plate linked to the fantail by an oval unit. In the second place the brooch lacks a headloop which is a typical feature on the British series. Other unusual features include the very delicate beading effects on the ribs and on the edges of the enamelled fields, and the fact that the upper bow and the fantail have a silvered band that frames the enamel. Silver decoration was applied to British brooches with fantails such as the Alcester type (Hull Type 162 – Bayley and Butcher 2004, 169) but the combination of the two decorative techniques is not common. Admittedly No. 6 has been very carefully conserved to exhibition standard and not all brooches have benefited from such treatment, but the beading effect would probably have been noticed before if it was common on British fantails.

On the continent this type of delicate beading is sometimes noted around the enamelled fields of the brooches with triangular or diamond-shaped bows and snake-headed feet (Exner 1939, Types I31 and I141). The detail is well illustrated on the examples from Augst (Riha 1979, 160 nos. 1413-7). An uncommon brooch form with a similar bow, a fantail foot and a central rounded rib or semi-circular enamelled unit is generally discussed alongside the snake-headed foot form where it occurs. Whilst these are not precisely identical to the Chester brooch as they have a triangular upper bow, No. 6 does appear to be more similar to this unusual continental form than it is to the British fantail family. Examples may be noted from Siscia (Sellye 1939. 73 pl. XI no. 11), from Saalburg (Exner 1939, 82 nos. 33 and 34) and from France (Feugère 1985, 357 no. 1875). If the continental scholars are correct in their linking of the fantail form with the snake's head form, then an origin for No 6 in the Rhineland or in the Danubian provinces would be likely. The brooch was unfortunately found unstratified but a late first and second century date is generally accepted for the snake-footed form (Riha 1979, 159).

These bow brooches provide an interesting sidelight on the garrison of the fortress. As will be discussed below with regard to the military equipment, there is good evidence that the quarry pits at Delamere Street were being infilled with rubbish from within the fortress. The geographic affiliations of the brooches other than the exotic No. 6 are all with the west midlands / Severn valley area, precisely as one would expect for a legion such as the Twentieth which had formerly been stationed at Wroxeter.

No. 10 is broken and distorted but is most likely to be a penannular brooch of Fowler (1960) Type D with the terminal folded back along the hoop. The fact that only one possible penannular brooch has been identified compared with six bow brooches is good evidence that the rubbish filling the pits is coming from within the fortress as penannular brooches were more favoured by the native population than by the military at this time.

The most interesting item amongst the other personal ornaments is the centre piece of a beaded torc (No. 11). It is now in two fragments and had clearly been broken prior to being disposed of as the pieces were found in different layers within pit 4041 with one piece coming from the lower fill. Both layers were assigned to Event 2. The illustration only shows a drawing of the conserved piece, but the outline of the second piece can clearly be seen in the X-radiograph. The ends have mortice-holes into which tenons on the ends of the plain bar that would have completed the torc would have fitted. The severe corrosion the pieces have been subjected to makes it difficult to ascertain the full decoration on the beads but at least some of the ribs were themselves beaded.

Beaded torcs were first studied as a complete group by MacGregor who showed there were two basic types, those where the beads were separate and threaded onto a central bar and those where the beaded section was cast in one as here (1976, 97-101, Map 15). For the latter the associations generally indicated a later first to second century date though the quality of the dating was not always good, so the recovery of this securely dated example is most welcome. The distribution of all the beaded torcs then known lay north of the Mersey/Humber line up to the Firth of Forth and she suggested they were the products of Brigantian smiths (MacGregor 1976, 97-101, Map 15).

More recently Hunter has re-assessed the type drawing on the new finds that have been made especially via metal detecting (Hunter 2010). It is possible to see that the examples where the beads are cast in one (Hunter's type B) are now strongly represented in the English midlands and a growing number are coming to light in the south west. Given the dating, beaded torcs of Hunter's type B are clearly not late Iron Age artefacts living an after-life in the Roman period, but a Romano-British development. Hunter has argued for them being a recreation of the earlier elite torc tradition within the new context of Roman Britain and suggests that wearing a torc 'became part of being Romano-British, for at least some people in at least some places'.

Currently the majority of beaded torcs like the Chester piece have been found on sites where there is no obvious military presence, but the Chester example is not alone in coming from a military site as two have been found at Carlisle (MacGregor 1976 nos. 199 and 208). One was found at Rocester in a context dated to c. AD 95 – 130 (White forthcoming). It is unclear whether this should be related to the first fort at Rocester or to the early stages of the second, but the military context is undoubted. We may also note, cautiously, a detectorist find from Wall, Staffordshire[1] where there was a military presence at the time when the torcs would have been in use. The presence of these torcs within Roman military milieus as at Chester is interesting given that torcs were symbols of power and authority within the late Iron Age British communities and, as Hunter has argued, the later neck ornament such as beaded torcs may have come to signify some new Romano-British identity. Torcs were also one of the lesser Roman military decorations (Maxfield 1981, 86-8) so had significance to soldiers. The question needs to be asked what do pieces like No. 11 mean when they are found on military sites. Are they the spoils of war acquired during the conquest period in the north? This is a possible though perhaps over-dramatic interpretation. Are they trinkets acquired as curios? Or should we be considering the possibility that they are the possessions of native British men serving in the army? These are questions that may be asked, but at present not answered.

An iron finger ring is represented by No. 12. Only a highly corroded scrap of the bezel survives in which a damaged glass intaglio is set. The ring would almost certainly have been of Henig's (1974) Class II or III style with a simple expanded bezel, the typical signet ring of the first to early third century. The fact that it was made of iron would also have been appropriate for a legionary fortress as iron was the correct metal for an ordinary citizen, i.e. one who was not of Equestrian or Senatorial rank. The other finger ring (No. 13) is an expanding form with outward spiralling loops. The type has a long history in both bracelet and finger ring size, but examples from Britain have generally come from late Roman contexts so the early dating of this piece is a useful addition (Cool 1983, 130-5, 221-4).

The recovery of the pentagonal peacock glass bead (No. 14) in pit 2110 is also a valuable addition to the dating of the type. Previously the main focus of the dating has suggested that they appeared first in the later second century, such as at Causeway Lane Leicester (Cooper 1999, 260 no.76). The only earlier example was from the fort rampart at Malton (Price and Cottam 1997, 130 no. 61) which indicated a Trajanic date. The pit that No. 14 came from had a large pottery assemblage which included Hadrianic or Hadrianic-Antonine samian and approximately 10% by weight of the pottery was Dorset BB1. This would suggest that the fill of the pit was not as early as the Malton rampart but it certainly provides evidence for the use of the type in the first half of the second century. Most pentagonal beads such as these use various shades of green glass to case an opaque yellow core. This example though is made of peacock glass throughout. The aim of all these beads was probably to simulate beryl emeralds which naturally form hexagonal-sectioned crystals and were common on Roman gold jewellery.

The presence of No. 14 indicates a female presence in the area generating the rubbish that was going into the pits. The earring No. 15 may do as well though it was recovered from one of the evaluation test pits and not from one of the well stratified contexts. Penannular rings such as these were the commonest form of Roman earring (Allason-Jones 1989, 2 Type 1). In the Roman world earrings were worn by females and not by men.

[1] PAS database no. WMID5241.

Frit melon beads are normally catalogued as part of personal ornaments though they are likely to have had other functions, such as the decoration of horse harness, as well. They are generally common on first to mid second century military sites so the presence of six (nos. 16-21) at Delamere Street is not surprising. They have frequently been noted within the fortress before. Of those that have been published we may note eleven from the Princess Street area (Newstead and Droop 1940, 41), five from St Michael's Street (Newstead 1928, 136), three from Bridge Street (Cool 2008, 302 nos. 1, 2, 4) and single examples from the Deanery Field (Newstead 1924, 80 no. 6) and the amphitheatre (Newstead and Droop 1932, 35).

The final item in this section (No. 22) consists of part of a shoe sole represented by the highly corroded remains of seven hobnails.

Catalogue

1. Nauheim Derivative brooch. Copper alloy. Oval-sectioned bow with spring of one turn on either side of bow and cord passing below the bow; rectangular-sectioned pin tapering to point; most of bow missing; triangular catch plate with slightly projecting foot. Length of pin 42mm. Sf 116 : 4007, Pit 4002, Group 112, Area C, Event 3. (ID 2). Not illustrated.
2. Polden Hill brooch. Copper alloy. Semi-cylindrical spring cover with closed ends, one retaining clear evidence for bar that passed through the spring of ten turns; head of bow has forward facing hook holding cord; two pairs of vertical grooves with rib between on each wing; semi-circular moulding on either side of head, bow tapering to foot; triangular catch plate with triangular perforation. Tip of pin missing. Length 62mm, width of wings 23.5mm. 5012, Pit 5013, Group 112, Area E, SF 149, Event 2. (ID 17). (Fig. 2.28.2)
3. Polden Hill brooch. Copper alloy. Semi-cylindrical spring cover with plates at end holding bar that runs through spring of c. 12 turns; cord of spring held by forward facing hook on head of bow; ends of wings have five vertical ribs – second from end has alternating nicks to produce a zig-zag effect, innermost has transverse punchmarks to produce a beaded effect. Semi-circular moulding on either side of head, D-sectioned tapering bow with central button consisting of wide rib with narrow beaded rib above and below, channel either side of this unit, a pair of narrow ribs above and below this central unit may originally have had a beaded effect; lower bow tapering with foot missing and empty triangular cell below button. Upper part of catchplate with angular fretwork perforation. Lower part of pin also missing. The whole much corroded and obscured. Present length 53mm, width of wings 34mm. 5002, Pit 5004, Group 112, Area E, Event 3. (ID 18). (Fig. 2.28.3)
4. Hinged T-shape brooch. Copper alloy. Short wings bent over to form cylinder with seam behind and vertical groove at end; remains of iron hinge pin within; triangular-sectioned tapering broken bow; small vertical ridge on head with transverse ridges. Present length 24mm, width wings 25mm.: 4016, Pit 4041, Group 112, Area C, SF57, Event 3. (ID 103) (Fig. 2.28.4)
5. Trumpet brooch. Small trumpet head with central lug on back; small knob on top of head that would have kept the missing separate head loop in place; D-sectioned bow expanding to central button consisting of wide channel either side of button consisting of wide rib between narrow ribs, shallow lentoid mouldings above and below; bow tapering to foot; trapezoidal catch plate. Head bent down. Traces of tinning or silvering on back. Length 44mm, width of head 9mm.: 4030, Pit 4041, Group 112, Area C, SF65, Event 2. (ID 164). (Fig. 2.28.5)
6. Fantail brooch. Copper alloy. Short cylindrical open-ended hinge cover retaining scrap of pin, a beaded rib on head separating hinge cover from bow; upper bow consists of asymmetrical diamond plate with transverse projecting beaded rib above; centrally an oval unit with transverse beaded rib below; short fantail with broken catchplate behind. Edges of diamond plate and fantail have border of transverse grooves producing a beaded effect similar to the transverse ribs; all three units have triangular cells, enamel now much degraded and colour uncertain. Diamond plate and fantail have narrow framing strip between beaded edges and cells which retains traces of white metal coating identified as tin during conservation. Length 41mm, hinge width 18mm. SF21, unstratified (Fig. 2.28.6).
7. Brooch spring (?) Copper alloy. Circular-sectioned bent pin retaining one and a half turns of a spring. Length 31mm, section 2mm.: 5014, Pit 5013, Group 112, Area E, SF157, Event 2. (ID 61). Not illustrated.
8. Brooch chain. Copper alloy. Small penannular wire ring, length of double loop-in-loop chain in very fine gauge wire. Diameter of ring 8mm, ring section 2mm, length of chain 17mm, chain section 3 x 2.5mm.: 4030, Pit 4041, Group 112, SF75, Event 2 . (ID 28). (Fig. 2.28.8)
9. Brooch chain. Copper alloy. Length of double loop-in-loop chain broken into three pieces. Rectangular section. Length 48mm, section 3.5 x 2.5mm. 1008, Pit 1019, Group 112, Trench 1, Area C, SF1, Event 3. (ID 153). Not illustrated.
10. Penannular brooch. Copper alloy. Circular-sectioned hoop, one terminal bent back along hoop, other end broken and twisted out of shape. Diameter 29mm, hoop section 2.5mm. 5014, Pit 5013, Group 112, Area E, SF159, Event 2. (ID 13). Not illustrated.
11. Beaded torc. Copper alloy. Front section consisting of nine beads; central beads consist of five

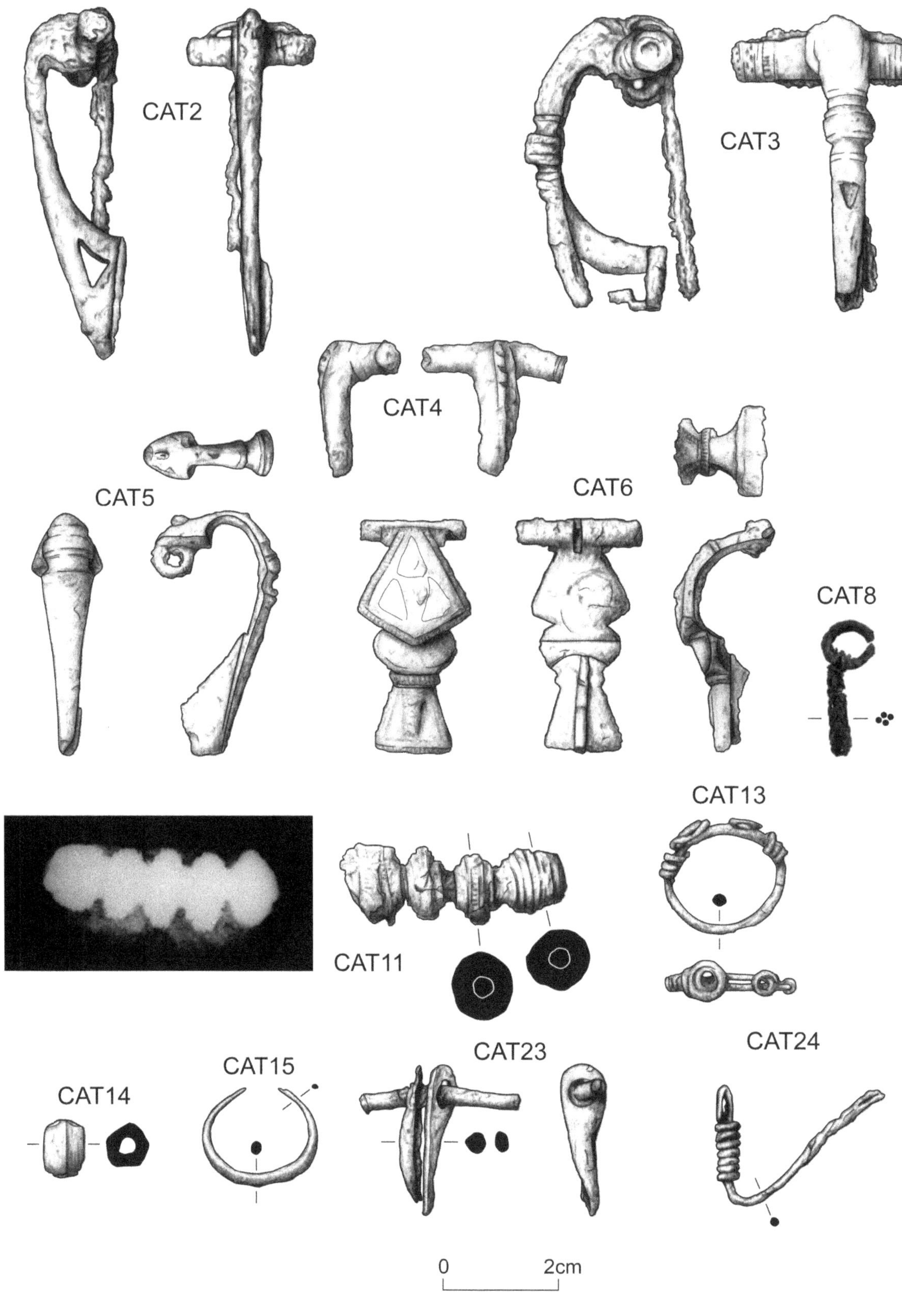

FIGURE 2.28 SMALL FINDS CAT2-24.

graduated ribs central one beaded and outer ones also retain traces of beading; hollow tapering terminal with seven ribs with traces of beading. Length 90mm, diameter of bead 14mm. 4030 and 4046, Pit 4041, Group 112, Area C, Event 2. (ID 112). (Fig. 2.28.11)

12. Finger ring. Corroded iron bezel retaining broken 'black' glass intaglio, flat oval with bevelled edge – possible standing figure motif. Dimensions of intaglio 13 x 9mm, thickness 3mm. 4030, Pit 4041, Group 112, Area C, SF93, Event 2 (ID 191). Not illustrated.
13. Finger ring. Copper alloy. Circular-sectioned wire, expanding junction with rosette of two and a half turns on each end before the spare end wrapping around hoop. Diameter 20mm, hoop section 2mm. 3007, Pit 3012, Group 112, Area B, SF 37, Event 3. (ID 31). (Fig. 2.28.13).
14. Pentagonal cylindrical bead. Translucent peacock glass. Scar around one end and small collar around other end. Diameter 8mm, length 11mm, perforation diameter 3mm.: 2111, Pit 2110, Group 112, Area A, SF11, Event 3. (Fig. 2.28.14).
15. Earring. Copper alloy. Oval-sectioned penannular hoop tapering to points. Diameter 20 x 17mm, maximum section 4.5 x 2.5mm. E603 Test Pit 3. (ID 190). (Fig. 2.28.15).
16. Melon bead. Turquoise frit retaining glaze. Well formed gadroons. Diameter 13.5 x 12mm, length 10.5mm, perforation diameter 5.5. : 4030, Pit 4041, Group 112, Area C, SF74, Event 2. (ID G237). Not illustrated.
17. Melon bead. Turquoise frit retaining glaze. Shallow gadroons around centre; one side lower than other. Diameter 12 x 10.5mm, length 10 - 7mm, perforation diameter 5m. 4016, Pit 4041, Group 112, Area C, SF59, Event 3. (ID G238). Not illustrated.
18. Melon bead. Turquoise frit with traces of glaze. Slightly irregular gadroons. Diameter 12mm, length 10.5mm, perforation diameter 5.5mm. 4009, Pit 4002, Group 112, Area C, Event 3. (IDG 217). Not illustrated.
19. Melon bead. Turquoise frit retaining some of the glaze. Shallow gadroons. Diameter 13 x 11mm, length 10.5mm, perforation diameter 5mm.: 5005, Pit 5007, Group 112, Area E, SF122, Event 3. (ID G220). Not illustrated.
20. Melon bead. Frit now appearing grey. Irregular gadroons, narrowing on one side. Diameter 11mm, length 9-5mm, perforation diameter 5mm. 2283, Pit 2292, Group 112, Area A, Event 3. (ID G236). Not illustrated.
21. Melon bead; approximately half extant. Turquoise frit with some of original glaze. Well formed gadroons. Diameter 19mm, length 15.5mm, perforation diameter c. 7mm. Area E, unstratified, (ID G235). Not illustrated.
22. Hobnails. Iron. Seven hobnails corroded together. Dimensions 37 x 32mm, diameter of individual head 9mm. Unstratified (ID 186). Not illustrated.

Toilet equipment

An interesting group of toilet equipment was found. At least two different chatelaines were represented. No. 23 consists of only the bar and simple perforated disc heads of two implements, and so does not retain sufficient information to characterise it further. The other two chatelaine tools are both formed from a wire shank with the top bent over to form a loop with the end spiralling around the shank. No. 24 is clearly a nail cleaner, No. 25 lacks the tool end. This style is discussed by Eckardt and Crummy (2008, 128 and 146) who noted a divide between those where the twist is low down the shank and those where it is high. Nos. 24 and 25 fall into their Sub-group B where it is high. Interestingly this sub-group has a more western and northern distribution than many groups. A small spoon of this type from within the fortress at Caerleon (scott in Mason and MacDonald 2010, 80 no. 9 fig. 7) and this pair from Chester adds to this bias in the distribution. Three of the nail cleaners of this sub-group come from forts and this they consider unusual as it is argued that nail cleaners are much more a native fashion (Eckardt and Crummy 2008, 97 fig. 44). Nos. 24 and 25 add to this military bias as all the evidence points to the material in these pits coming from the interior of the fortress. In their sub-group as a whole the number of turns ranges from two to seven. Both of these have numbers of turns at the higher end of this range as do the instruments of this type on a chatelaine at Holt (Grimes 1930, 128 no. 19) and the small spoon from Caerleon. So it is possible that we might be looking at a distinctive western military style.

In addition to these personal grooming tools there were also two fragmentary implements that may have been ligulae (Nos. 26-7), long-handled tools that had a use as toilet implements and in pharmacy and medicine. No. 26 retains very slight indications that there may have been a circular perforation in the blade. This would be an unusual feature in a ligula. Another broken implement with this feature has been found within a legionary setting at York (Cool *et al* 1995, 1557 no. 6346) and so it is possible that this item served a different purpose.

Catalogue

23. Chatelaine. Copper alloy. Slightly facetted bar with burred ends; perforated disc ends of two instruments threaded on this; tool end missing. Length of bar 28mm, maximum section of terminal 9 x 1.5mm. 2283, Pit 2292, Group 112, Area A, SF32, Event 3. (ID 33). (Fig. 2.28.23).
24. Nail cleaner. Copper alloy. Wire bent into loop and wrapped seven and a half turns around shank; lower end flattening and expanding where much corroded; traces of vertical groove. Implement bent almost in two. Original length 55mm, shank section 1.5mm. 4016, Pit 4041, Group 112, Area C, SF 63,

Event 3. (ID 10). (Fig. 2.28.24)

25. Chatelaine tool. Copper alloy. Rectangular-sectioned shank lacking tool end; other end tapering to wire and forming a closed loop with end wrapped around shank five times. Present length 48mm, shank section 2 x 1mm.: 5002, Pit 5004, Group 112, Area E, SF128, Event 3. (ID 73). Not illustrated.
26. Ligula. Copper alloy. Faceted circular-sectioned shank with rounded end, other end broken across angled blade that may have been perforated. Bent shank. Present length 73mm, shank diameter 2.5mm. 4030, Pit 4041, Group 112, Area C, SF75, Event 2 (ID 29). (Fig. 2.29.26).
27. Ligula; fragment. Copper alloy. Circular-sectioned handle, one end broken, other broken across circular flat spoon blade. Present length 56mm, section handle 3mm. 2235, Pit 2234, Group 112, Area A, SF12, Event 3. (ID 137). Not illustrated.

Textile Equipment

In shape No. 28 is an example of a common Roman needle form (Crummy 1983, 65 Type 2) known to have been in use as early as the 1st century.

Catalogue

28. Needle. Copper alloy. Rectangular-sectioned shank tapering towards missing point, other end expanding to flat end; rectangular eye with transverse groove above and below on both faces. Present length 51mm, maximum section of needle 3 x 1mm, eye length 3.5mm. 4004, Pit 4002, Group 112, Area C, SF82, Event 3. (ID 178). (Fig. 2.29.28)

Household Equipment

No. 29 is part of a swing handle from a large copper alloy bowl. The tenon at the terminal end would have fitted into an attachment on the sides of the vessel. At Pompeii such handles occur on a few large shallow basins and on truncated conical bowls with wide rims which could have flat lids to which a similar swing handle was attached (Tassinari 1993, S1110-S1130 series, Type T). In her illustrations of the handles on these there does not appear to be the transverse rib between the tenon and the handle as here though they do occur on handles at the Augustan site of Magdalensberg which are believed to be associated with shallow bowls which had vine leaf escutcheons with tubes at the top into which the tennons fitted (Sedlmayer 1999, 55, 58 Taf. 23 nos. 5-6). These date from the first century BC to the early first century AD.

Swing handles appear to be very rare in Britain. The only ones known to me are two from the Sheepen site at Colchester, one of which was from a pre-Boudiccan context (Hawkes and Hull 1947, 332 nos. 9 and 10). In comparison the handles from basins with fixed handles that arch upwards and are either attached to the side (den Boesterd 1956, Type 172; Tassinari 1993, S4000 series) or to the rim (den Boesterd 1956, Type 185; Tassinari 1993, S3000 series) have been found regularly even though they are not numerous (Cool 2006, 138-40 especially table 15.3). The rarity of swing handles on British sites suggests that Den Boesterd's (1956, 52) date range for them from the first century BC to second century AD may not reflect accurately when they were generally in use. From the mid first century the bowls with fixed handles were becoming very common. Tassinari's monumental 1993 catalogue of the copper alloy vessels includes 15 vessels with swing handles together with seven individual handles. In comparison she lists 66 vessels and 25 individual handles for the S3000 and S4000 series with fixed handles. Clearly by the time of the eruption in AD 79 it was the fixed handle form that was most common. It may well be that swing-handled bowl form was going out of general use about the middle of the first century, and the bowl from which No. 29 came from would already have been old by the time the handle fragment came to be deposited in the Event 2 context it was found in. The changing fashion would explain why swing handles are so rare in Britain compared with the bowls with fixed handles.

The large bowls these handles came from were used as basins for hand-washing (Koster 1997, 86). They would have been used not only for normal hygiene but also in more ceremonial circumstances such as before making a sacrifice or within a polite dining environment. No. 29 thus hints that some of the rubbish being deposited at Delemere Street was indeed coming from officer's quarters as suggested by Wild (*see Chapter 3*). It has to be said though that it is the only item in the small find and vessel glass assemblage (*Cool see above*) that does suggest this.

Catalogue

29. Vessel handle. Copper alloy. Diamond-sectioned curved bar tapering to terminal; upper face has deep channel parallel to each side, rib between has zig-zag moulding; one end broken, terminal with transverse rib and disc with tenon. Curve probably elliptical but upper part has a diameter of c. 40mm, section 6mm, disc diameter 9mm. 4033, Pit 4041, Group 112, Area C, SF65, Event 2. (ID 125). (Fig. 2.29.29).

Tools

Fragments of two whetstones were recovered. No. 30 has the typical shape and probable lithology of Roman whetstones. In my experience No. 31 is unusual in both shape and lithology for a Roman whetstone, as it is much longer than normal and the very fine-grained stone it is made of is also unusual. It is possible that it is intrusive in this context which was a ditch fill heavily truncated by subsequent activity.

Two tools with sharp edges were recovered, one flaked from a piece of flint (No. 32) and the other from a body fragment of a blue--green glass bottle (*see Cool above*). Glass re-used in this way is a regular feature of Roman vessel glass assemblages (Price and Cottam 1998, 9). The use of flint to form sharp edged blades on Roman sites is less frequently commented on presumably because it is automatically assumed to be residual prehistoric material. Young and Humphrey (1999) have shown that the use of flint continued throughout the prehistoric period into the Iron Age and there is no reason why this utilitarian use of the material should not have continued into the Roman period and a flint knapping industry of that date was recognised over 30 years ago at the villa at Seaton (Miles 1977, 142-3). Legionary exploitation of the material has also been observed within the fortress in York (Woolrich in Cool *et al* 1995, 1596).

Iron tools are scarce in this assemblage, possibly as a result of the poor preservation of all the metalwork. No. 34 has not undergone investigative conservation but the blade does not appear to have a sharp edge. In general shape it is similar to an item of unknown function from Hod Hill which also appears to have had a blunt blade and thin tang (Manning 1985, 144 no. S136). Another tanged tool is represented by No. 35 but here only the tang survives.

The copper alloy assemblage also contains a small amount of evidence for the working of copper alloy in the form of two fragments of copper alloy casting waste (not catalogued here) from the fill of ditch [2200] (Event 3) and pit [3038] (Event 2).

Catalogue

30. Whetstone. Pale brown, indurated micaceous Palaeozoic sandstone, probably Devonian. Probably a secondary hone. Half of rectangular-sectioned, rectangular bar. Wear in centre producing an oval section. Present length 64mm, maximum section 30 x 18mm. 4030, Pit 4041, Group 112, SF72, Event 2. (ID 180). (Fig. 2.29.30)
31. Whetstone. Dark grey, fine-grained, indurated gritstone / "slate". Possibly Welsh. Primary hone. Rectangular-sectioned rectangular bar with waisted centre caused by use; many sharpening scratches. Length 168mm; maximum section 30 x 24mm. 2335, Pit 2234, Group 112, Area A, SF25, Event 3 . (ID 175). Not illustrated.
32. Blade. Flint flake. Approximate heart-shaped, long curved edge has re-touch to make it sharp. Length 52mm, width 40mm. 3028, Pit 3027, Group 112, Area B, Event 2. (ID 174). (Fig. 2.29.32).
33. Tool. Blue/green glass. Prismatic bottle body fragment; grozed on three sides to form a rectangular shape; one corner flaked to a sharp edge. Dimensions 45 x 22mm, thickness 6.5mm. 2164, Pit 2165, Group 112, Area A, Event 2. (ID G218). (Fig. 2.29.33).
34. Tool? Iron. Triangular blade with narrow broken tang. Length 117mm, maximum width blade 45mm width tang 10mm. 5002, Pit 5004, Group 112, Area E, Event 3. (ID 148). Not illustrated.
35. Tang. Iron. Tang tapering to point, blade expanding and broken. Present length 65mm, maximum width of tang 9mm. 5005, Pit 5007, Group 112, Area E, Event 3. (ID 48). Not illustrated.

Fasteners and fittings

As normal with Roman military assemblages this category of items is well represented. There were three items concerned with security. Two of these were the bolts from slide locks (Nos. 36-7), the complete example coming from a post medieval pit. The outline of an iron key (No. 38) of the type that would have been used with bolts such as these (Manning 1985, fig. 25 nos. 4-5) is visible via an X-radiograph within a large lump of iron corrosion found residually in a Phase 4 context. Fittings such as these for slide locks are frequent finds in Roman military assemblages. Similar lock bolts, for example, have previously been recorded from Linenhall Street (Thompson 1969, 15 fig. 5.6) and Godstall Lane (Newstead 1901, 85 fig. 3c) and two iron keys were noted as occurring residually at Bridge Street (Mould 2008, 317).

Most of the rest of the items require little additional comment. There is the normal range of nails and studs, a suspension loop (No. 48) which may have been part of the fittings of a steelyard and lengths of a chain (Nos. 46-7) that would have been of a suitable gauge for acting as a suspension chain for something like a bath flask or a lamp.

The hinged mount No. 51 is an unusual find. It may have been enamelled but the state of the surface even after careful conservation means that it is difficult to be certain that the radiating depressions are indeed deliberate cells. The two long rivets projecting below two of the lugs indicate that it was fastened to something substantial such as a wooden panel as they are longer than was usual for when mounts were attached to leather. They also lack the burred ends that are frequent on items intended for use with that material (see for example No. 53), and give the impression that they were designed to be driven into, rather than through, something. The lugs on one side indicate only a small hinge, and would only have been appropriate for something light, probably just a pendant fitting.

Also worthy of note is the stud with a pelta-shaped head (No. 53). It is inlaid with silver. Though the soldiers of the 1st century army liked shiny trappings, it was more usual to inlay patterns with black niello against a silvered or tinned coating, rather than use silver strips on copper alloy as on No. 54. As the piece is unstratified a Roman date is not certain and stylistically the inlay pattern would be unusual too for a Roman item.

Catalogue

36. Slide lock bolt. Copper alloy. Rectangular bar, four square perforations and two large ones in shape of a quarter circle. Length 60mm, section 12 x 3mm.

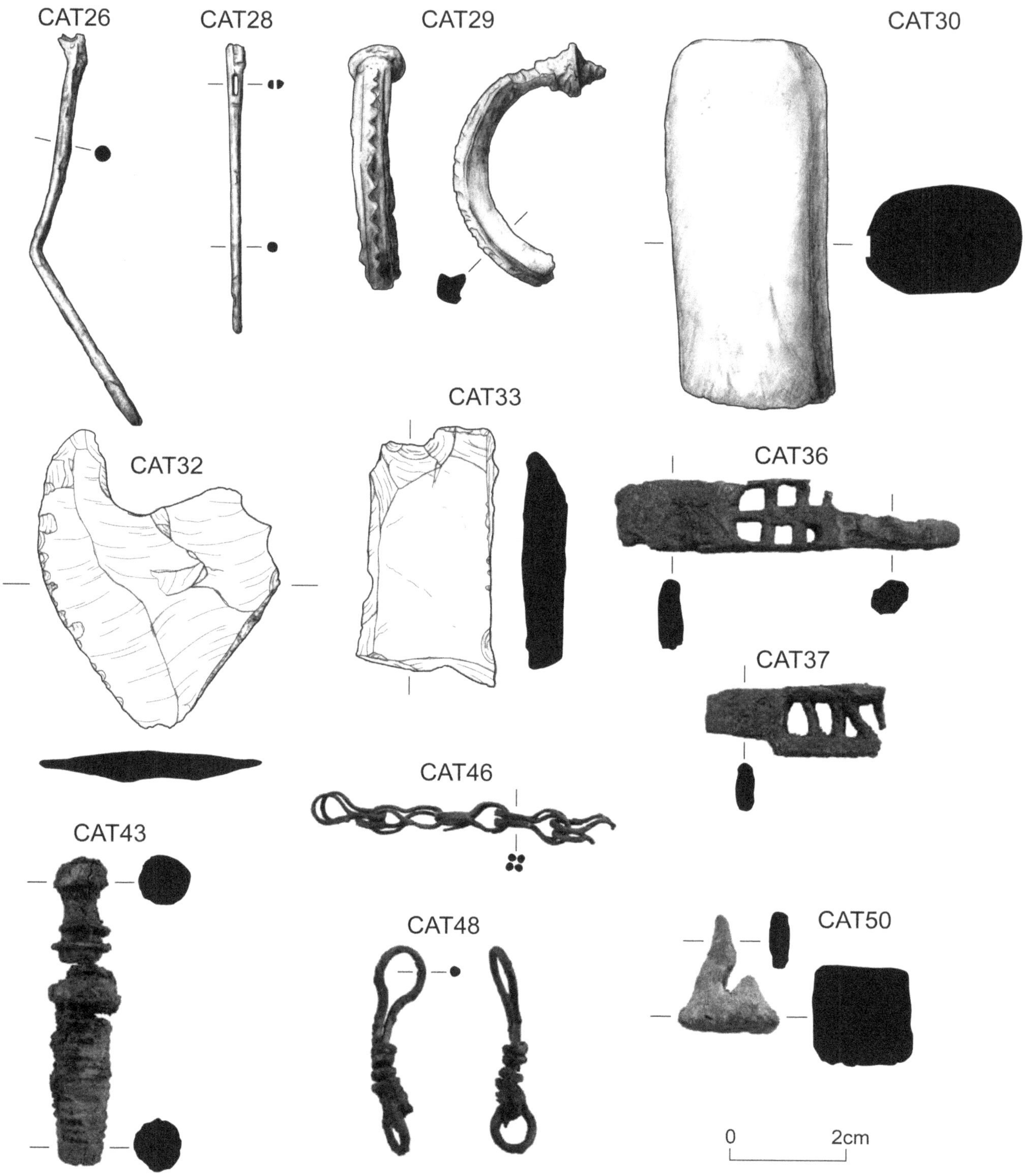

FIGURE 2.29 SMALL FINDS CAT26-50.

1021, Pit 1020, Group 112, Trench 1, Area C, Phase 4 . (ID 87). (Fig. 2.29.36).

37. Slide lock bolt. Copper alloy. Rectangular bar,one end broken across perforations; two rectangular and two triangular perforations. Present length 27mm, section 11 x 5.5mm. 3029, Pit 3030, Group 112, Area B, SF106, Event 2. (ID 107). (Fig. 2.29.37).

38. Slide lock key. Iron. Catalogued from X-radiograph. Broken loop terminal; rectangular block handle; wards projecting to one side. Length 80mm, width of handle 20mm. 2194, Pit 2193, Group 106, Area A, Phase 4. (ID38). Not illustrated.

39. Drop handle fragment? Copper alloy. Circular-sectioned wire bent in 'U'-shaped hook. Section 2mm, length 25mm. 5006, Pit 5007, Group 112, Area E, SF152, Event 2. (ID 59). Not illustrated.

40. Collar. Copper alloy. Sheet with three transverse ribs, Diameter 10mm, depth 8mm. 4006, Area C,

SF45, unphased. (ID 6). Not illustrated.

41. Ferrule. Copper alloy. Sheet bent into conical ferrule with seam up back; end bent to one side. Length 65mm, maximum section 8.5mm. 4006, Area C, SF46, unphased. (ID 170). Not illustrated.

42. Ferrule. Copper alloy. Rectangular-sectioned strip with two ribs on exterior; approximately one-third extant. Diameter 25mm, section 11.5 x 2mm. 2175, Pit 2209, Group 112, Area A, SF7, Event 3. (ID 146). Not illustrated.

43. Terminal. Copper alloy. Hemispherical knob terminal, concave neck, two narrow ribs; expanding ribbed rod below with collar at base before break; Present length 54mm, maximum diameter 12mm. 4033, Pit 4041, Group 112, Area C, SF79, Event 2. (ID 57). (Fig. 2.29.43).

44. Binding. Copper alloy. Rectangular strip bent along long axis and flattened. Length 50mm, depth 7.5mm. 2192, Pit 2191, Group 112, Area A, SF8, Event 3. (ID 149). Not illustrated.

45. Binding. Copper alloy. Rectangular strip bent along long axis and flattened. Present length 42mm, depth 6mm. 5002, Pit 5004, Group 112, Area E, Event 3. (ID 93). Not illustrated.

46. Chain. Copper alloy. Wire links formed from an oval loop with centre pinched together, bent in two with centre passing through paired loops of previous link. Five conjoined links with one separated one. Length of link 13mm, width link 6mm. 5005, Pit 5007, Group 112, Area E, SF130, Event 3. (ID 27). (Fig. 2.29.46).

47. Chain. Copper alloy. Links of rectangular-sectioned sheet with a ring at either end of a central bar; each link bent in half with bar passing through the paired rings of the previous link. Two fragment of two links each. Length of folded link 17mm, width 6mm. 1008, Pit 1019, Group 112, Trench 1, Area C, SF3, Event 3. (ID 144). Not illustrated.

48. Suspension loop. Copper alloy. Wire loop bent into circular loop at one end and wrapped around shank one and a half times; forming large circular loop at the bottom, other end runs parallel to shank and then wrapped around shank twice. Length 44mm, diameter of larger loop 11.5mm. 4030, Pit 4041, Group 112, Area C, SF78, Event 2. (ID 16). (Fig. 2.29.48).

49. Split pin. Copper alloy. Rectangular-sectioned strip tapering to point, bent in two with circular loop at top with arms parallel and then bent at 90 degrees, tip of one arm broken. Length 24mm, maximum section 4 x 2mm. 5005, Pit 5007, Group 112, SF132, Area E, Event 3. (ID 67). Not illustrated.

50. Slotted base. Copper alloy. Square flat-based pyramid with one projecting arm and slot at base. Base dimensions 17mm, height 22mm.: 5002, Pit 5004, Group 112, Area E, SF123, Event 3. (ID 81). (Fig. 2.29.50).

51. Hinged mount. Copper alloy. Hollow hemisphere with two semi-circular lugs for a hinge projecting below; three projecting circular projecting lugs spaced around the edge, the two on either side of the hinge have long circular-sectioned shanks projecting on back. The upper face of the hemisphere appears to have narrow triangular cells radiating out from a small central cell, but the piece was highly corroded and details are obscure. Diameter of hemisphere 18mm, maximum length 24mm, length of shanks 7mm. 1008, Pit 1019, Group 112, Trench 1, Area C, Event 3. (ID 39). (Fig. 2. 30.51).

52. Mount. Copper alloy. Rectangular mount with two integral rivets with burred ends; extant rounded end has large circular perforation with triangular notch on either side; other end broken at notch. Present length 50mm, section 9.5 x 2mm, length rivets 6mm. 4030, Pit 4041, Group 112, Area C, Event 2. (ID 163). (Fig. 2. 30.52).

53. Stud. Copper alloy. Peltate head inlaid with silver (identified during conservation) outlining edges and with a rayed pattern along the lower part; central square-sectioned tapering shank. Dimensions head 19 x 17mm, length 10mm, shank section 3mm. SF43, 4015, Pit 4041, Area C, Event 3 (ID 114). (Fig. 2. 30.53).

54. Nail. Copper alloy, Spherical knob with circular-sectioned shank inserted into a broken plate. Length 16mm, head diameter 5.5mm, shank section 2.5mm. : 4030, Pit 4041, Group 112, Area C, SF92, Event 2. (ID 104). Not illustrated.

55. Nail. Copper alloy. Hemispherical knob head; square-sectioned shank tapering to point. Length 21mm, head diameter 9mm, shank section 4mm. 5002, Pit 5004, Group 112, Area E, SF118, Event 3. (ID 145). (Fig. 2. 30.55).

56. Nail. Copper alloy. Hemispherical knob head; broken circular-sectioned shank. Present length 15.5mm, head section 11 x 10.5mm, shank section 3.5mm.: 5002, Pit 5004, Group 112, Area E, SF120, Event 3. (ID 128). Not illustrated.

57. Nail. Copper alloy. Oval knob head, broken circular-sectioned shank section. Present length 14.5mm, head section 6mm, shank section 2.5mm.: 3033, Pit 3031, Group 112, Area B, SF105, Event 2. (ID 172). Not illustrated.

58. Rivet. Copper alloy. Rectangular-sectioned shank tapering to bevelled point; slightly expanded flat-topped head. Length 21mm, head section 3.5mm, shank section 2.5 x 2mm. 5002, Pit 5004, Group 112, Area E, SF138, Event 3. (ID 74). Not illustrated.

59. Stud. Copper alloy. Large flat head with broken edges; square-sectioned shank tapering to blunt end and bent to one side. Diameter 31mm, length 8mm, shank section 3.5mm. 4030, Pit 4041, Group 112, Area C, SF119, Event 2. (ID 8). (Fig. 2.30.59).

60. Stud. Copper alloy. Flat broken disc head with edges bent over; short square shank with burred end. Length 4.5mm, head diameter 18mm, shank

section 2.5mm. 5005, Pit 5007, Group 112, Area E, SF127, Event 3. (ID 185). Not illustrated.

61. Stud. Copper alloy. Flat disc head with chipped edges bent over; short rectangular-sectioned shank tapering to point. Length 6mm, diameter head 12mm, shank section 1.5 x 1mm. 4015, Pit 4041, Group 112, Area C, SF60, Event 3. (ID 69). Not illustrated.
62. Stud. Copper alloy. Broken flat circular head; slender rectangular-sectioned shank bent sideways. Head diameter 12mm, shank section 3.5 x 1.5mm, length 13mm. 4030, Pit 4041, Group 112, Area C, SF89, Event 2. (ID 122). Not illustrated.
63. Stud. Iron. Flat disc head with stump of shank. Diameter head 23mm. 4003, Pit 4012, Group 112, Area C, SF34, Event 3 (ID 159). Not illustrated.
64. Stud. Copper alloy. Circular-sectioned tapering shank with broken tip; fragment of flat head. The whole bent out of shape. Length c. 20mm. 4003, Pit 4012, Group 112, Area C, SF120, Event 3. (ID 131). Not illustrated.
65. Stud. Copper alloy. Circular-sectioned shank tapering to point; broken flat head. Length 11.5mm. 4006, Area C, SF44, unphased. (ID 119). Not illustrated.
66. Stud. Copper alloy. Square-sectioned shank broken at tip; broken flat head. Length 13mm, section shank 2mm. 4030, Pit 4041, Group 112, Area C, SF73, Event 2. (ID 171). Not illustrated.
67. Stud. Copper alloy. Flat head fragments. Head diameter 6.5mm. 2111, Pit 2110, Group 112, Area A, SF9, Event 3. (ID 132). Not illustrated.
68. Stud. Copper alloy. Hollow hemispherical head, broken circular-sectioned shank. Head diameter 12mm, depth of head 8.5mm. 4006, Area C, SF45, unphased. (ID 5). Not illustrated.
69. Stud. Copper alloy. Hollow domed head; shank bent sideways. Diameter 9mm. 4007, Pit 4002, Group 112, Area C, SF117, Event 3. (ID 165). Not illustrated.
70. Stud. Copper alloy. Hollow domed broken head; square-sectioned tapering shank. Length 13mm, head diameter 5.5mm, shank section 1.5mm. 5002, Pit 5004, Group 112, Area E, SF144, Event 3. (ID 91). Not illustrated.
71. Rivetted strip. Copper alloy. Two fragments of strip, rivetted together with dome-headed stud. Dimensions 14 x 8mm, diameter of stud head 4.5mm. 5002, Pit 5004, Group 112, Area E, Event 3. (ID 89). Not illustrated.
72. Rivetted sheet. Copper alloy. Fragment retaining dome-headed stud. Dimensions 21 x 14mm, diameter of stud 8mm. 5012, Pit 5013, Group 112, Area E, SF155, Event2 (ID 58). Not illustrated.

Military Equipment

A wide range of military fittings were recovered including belt equipment, fragments of armour and harness fittings, providing very strong evidence that this was indeed rubbish being derived from inside the fortress. Several of the studs catalogued in the previous section could also have come from military items but only those that regularly occur on military sites but are rare on civilian ones have been included here. Military equipment of the 1st century was quite often coated with white metal with contrasting black niello inlays. As metal does not survive well at this site and most of the surfaces show great erosion, any decorative details such as those have been destroyed.

Nos. 73-4 are quadrilateral buckle frames from narrow belts. In No. 73 the buckle pin could have hinged on the crossbar, but on No. 74 the finished edge with an internal volute indicates it must have been hinged on the bar that ran through the hinge linking it to the first belt plate. The presence of the quadrilateral form in these contexts is interesting as the commonest military belt buckles of the time were D-shaped with internal volute patterns such as the example made of ivory from the *principia* within the fortress found during the excavations at Watergate Street (Lloyd Morgan 1988), and the examples found during the excavation of the amphitheatre (Thompson 1976, 190 nos. 12 and 17). The quadrilateral form is known in the first century but is more typical of the second century (Bishop and Coulston 2006, 107 figs. 62.14, 88.8). The internal volute on No. 74 is reminiscent of the volutes seen on the wider D-shaped buckles. These normally have a small bar between them for hinging the pin, but occasionally this is lacking and the pin was hinged on the cross bar running through the hinge loops. An example from Hod Hill shows this clearly (Brailsford 1962 fig. 4 no. A78). No. 74 may thus be demonstrating an early variant for the quadrilateral form. The context date would fit this interpretation.

There are numerous fittings from the form of armour made of bands of iron plate known as *lorica segmentata*. The Corbridge form of this armour, in use during the first century and into the early second century, relied on hinged copper alloy buckles with square plates, lobate hinges, and rectangular plates with looped ends to fasten all the pieces together (Bishop 2002, 31-44). Pieces that definitely belong to this style include the buckle frame No. 75, the hinge plate No. 76 and the tie plate No. 78. The hinge plate No. 77 probably also belongs to this type. It is shorter than is normally to be expected but that may be because it has been damaged. No. 80 is very badly corroded and fragmented but the extant features would agree with it being a Corbridge style tie plate. No. 79 has been included as a possible third example but is much smaller than is to be normally expected. The type of cuirass that replaced the Corbridge style is known as the Newstead style and this was developing in the second quarter of the second century and was dominant by the middle part of the century (Bishop 2002, 46-61). This type of armour had different types of copper alloy fittings which did away with the need of so many fiddly hinges and buckles that had probably been in need of constant repair. Key amongst the improvements was the substitution of tie rings for the tie loops. No. 81 is an example of one of these tie rings. The fact that only one fitting from a Newstead cuirass can be identified at

Delamere Street compared to the numerous items from the Corbridge style is to be expected given that this is a late first to early second century assemblage.

In addition to these copper alloy fittings there were three fragments of iron plate with copper alloy (Nos. 82-4). The first two can certainly be identified as small parts of girdle plates with attached copper alloy fittings. The features they preserve would be consistent with coming from a Corbridge style cuirass. No. 84 is too corroded for the identification to be made with certainty. Finally a copper alloy D-shaped loop with an iron plate may be noted (No. 85). It is possible that this was an *ad hoc* repair for a cuirass.

At Chester fittings for the Corbridge style of cuirass have been found before at King's Buildings. From the description several elements of the cuirass may have been found still connected together, though the initial identification described the item as a box or casket (Newstead 1928, 91 nos. 3-8, pl IX). Tie loops for this style are also recorded from the amphitheatre (Thompson 1976, 192 no. 20) and Princess Street (Newstead and Droop 1940, 40 no. 11 pl. X), while a buckle and plate similar to Nos. 75 and 76 was found at Deanery Fields (Newstead 1924 pl. VI no. 10). The Newstead style of cuirass is also represented by a tie ring (Newstead 1939, 59 no. 5). The presence of all these pieces of *lorica segmentata* is not surprising given that it was the armour form associated with legionaries (Bishop 2002, 91). There is evidence from within the fortress at Bridge Street that the legionaries were also wearing scale armour (Cool 2008, 309 nos. 24-7) and mail armour in both copper alloy (Cool 2008, 309 nos. 28-32) and iron (Mould 2008, 317-8). A fastener of the type associated with mail armour was also recovered at King's Buildings (Newstead 1928, 90 no. 2) whilst armour scales came from two sites on Hunter Street (Newstead 1928, 73 no. 8). Neither of these types are represented at Delamere Street but given the soil conditions it is unlikely that the latter would have survived in a recognisable form, and the same might be suspected for the former.

The strap terminal No. 86 belongs to a style of harness fitting called the 'spectacle type' (Bishop 1998, 72 nos. 270-7) which was well established in the later first century and continued into the Antonine period. The clip at the back of No. 86 shows that it was a terminal plate for a narrow strap as the end of the plate would have fitted into it. Five strap terminals with similar clips were found amongst the harness fittings found in a Claudio-Neronian pit at Canterbury which confirmed that these were horse equipment (Lawson 1995, 991 nos. 165-9). Such fittings would have been appropriate for the narrow decorative straps that hung down over the legs of the horse (see for example Jenkins 1985, figs. 1 and 15).

Pendants such as No. 87 were most probably part of military horse harness though the possibility that they may have been part of legionary aprons is sometimes advanced. Its fragmentary state means that the method of attachment is unclear, it could either have been an extension bent over to form an open loop or a perforated lug. The latter is probably more likely. It seems very probably that the piece had an openwork pattern, though it is likely that it was a relatively simple 'T' shape or the like. Oldenstein (1977, 126) suggests perforated pendants such as these are of second century date which would agree with the other material from the context. No. 88 probably served a similar purpose. No precise Roman parallel has been identified and the context is undated, but within the context of this excavation where post Roman metalwork is rare, a Roman date seems likely and the peltate terminal is a typical Roman motif.

There are also a small number of items whose precise function cannot be suggested but which are regularly found on military sites. The stud No. 89 has the typical dished centre seen on military fittings. It is generally accepted that ribbed collars such as No. 90 had some role to play within military equipment though no-one is quite sure what (Bishop 1998, 91 no. 343). Writing about a large example found at Blake Street, York, Lloyd-Morgan drew attention to the first to second century military contexts they were often found in (Cool *et al* 1995, 1536) and cited two published examples from Chester. One from Newgate Street associated with the Trajanic construction of a rampart building (Thompson and Tobias 1957, fig. 4 no. 16) and one found with first and second century pottery at Princess Street (Newstead and Droop 1940, 40 no. 15). She describes a 'good number' as having been found at Chester, presumably coming from the many still unpublished reports she wrote on the finds from the sites excavated in the third quarter of the 20th century within the town. No. 91 is an example of a dumbell button (MacGregor 1976, 134). Again these are regularly found in 1st and 2nd century military contexts. Examples from Castleford in both copper alloy and bone were found in contexts associated with the early Flavian fort (Cool and Philo 1998, 116 nos. 782-4, 281 no. 161), whilst a bone example was included in the chest of military fittings found in a probable Hadrianic context at Corbridge (Allason-Jones and Bishop 1988, 83 no. 285).

Finally the button and loop fastener No. 92 may be included with the other military fittings. They obviously had a use in both military and civilian milieus but again are regular finds on military sites and are generally included with the other military equipment. No. 92 belongs to Wild's Class Vb with circular enamelled heads for which a second century date is suggested (Wild 1970, 140). The context of this example is a useful early instance of the type as a Trajanic date is suggested for this pit fill. Button and loop fasteners have occasionally been noted at Chester before. Two circular examples with ribbed faces (Wild 1970 Class Va) came from excavations at the amphitheatre; one from a probable late first century context (Newstead and Droop 1932, 35 no. 5), whilst the other came from a post-medieval one (Thompson 1976, 190 no. 6).

Catalogue

73. Belt buckle frame. Copper alloy. Square with concave sides and front edge, straight crossbar with one broken hinge loop. The whole much corroded and eroded. Length 26mm, width crossbar 26mm. 3033, Pit 3031, Group 112, Area B, SF103, Event 2. (ID 136). (Fig. 2. 30.73).
74. Belt buckle frame. Copper alloy. Side of square frame, concave sides and part of front edge, part of straight crossbar with broken hinge loop. Transverse beaded rib on cross bar which terminates in roundel; central crossbar probably with gap for pin. Length 26mm, width crossbar 29mm. 2175, Pit 2209, Group 112, Area A, Event 3. (ID 94). (Fig. 2. 30.74).
75. Cuirass buckle frame. Copper alloy. Square-sectioned D-shaped frame with broken expanded terminals for missing cross-bar. Width 26mm, section 3mm. 4004, Pit 4002, Group 112, Area C, SF98, Event 3. (ID 118). (Fig. 2. 30.75).
76. Cuirass hinge plate. Copper alloy. Rectangular sheet strip with diagonal corners bent in half around an iron hinge bar and fastened with two dome-headed rivets. Length 22.5mm, width 20mm. 5006, Pit 5007, Group 112, Area E, SF151, Event 2. (ID 78). (Fig. 2. 30.76).
77. Cuirass hinge plate. Copper alloy. Sheet bent in half with tubular hinge with rectangular cut-out, outer edge irregular. Width 22mm, current length 11mm. 4030, Pit 4041, Group 112, Area C, SF90, Event 2. (ID 9). (Fig. 2. 30.77).
78. Cuirass girdle tie plate. Copper alloy. Irregular slightly waisted plate with two rivets and rectangular-sectioned projection with bevelled end. Length 63mm, maximum section of plate 16 x 1.5mm., section of loop 4.5 x 3mm. 4003, Pit 4012, Group 112, Area C, SF107, Event 3. (ID 52). (Fig. 2.30.78).
79. Cuirass girdle tie plate ?. Copper alloy. Rectangular sheet plate with rounded end, other end tapering and curved back to form loop. Unusual corrosion products on underside, possibly the remains of mineralised organics. Length 25mm, section plate 6.5 x 1mm. 5005, Pit 5007, Group 112, Area E, SF148, Event 3. (ID 80). Not illustrated.
80. Cuirass girdle tie plate ? Copper alloy. Highly corroded sheet, broken plate with tapering loop. Present length 23mm, width 10.5mm. 4030, Pit 4041, Group 112, Area C, SF64, Event 2. (ID 158). Not illustrated.
81. Cuirass tie ring. Copper alloy. Square-sectioned ring with cross bar; circular sectioned shank. Length 24mm, diameter of ring 15.5mm, thickness 2.5mm. 5002, Pit 5004, Group 112, Area E, SF134, Event 3. (ID 123). (Fig. 2.29.81).
82. Cuirass girdle plate fragment. Two overlapping iron plate fragment with copper alloy sheet on one face. Dimensions 31 x 22.5mm. 2311, Pit 2293, Group 112, Area A, SF31, Event 3. (ID 12). Not illustrated.
83. Cuirass girdle plate fragment. Fragment of iron sheet with copper alloy sheet over fastened by dome-headed stud. Dimensions 12.5 x 15mm, stud head diameter 6mm. 5002, Pit 5004, Group 112, Area E, SF135, Event 3. (ID 76). Not illustrated.
84. Cuirass girdle plate fragment? Iron with fragment of copper alloy fitting. Much corroded. Dimensions 16 x 13mm. 2311, Pit 2293, Group 112, Area A, SF30, Event 3. (ID 154). Not illustrated.
85. Strap fitting. Rectangular-sectioned approximately 'D'-shaped copper alloy loop; cross bar enclosed by iron sheet. Length loop 18.5mm, loop width 25mm, section 4 x 1mm.: 5014, Pit 5013, Group 112, Area E, SF158, Event 2. (ID 32). Not illustrated.
86. Strap terminal. Copper alloy. Flat backing plate with two circular expansions at top and a tapering terminal consisting of a square unit above a 'D'-shaped one; each circular expansion has remains of a dome-headed stud; on the back a projecting triangular clip to take the end of the strap, Length 42mm, diameter of stud 10mm, thickness plate 1mm.4033, Pit 4041, Group 112, Area C, SF77, Event 2 (ID 111). (Fig. 2.30.86).
87. Harness pendant. Copper alloy. Fragmentary heart-shaped pendant with flat-backed baluster terminal; broken top edge of pendant slightly thickened; broken central part retains smooth edges in places showing that it originally had an openwork pattern. Traces of punched dots around upper edges. Present length 57mm, original maximum width c. 35mm. 1008, Pit 1019, Group 112, Trench 1, Area C, Event 3. (ID45). (Fig. 2.31.87).
88. Harness pendant. Copper alloy. 'D'-sectioned terminal bent over to form hook with moulded end; flat disc with central hollow-backed hemispherical boss with wide flange with two concentric grooves, two lugs centrally with pair of lugs at hook/disc junction; bottom has broken flat pelta terminal. Present length 60mm, maximum width 39mm. SF67, 4011, Pit 4012, Area C, Event 3, (ID 105). (Fig. 2.31.88).
89. Stud. Copper alloy. Circular head with central dishing and edges bent over; thick oval-sectioned shank with burred end. Differential corrosion in ring around dishing, possibly indicative of inlay. Diameter 16mm, length 8mm, shank section 5.5 x 4.5mm. 5015, Pit 5013, Group 112, Area E, SF153, Event 2. (ID 72). (Fig. XX.89).
90. Collar. Copper alloy. Sheet strip bent into a cylinder and now flattened; thick rib at either end with more shallow ribbing centrally. Length 21.5mm, present diameter 10.5 x 7mm. 4016, Pit 4041, Group 112, Area C, SF56, Event 3. (ID 108). (Fig. 2.31.90).
91. Dumbell button. Copper alloy. Saddle between two spheres with ridge on either side. Length 20mm, maximum diameter 10mm. 3049, Pit 3058, Group 112, Area B, SF118, Event 2. (ID 126). (Fig. 2.31.91).
92. Button and loop fastener. Copper alloy. Circular disc with upper face divided into six triangular

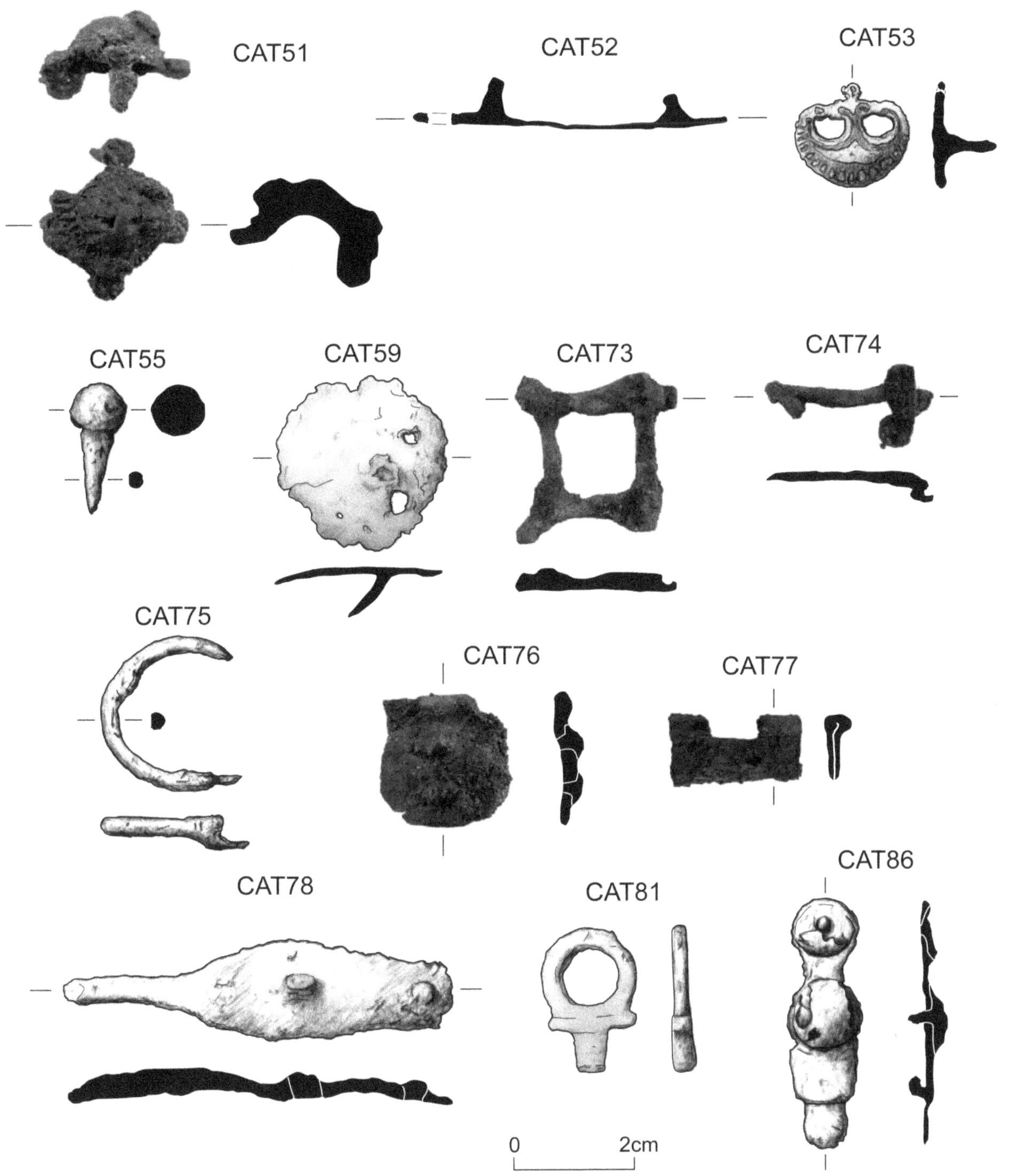

FIGURE 2.30 SMALL FINDS CAT51-86.

cells, ribs between cells have feathered edges, small central circular cell centrally; triangular cells enamelled alternately dark blue and now decayed green, central dot also decayed green. Stump of broken loop below. Diameter 24mm. 4015, Pit 4041, Group 112, Area C, SF42, Event 3. (ID 113). (Fig. 2.31.92).

Miscellaneous

Fragments of sheet, wire and the other items that have been reduced to unidentifiable lumps have not been catalogued here but are available in the archive listings. Of the items catalogued here only a few call for any special comment.

There are two enamelled fragments. The larger No. 93 might come from a belt plate but enamelled belt plates often have a central rivet at the back or the scar of one as can be seen on examples from Caerleon (Fox 1940. fig. 6.10; Webster 1992, 123 no. 88) and this piece does not show this feature. It may also be noted that the quatrefoil design would not be a common one for belt plates of this type. (The Caerleon examples show two of the most common designs). Enamelled rectangular strips are

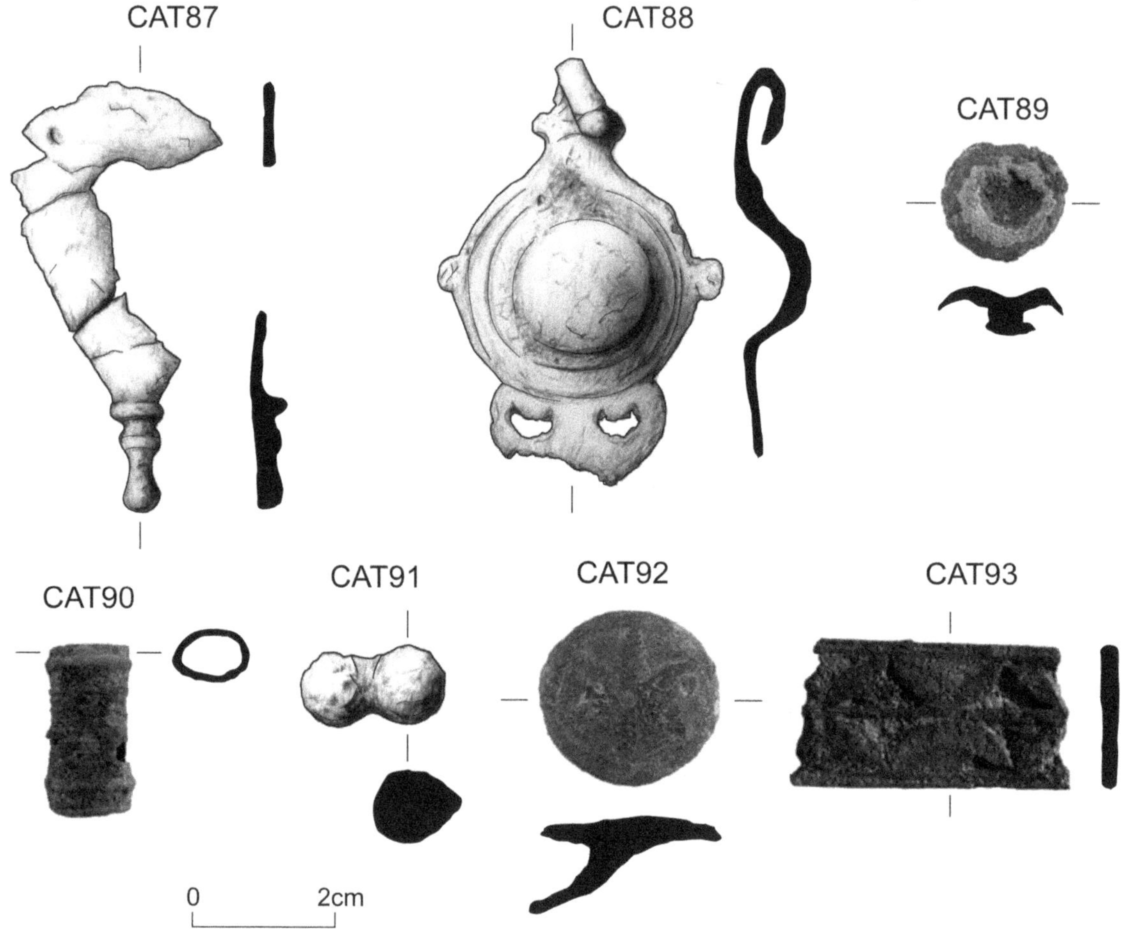

FIGURE 2.31 SMALL FINDS CAT 87-93.

found from time to time such as that from South Shields (Allason-Jones and Miket 1984, 96 no. 3.12) which is of the same width as No. 93. These might just be decorative inlays but the possibility exists that they could be elements of composite enamelled vessels. The hexagonal conical flask from a burial at Corbridge, for example, has six tapering side panels as well as a separate base and top (Casey and Hoffman 1995, 24 no. 6). Neither No. 93 nor the South Shield piece is tapering but it is clear that we are still long way from knowing the full range of the vessel forms that were produced in this technique. In support of this identification it may be noted that reserved quatrefoils such as those on No. 93 occurred regularly on the Castleford moulds for making enamelled vessels (Bayley and Budd 1998, 211 Patterns 3-6) which may be dated to within the AD 85 – 100 period. No. 93 is possibly not the only enigmatic enamelled strip to have been recovered from Chester. The end of a plate with a running vine leaf scroll pattern was found at the amphitheatre described as having been found 'at a low level below the seating bank' (Newstead and Droop 1932, 33). Unfortunately there is no record of whether there were any rivets at the back which would indicate it was a belt plate and the piece was subsequently destroyed. Again though, the pattern would be unusual for a belt plate but is one known to have been used on the metal vessels occurring on Patterns 9, 11, 13 and 20 of the Castleford moulds.

The other enamelled fragment consists merely of a single circular cell retaining enamel and part of the field around it. It probably did not come from a stud, as unless the stud was very large with a ring of such cells, it might be expected that the extant cell would have been the central one. In which case there ought to be a shank on the back and there is no sign of one. It is thus more likely that the fragment came from a brooch or a seal box lid, both of which could have flat rear surfaces at the centre.

Catalogue

93. Enamelled strip. Copper alloy. Rectangular tapering very slightly; both ends broken. Front divided into two fields by central strip with pointed oval blocks in zig-zag either side of it forming a reserved quatrefoil pattern; groove parallel to each edge. Recessed cells have traces of blue, orange and red enamel together with decayed enamel appearing green. Present length 38mm, width 9-10mm, thickness 2.5mm. 5005, Pit 5007, Group 112, Area E, SF147 Event 3. (ID 71). (Fig. 2.31.93).

94. Enamelled fragment. Copper alloy. Fragment of plate with circular cell containing orange enamel. Dimensions 13 x 12mm, diameter of cell 7.5mm, 4004, Pit 4002, Group 112, Area C, SF101, Event 3. (ID 130). Not illustrated.

95. Disc. Lead alloy. Thick sheet cut into circular disc. Edges burred. One face has dot at centre of two finely scored lines locating the centre. More random scratch marks on under side. Diameter 22.5 mm, thickness 3mm. Weight 13g. 4001, Area C, SF27, unphased. (ID 162). Not illustrated.
96. Disc. Lead alloy. Plano-convex. Diameter 75mm, height 25mm, weight 663g. 4009, Pit 4002, Group 112, Area C, Event 3. (ID 187). Not illustrated.
97. Disc. Lead alloy. Circular disc with straight rounded ridge across one face close to edge. Diameter 42 mm, thickness 6mm. Weight 65g. 3049, Pit 3058, Group 112, Area B, SF102, Event 2. (ID 109). Not illustrated.
98. Bar. Lead alloy. Rectangular-section bar with grooves on three faces; broken ends. Length 49mm, section 12 x 9mm, weight 47mm. 4016, Pit 4041, Group 112, Area C, SF55, Event 3. (ID 161). Not illustrated.
99. Spatula (?). Lead alloy. Square-sectioned shank expanding slightly to oval bowl; bowl bent back over part of shank. Length c. 140, shank section 7.5mm, width bowl 50mm. 4003, Pit 4012, Group 112, Area C, SF51, Event 3. (ID 53). Not illustrated.
100. Ring. Copper alloy. Rounded square section. Diameter 20mm, section 3mm. 5002, Pit 5004, Group 112, Area E, SF125, Event 3. (ID 140). Not illustrated.
101. Ring. Copper alloy. Oval-sectioned. Diameter 25.5mm, section 4 x 3mm. 4009, Pit 4002, Group 112, Area E, Event 3. (ID 51). Not illustrated.
102. Penannular ring. Copper alloy. D-sectioned of varying thickness, bent into an approximate ring shape. Diameter 26 x 24mm, maximum section 3 x 2.5mm.: 5012, Pit 5013, Group 112, Area E, SF160, Event 2. (ID 54). Not illustrated.
103. Ring. Iron. Much fragmented. Diameter 50mm. 3039, Pit 3038, Group 112, Area B, SF95, Event 2. (ID 47). Not illustrated.
104. Ring. Copper alloy. Rectangular-sectioned, Diameter 18.5mm, section 3 x 2mm. 5005, Pit 5007, Group 112, Area E, SF142, Event 3. (ID 70). Not illustrated.

Discussion of the Roman finds

This group of finds has provided an interesting insight into the legionary community at Chester. The former stationing of the legion at Wroxeter can be seen from the geographical affiliations of the majority of the brooches which came from the west Midlands / Severn valley area rather than being the types found on military sites further to the north. The possibility that some of these individuals might have been of native British origin is an intriguing possibility given the presence of the nail cleaner, also a local type, and the beaded torc. The possibility that some of the soldiers might have been relatively recently arrived from the continental frontiers though, is suggested by the unusual fantail brooch No. 6.

On the whole this is a fairly utilitarian legionary assemblage. Much of the military equipment appears to be damaged in some way; in as far as it is possible to tell given the state of the metalwork. It consists primarily of armour and belt equipment and so fits well with the concentration on personal items seen in the other categories, not only the brooches and other ornaments but also the toilet and textile equipment. Generally the assemblage has a very 'domestic' feel. It is useful to compare the range of functional categories present (Table 2.18) to those originally defined by Crummy (1983). Tools are scarce, there is nothing that relates to weighing and measuring unless the suspension loop No. 48 comes from a steelyard. Other missing categories are writing equipment, transport equipment and recreation items. This is a very limited assemblage and not all the absences can be put down to the problems of survival on the site. Possibly what we are looking at is a rubbish source where a limited range of activities were being carried out, rather than the fact that these quarries and pits were a general dumping ground for the whole fortress. It is here that the lack of publication of such sites as Abbey Green, just inside the fortress to the

TABLE 2.19: A COMPARISON OF THE FINDS FROM DELAMERE STREET, CHESTER AND BLAKE STREET, YORK. (BLAKE STREET DATA FROM COOL ET AL 1995, TABLE 113 PLUS THE VESSEL FRAGMENT NO. 6351)

Function	Event 2	Event 3	Blake Street			
	Number	%age	Number	%age	Number	%age
Personal ornaments	8	25	11	22	13	23
Toilet equipment	1	3	4	8	1	2
Textile equipment	-	-	1	2	-	-
Household items	1	3	-	-	1	2
Recreational items	-	-	-	-	10	18
Writing equipment	-	-	-	-	4	7
Tools	3	9	3	6	8	14
Fasteners etc	11	34	19	38	14	25
Military equipment	8	25	12	24	5	9
Total	32		50		56	

south, becomes problematic as there are no useful large assemblages to compare the Delamere Street finds to. In the absence of Chester assemblages it can be compared to the assemblage from the buildings at Blake Street in the *praetentura* at York as is done in Table 2.19. The Delamere finds include only those stratified in the Event 2 and 3 contexts and the Blake Street material includes only that from the Phase 2 and 3 contexts which broadly date to the last two decades of the first century and the first two of the second (Hall 1997, 377) so the assemblages are contemporary. They should be directly comparable as only four items of the Delamere Street assemblage being considered came from samples, so the difference in digging regimes at the two sites should not have influenced finds retrieval to any great extent. The type of occupation at Blake Street at the time is not known as much of their plan was destroyed by the late second century stone buildings built on the site though the position would be correct for a barracks and the finds assemblage seemed to indicate that is what it was.

Given that the small finds do look as if they might be coming from a specialised source though possibly not from a barracks, it is useful to compare them to the other finds. It has been suggested that the samian pottery might indicate that the rubbish was coming from officer's quarters, and the ceramic building material also includes evidence that it was being derived from areas where heated rooms were present. In contrast to this the vessel glass shows no evidence of the quality of tablewares that might have been expected at an officer's table and is very much a kitchen assemblage. There is no obvious evidence that the small finds were coming from an officer's quarters other than perhaps the handle No. 29. It would have come from the sort of basin more likely to have been owned by an officer than by an ordinary soldier but as has been pointed out, the vessel it came from would have been quite old by the time this piece was lost. Whether this might indicate it was a second-hand battered item which had lost its elite connotations is an interesting question.

One possibility that might account for the high proportion of copper alloy items and the high proportion of generally damaged military equipment is that some of the rubbish derived from a *fabrica* that repaired armour and belt fittings and the like. Copper alloy casting waste is represented in both Event 2 and Event 3 contexts (*see Mackenzie Chapter 6*) as are numerous unidentifiable copper alloy fragments, which might support this hypothesis.

So though we can say that this material was undoubtedly derived from within the fortress and appears to have elements from a specialised source, identifying who would have used it more closely is not possible.

Table 2.19 allows a rapid comparison of the functional profile from both Events. As can be seen there is very little difference between the two. The only possible significant one being the higher proportion of toilet equipment in Event 3. As far as the dates of the items are concerned, the assemblages from both events are broadly contemporary and can be happily accommodated within the later first to early second century period. The military equipment is particularly useful here as it is a category of find subject to fashion changes and is present here in large numbers. There are items for which a second century date would normally be suggested as more likely than a late first century one. The quadrilateral buckles (Nos. 73-4) and the harness pendant (No. 87) being good examples. The buckles, however occur in both Event 2 and Event 3 contexts and so do not provide compelling evidence for differentiating the date of the events. What is very noticeable is the absence of the distinctive suite of military equipment that develops from the late Antonine period onwards (see for example Bishop and Coulston 2006, 128-48). The *lorica segmentata* is overwhelmingly of the Corbridge type. Had the assemblage been continuing to accumulate in the second half of the century then, given the amount of *lorica segmentata* fittings present, the Newstead form should have been better represented than it is. The brooches too tell the same story. Both in the contexts of Events 2 and 3 they are first century to early second century forms. The knee brooches and divided bow brooches typical of northern military sites in the second half of the second century are absent.

It would thus appear that the rubbish being disposed of during both Events 2 and 3 was broadly contemporary and to be dated to the later first to earlier second centuries. If, as suggested in Chapter 2, the Event 3 assemblages result from late second century midden clearances within the fort, then the small finds would strongly suggest that those middens could well have been 50 years or more old by that time.

THE COINS *by Roger White*

A total of 16 coins were recovered during the excavation, the majority from stratified contexts. Of these two are modern and are not considered further. The group is too small for any meaningful statistical analysis. Where the coins are not corroded, the preservation is excellent and the surfaces were very legible after conservation. Even the illegible coins can, however, be ascribed to a rough date range consistent with the identifiable coins in the group and it can be said with some confidence that these coins fall into the period from the foundation of the fortress at Chester towards the end of Vespasian's reign through to early in the reign of Hadrian (Mason 2001, 44-6). This conclusion can be refined further in that the coins that are clearly of Vespasian and Domitian are largely worn so that the faces are quite smooth and only some letters are visible. The illegible coins are probably extreme versions of these Vespasianic and Domitianic coins. In this context it is worth noting that only one of the coins is a *Sestertius*, four others are *Dupondii* while the rest are *Asses*. These last will have had a low value – the *As* is often translated from Latin in texts as a 'penny' – and will thus have received high circulation as small change (Reece 1987, 31-3), suffering in consequence from extreme wear. The high proportion of *Asses* of Domitian, at least three and probably five, probably derive from a large number of *Asses* minted in Rome and directly shipped to Britain in AD 85-6 (Burnett 1987, 57). The three later coins, two of Trajan (98-117) and one of the reign of Hadrian (117-138), are a distinct element in this group. Their preservation is very good with all showing scarcely any wear while the latest coin, a scarce issue of Hadrian's wife Sabina issued before her death in AD128, is very crisp, despite its corrosion.

Taken together, it seems clear that these coins could all have been circulating contemporaneously as a group, i.e. the group does not appear to represent an occupation lasting from the time of Vespasian until that of Hadrian but rather that these represent casual losses of low value coins at the same moment. If so, then that moment is likely to have been early in the reign of Hadrian.

Catalogue

All references are to *Roman Imperial Coinage* (RIC) volume II (Mattingley and Sydenham 1926).

Roman

1. Obv: [IMP C]AES VESPASIAN [AVG]COS [III] Head, laur. R.; Rev: AEQVITAS AVGVS[TI] Aequitas standing l. holding scales In field, SC; Die Axis: 12; Diam: 27mm; Wt: 9g; W/W (locally corroded); *As* of Vespasian, probably RIC II, 482 Consulship unclear but post AD71; 3049, Pit 3058, Area B, SF115, Event 2. Not illustrated.
2. Obv: [IMP CAES VESPA[SIAN AVG [COS --] Head, laur. R.; Rev: Standing figure [In field, SC]; Die Axis: 12; Diam: 27mm; Wt: 6g; Corr/Corr; *As* of Vespasian, AD69-79; 4033, Pit 4041, Area C, SF80, Event 2. Not illustrated.
3. Obv: IMP CAES DOMIT AVG GERM [COS – CENS PER PP] Bust, laur. R. ; Rev: VI[RTVTI] AVGVSTI Virtus standing r. holding spear & parazonium In field, SC; Die Axis: 11; Diam: 28mm; Wt: 14g; W/W ; *Dup* of Domitian, as RIC II, 331 or similar (Consulship unclear but probably postdates AD85) AD85-96; 4006, Area C, SF40. Not illustrated.
4. Obv: [IMP CAES DOMIT]AVG GER[M]COS XII CENS PER PP Bust, laur. R. with aegis; Rev: MO[NETA AVGVSTI] Moneta standing l. [holding scales and cornucopiae. In field, SC]; Die Axis: 1; Diam: 27mm; Wt: 11g; UW/UW (locally corroded); *Dup* of Domitian, RIC II, 327, AD86; 4016, Pit 4041, Area C, SF47, Event 3. Not illustrated.
5. Obv: IMP CAES DOMIT AVG GERM [COS -- CENS] PER PP Bust, laur. R. with aegis; Rev: Illegible [In field SC]; Die Axis: --; Diam: 27mm; Wt: 7g; W/W; *As* of Domitian, Consulship unclear but post AD88; AD88-96; 4006, Area C, SF39. Not illustrated.
6. Obv: IMP CAES NER TRAIANO OPTIMO AVG GER DAC PARTHICO PM TR P COS VI PP Radiate bust r. dr.; Rev: PROVIDENTIA AVGVSTI Providentia standing l, resting elbow on column and pointing to globe at feet. In field: SC; Die Axis: 12; Diam: 27mm; Wt: 12g; UW/UW (slightly corroded); *Dup* of Trajan, RIC II, 665, AD114-7; Area B, U/S (spoil heap) SF99, unphased. Not illustrated.
7. Obv: IMP CAES NERVAE TRAIANO AVG GER DAC PM TR PCOS V PP Laur bust, r. with aegis; Rev: SPQR OPTIMI [PRIN]CIPI Pax standing l. with branch and cornucopiae, r. foot placed on Dacian captive. In field SC; Die axis: 12; Diam: 32mm; Wt: 24g; SW/SW (local corrosion); *Sest* of Trajan, RIC II, 503, AD 103-111; 5005, Pit 5007, Area E, SF145, Event 3. Not illustrated.
8. Obv: [SABINA AVGVSTA] HADRIANI [AVG PP] Bust, dr. r.; triple tiara above brow with hair drawn up behind and elaborately piled; Rev: Vesta seated l. holding palladium and sceptre; Die axis: 12; Diam: 27mm; Wt: 12g; SW/SW (surface corroded); *Dup* of Sabina, RIC II, 1024, AD 122-8; 5005, Pit 5007, Area C, SF146, Event 3. Not illustrated.

Roman, illegible

9. Obv: Head l.; Rev: Illegible; *As*, Die Axis: --; Diam: 27mm; Wt: 5g; Corr/Corr; 4015, Pit 4041, Area C, SF61, Event 3. Not illustrated.
10. Obv: Head r.; Rev: Illegible; *As*, Die Axis: --; Diam: 28mm; Wt: 6g; Corr/Corr; 5002, Pit 5004, Area E, SF119, Event 3. Not illustrated.
11. Obv: Head r.; Rev: Illegible; Die Axis: --; Diam: 27mm; Wt: 6g; Corr/Corr; *As* Bust is perhaps that

of Vespasian. 2206, Pit 2204, Area A, SF20, Event 3. Not illustrated.

12. Obv: Head r.; Rev: Standing figure; Die Axis: --; Diam: 27mm; Wt: 8g; Corr/EW (Corr); *As*, Bust is perhaps Vespasian. 5002, Pit 5004, Area XX, SF133, Event 3. Not illustrated.
13. Obv: Head r.; Rev: Illegible; Die Axis: --; Diam: 29mm; Wt: 8g; Corr/Corr; *As*, Bust is perhaps that of Domitian; 4030, Pit 4041, Area XX, SF91, Event 2. Not illustrated.
14. Obv: Head r.; Rev: Illegible; Die Axis: --; Diam: 26mm; Wt: 6g; Corr/Corr; *As*, Bust is perhaps that of Domitian; 3049, Pit 3058, Area XX, SF111, Event 2. Not illustrated.

OTHER METALWORK *by Erica Macey-Bracken*

Nails

A total of 262 nails and fragments of nails were identified after x-radiography. The nails were identified in accordance with Manning's typology (Manning 1985, 134-7). All of the identifiable nails were of Manning's Type 1, the most common nail of the Roman period (Manning *et al* 1995, 291) with the exception of one possible Type 10 hobnail (5006). These nails have a square-sectioned, tapering stem and either a conical / pyramidal (Type 1a) or flat (Type 1b) head depending on the size of the nail. Nails of Type 1a are generally larger than 150mm in length, whilst Type 1b nails are smaller than this.

All of the nails from the site were identified as Type 1 nails (see Table 2.20). Only a small number of the nails were complete and could be identified by their length and head shape as Type 1b. A near-complete example from pit [5007] (5005) was thought more likely, due to its length of 155mm, to be of Type 1a. The smaller Type 1b nails (normally smaller than 150mm) are often found in greater quantities than the larger Type 1a nails, reflecting their versatility in all areas of construction, and also the fact that the larger Type 1a nails were a more convenient source of re-usable iron than their smaller counterparts (Manning 1985, 134-5). This pattern is noted at other forts such as Inchtuthil in Scotland (*ibid*).

Table 2.20: Nails by Event, length and type

Phase/Event	Context	Length (mm)	Type
Event 2	3029	61	1
Event 3	1008	86	1
Event 3	1008	76	1
Event 3	2111	51	1
Event 3	2111	59	1
	2248	50	1
Event 3	3007	58	1
Phase 6	1003	78	Modern, wire drawn

The majority of the assemblage consisted of fragmentary nails. Of these 144 nails (55%) had surviving heads, whilst the remainder were fragments from shafts. The largest incomplete fragment measured 145mm, with most of the remainder of the assemblage being considerably smaller. This may suggest that the majority of the nails were Type 1b but it would be impossible to say given the fragmentary state of the assemblage.

With the exception of contexts (2005, 2094, 2137, 2139, 2205, 2331, 4024), all the nails were from securely stratified Roman contexts, but these nails appear to be residual Roman nails of Type 1 (see Table 2.21). The only exception to this is a machine-drawn wire nail of modern date (1003, Phase 6).

Table 2.21: Nail quantities recovered by Event:

Event/phase	Total number of nails / nail fragments
Event 2	66
Event 3	133
Phase 3	18
Phase 4	4
Phase 5	3
Phase 6	1
U/S or unphased	37
TOTAL	**262**

The majority of the nails came from Event 3 (rubbish disposal) which closely corresponds with a period of clearance within the fortress, when the earlier wooden fort began to be rebuilt in stone. These may derive from the demolition of the old wooden buildings and they may have still been deposited with the wooden remains which were not preserved.

Other iron

Identification of the iron objects was made followed x-radiography. Preservation of the iron objects was poor and as such only a few of the items could be identified by form. These items were a U-shaped piece of iron, resembling a split-pin staple (2111), an iron loop (4033) and a metalworking punch (Manning *et al* 1995, 247 – 249). The reminder of the assemblage was composed of unidentifiable scraps of iron rod or sheet (Table 2.22).

Table 2.22: Unidentifiable scrap, sheet and rod metal by Event and phase

Phase	Total numbers of iron fragments
Event 2	15
Event 3	36
Phase 3	4
Phase 4	3
Unphased	9
TOTAL	**69**

Catalogue

1. U-shaped piece of iron, heavily covered with corrosion products, resembling a split-pin staple. Length 63mm; width 30mm at widest point. 2111, Pit 2110, Group 112, Area A, Event 3. Not illustrated.
2. Iron loop, heavily covered with corrosion products. Length 36mm; width 17mm at widest point. 4033, Pit 4041, Group 112, Area C, Event 2. Not illustrated.
3. Metalworking punch, partially obscured with corrosion products. Length 103mm; width 21mm at widest point. 5005, Pit 5007, Group 112, Area E, Event 3. Not illustrated.

Lead

The bulk of the lead assemblage was composed of droplets of lead waste (see Table 2.23). This material is likely to have come from casting and other metal spills during manufacture and shows patterns of floor clearance and rubbish disposal from within the Fortress. Parallels exist from the fort at Segontium at Caernarfon in Wales where pit and drain fills containing rubbish also contained lead working waste (Heyworth 1993, 211). Given the presence of both ferrous and non-ferrous (cu alloy) working residues within the waste material (see Mackenzie below) the likelihood is that the lead droplets are part of general metalworking waste associated with the *fabrica* within the Fortress or metal-working within the *canabae*.

TABLE 2.23: LEAD DROPLET WASTE BY EVENT AND PHASE

Phase	Quantity of lead droplets
Event 2	14
Event 3	35
3	1
4	1
U/S or unphased	22
TOTAL	**73**

A few pieces of lead rod were also noted (1008, 2111, 3029, 4033), as well as some scraps of lead sheet (2311, 4009, 4016, 4030, 4037, 4046, Area B U/S MD) and a larger piece of lead sheet that had been rolled into a tube-like shape (4015). A fragment of lead pipe was also recovered (1008). Unalloyed lead was often used for large items, particularly those associated with construction and water supply, such as pipes and tanks (Crummy 2004, 136).

Catalogue

4. L ead pipe fragment, crushed almost flat. Length 137mm; width 64mm; thick. 22mm. 1008, Pit 1019, Group 112, Area C, Event 3.
5. Fragment of lead sheet, rolled into tube shape. Length 97mm; width 43mm, thick. 39mm. 4015, Pit 4041, Group 112, Area C, Event 3.

THE METALLURGICAL RESIDUES

by Rod Mackenzie

Introduction

Various residues were recovered from the excavations that were thought to relate to the production or use of metals. Post-excavation assessment of these residues identified significant amounts of material relating to the smithing of iron, as well as possible evidence of non-ferrous metal production. A high proportion of the metalliferous production residues were recovered from archaeological contexts dating from the Roman period; although some were also recovered from post-medieval contexts.

The initial post-excavation assessment report recommended that selected fragments of corroded metal including the possible chain mail rings were X-rayed to help determine whether the items were 'scraps' or artefacts (Table 2.24). Examination of the X-rays revealed that most items were undiagnostic and as there did not appear to any benefit from metallographic analysis no further work on these has been carried out. A small number of objects X-rayed did appear to be discrete artefacts and it was recommended that the X-rays of them be passed to a Roman finds specialist for identification of the artefacts. The fragment of ferrous metal from context (4016) was recommended for further analysis because its fracture surface was suggestive of cast steel, which would be highly unusual for the Roman period.

The production residues assemblage contained some unusual clay-like residues from the Roman layer (2156) and further work to identify these was recommended. However further detailed investigations have failed to identify the nature or source of the material, but it is thought that it is highly unlikely to relate to the production of metals.

The following sections describes the chemical and metallographic analysis of the possible cast metal item from [4016] and small selection of metal production residues from context (4030) which were both infilling the large quarry pit [4041].

An interpretation of the results of the scientific analyses are given, together with an overall interpretation of the 'production residues' assemblage – and how it might relate to metal finds.

Methodology

To investigate the range of metal production represented in the assemblage, three pieces of slag were selected from context (4030). The pieces chosen each had a slightly different morphology and were typical examples of the range of slag types present. Metallography, using a combination of optical and electron microscopy, was used to analyse the selected specimens. The Scanning Electron Microscope used for the analysis was equipped with an

Energy Dispersive Spectrometer which was used to determine the chemical composition of the slag specimens.

Analytical Methods

Specimens from the samples of 'bulk' slag were removed by a combination of fracturing and diamond blade cold disc cutting. The specimen of metal was removed using a jeweller's saw. All specimens were mounted in cold setting resin, before being ground and polished in the conventional manner (Van der Voort 1999); the final stages of polishing were carried out using 1µm diamond paste. The metal specimen was etched in a 2% Nital solution to reveal the microstructure. Photomicrographs were taken using a Nikon Optiphot 2 reflected light microscope fitted with a digital camera.

The polished specimens of slag were examined using reflected light microscopy, before being carbon coated for further examination and analysis using an analytical Scanning Electron Microscope with Energy Dispersive Spectrometer (SEM-EDS). Images of the microstructure of the specimens were recorded as back-scattered images (BEI); in this type of image, the various shades of grey recorded are based upon the mean atomic mass of the components present. Components with a high atomic mass, such as iron, appear brighter than those with a lower atomic mass.

The bulk composition of the slag specimens was determined by taking the average of three scans of areas approximately 500 by 330 microns. Spot analyses of individual mineral phases were also carried out to aid identification of each phase present.

The Material

Specimen 4016.1, Pit 4041, Group 112, Area C, Event 3.

This specimen was removed from a small corroded fragment of metal whose fracture surface was suggestive of cast iron or steel. The fracture appeared to be very fine grained and free from non-metallic inclusions or voids. For an artefact to be made from cast ferrous metal during this period would be extremely unusual and metallographic analysis was recommended to identify the metal. The fragment measured approximately 5mm x 5mm in cross-section and approximately 7mm long, possibly tapering toward one end. The small size and fragmentary nature of the piece make it difficult to determine the function of the object. The specimen was mounted to reveal transverse and longitudinal sections of the microstructure.

Specimen 4030.1, Pit 4041, Group 112, Area C, Event 2.

This specimen was removed from a fragment of possible smithing slag. The clean fracture surface is dark-graphite grey in colour with moderate vesicularity; individual vesicles range in diameter from <1mm to approximately 6mm diameter. Corrosion products within some vesicles and relative weight of the slag suggest that it is relatively iron rich.

Specimen 4030.2, Pit 4041, Group 112, Area C, Event 2.

The specimen was taken from a fragment of possible iron smithing or smelting slag. The surface texture and shape of the slag suggests that it had flowed or dribbled as a viscous liquid before solidifying. The clean fracture surface of the slag is mid-graphite grey in colour and it is moderately vesicular; vesicles range from <1mm to approximately 4mm diameter.

Sample 4030.3, Pit 4041, Group 112, Area C, Event 2.

The specimen was removed from a small fragment of possible slag or slagged refractory material, such as hearth lining. One small (<1.5%) oxidised copper rich inclusion could be seen in one of the weathered fracture surfaces of the piece, suggesting that it may relate to non-ferrous metal production.

TABLE 2.24: RESULTS OF SEM-EDS ANALYSES

	Oxide	MgO	Na2O	Al2O3	SiO2	P2O5	SO	CaO	MnO	K2O	TiO2	FeO	CuO	PbO
Specimen 1 - area 1*		-	2.62	21.26	46.82			3.89		6.51		18.90		
Specimen 1 - area 2		0.54	1.40	7.87	36.45			1.33		1.47		50.93		
Specimen 1 - area 3		-	1.59	11.11	34.37			1.60		2.33		48.20		
Average		**0.27**	**1.50**	**9.49**	**35.41**			**1.47**		**1.90**		**49.57**		
SD		*0.38*	*0.13*	*2.29*	*1.47*			*0.19*		*0.61*		*1.93*		
Specimen 2 - area 1		0.39	0.58	4.04	50.32	0.42		2.78		1.85	0.27	39.28		
Specimen 2 - area 2		0.30	0.56	4.27	50.38	-		2.91		1.81	-	39.37		
Specimen 2 - area 3		0.36	0.52	4.18	50.38	-		3.04		1.72	-	39.08		
Average		**0.35**	**0.55**	**4.16**	**50.36**	**0.14**		**2.91**		**1.79**	**0.09**	**39.24**		
SD		*0.05*	*0.03*	*0.12*	*0.03*	*0.24*		*0.13*		*0.07*	*0.16*	*0.15*		
Specimen 3 - area 1*					100.00									
Specimen 3 - area 2		1.09		7.93	74.98	1.23		0.58		1.89	0.42	1.98	8.47	0.31
Specimen 3 - area 3		0.63		7.34	81.57	0.74		-		2.74	-	1.77	4.34	-
Average		**0.86**		**7.64**	**78.28**	**0.99**		**0.29**		**2.32**	**0.21**	**1.88**	**6.41**	**0.16**
SD		*0.33*		*0.42*	*4.66*	*0.35*		*0.41*		*0.60*	*0.30*	*0.15*	*2.92*	*0.22*
Specimen 3 - point 1					0.5	2.58	1.58	0.51				1.41	91.01	0.28

Results of analysis

The results of SEM-EDS analyses are listed in Table 2.24.

Specimen 4016.1, Pit 4041, Group 112, Area C, Event 3.

The microstructure is composed of widmanstatten ferrite and pearlite. The carbon content of the specimen is variable and it ranges from approximately 0.3% to 0.6% carbon. Plates 2.15 and Plate 2.16 show two areas of microstructure and illustrate the variation in carbon content. Non-metallic (slag) inclusions are present in the microstructure and their size and shape are typical of those found in iron and steel produced by bloomery furnaces during the Romano- British period. No obvious weld lines were observed in the microstructure of the specimen.

Specimen 4030.1, Pit 4041, Group 112, Area C, Event 2.

The microstructure is predominantly composed of fayalite and wustite within a matrix of leucite (see Plate 2.17). Hercynite is also present, primarily along the edges of the fayalite laths and, in some areas fine dendritic wustite appears to be precipitating from the leucite matrix. The bulk chemical composition of the specimen approximates that of fayalite, although the relative amounts of silica and alumina are both slightly elevated. Minor amounts of alkalis and lime were also detected (see Table 2.24).

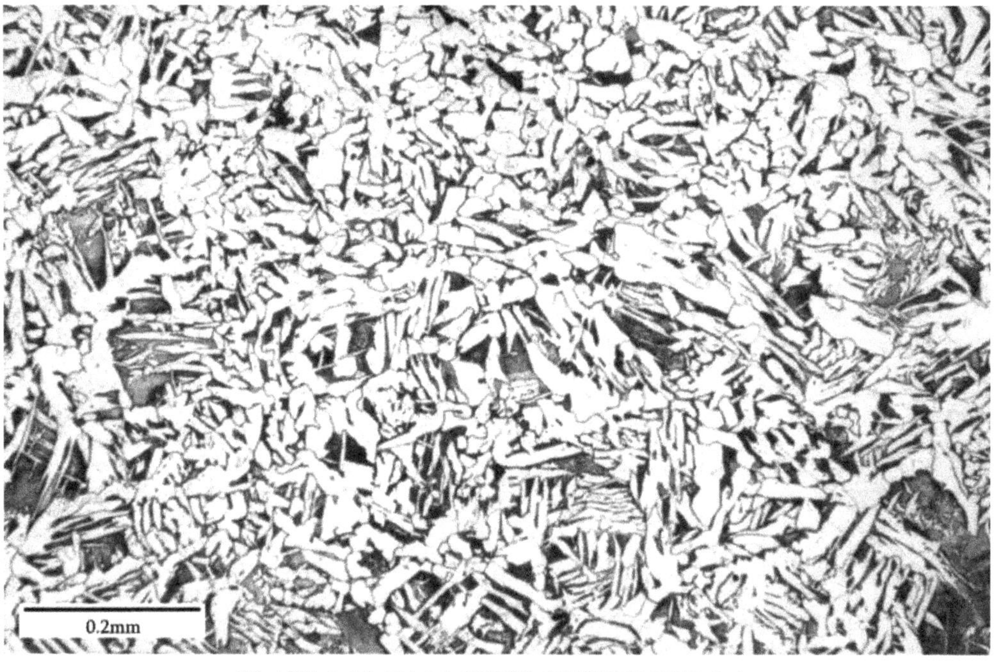

PLATE 2.15 4016.1 STEEL PHOTOMICRO 1.

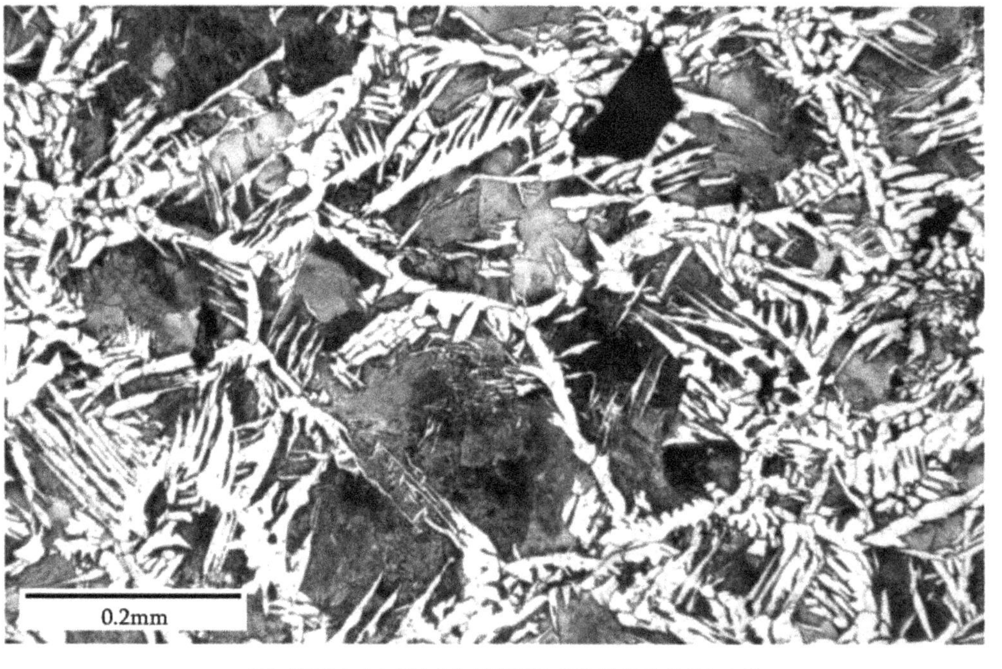

PLATE 2.16 4016.1 STEEL PHOTOMICRO 2.

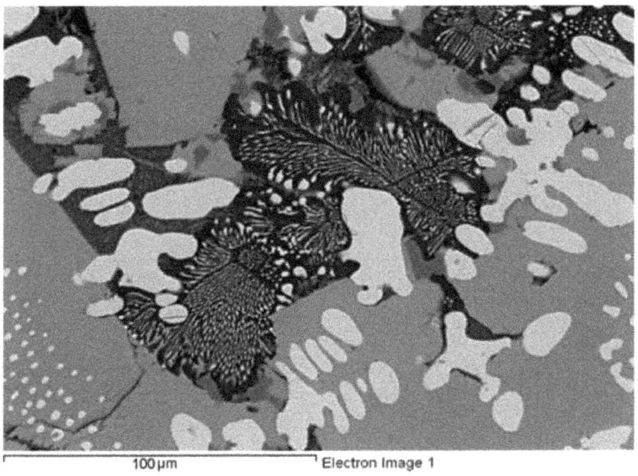

PLATE 2.17 4030.1 BEI PHOTOMICRO OF SLAG SPECIMEN.

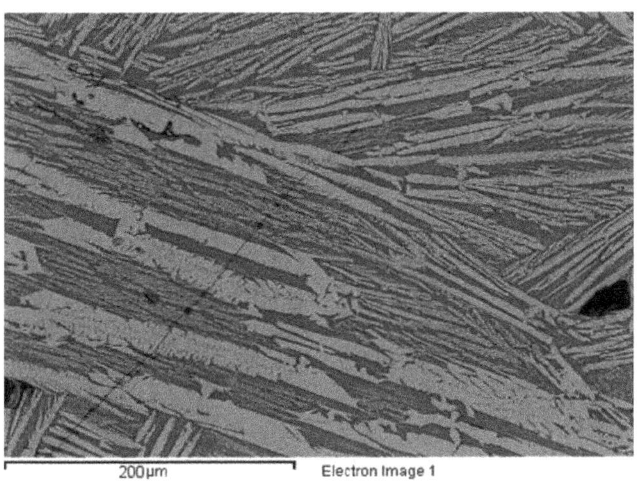

PLATE 2.18 4030.1 BEI PHOTOMICRO OF SLAG SPECIMEN.

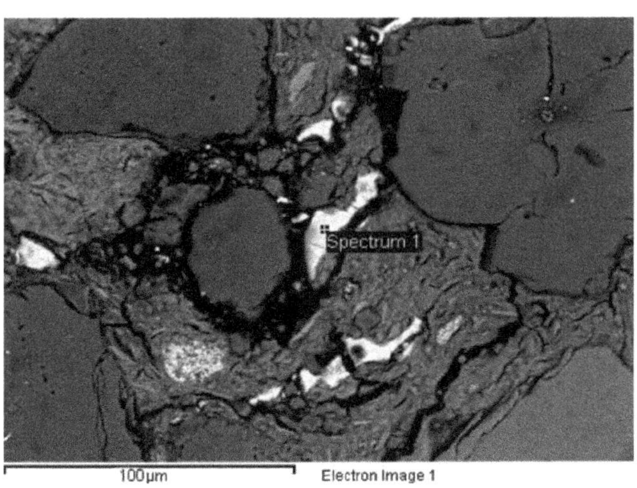

PLATE 2.19 4030.1 BEI PHOTOMICRO OF SLAG SPECIMEN.

Specimen 4030.2, Pit 4041, Group 112, Area C, Event 2.

The phases present in the specimen are fayalite and glass (see Plate 2.18). The fayalite takes the form of laths and 'feather edged' crystals. Area EDS scans of the specimen gave an average composition of approximately 40% FeO and 50% SiO_2, with the remaining 10% largely made up of Al_2O_3, CaO and K_2O. One bleb of metallic iron is present in the slag.

Specimen 4030.3, Pit 4041, Group 112, Area C, Event 2.

The microstructure of the specimen is composed of numerous quartz inclusions within a clay matrix. Metallic blebs are present within the clay (see Plate 2.19) and these were found to be largely composed of copper. Significant amounts of zinc, phosphorus and sulphur were also detected in the blebs.

Interpretation and Discussion of Results

The piece of metal 4016.1 is interesting because of its comparatively high carbon content, effectively that of medium carbon steel; this type of steel would be suitable for tools or weapons requiring a hard cutting edge. However, the type of microstructure present in the specimen would have made the steel comparatively brittle, which may explain why the original parent object had fragmented. Brittleness in steels can be reduced by heat treatments, and evidence of this would be visible in the microstructure. There is some evidence of forging in the microstructure, but none of deliberate heat treatment. It should be noted however, that if the object was a single use type weapon, such as a projectile point, brittleness may not have been problem, and in fact, may have been desirable. The type of slag inclusions present in the microstructure are typical of those found in iron and steel produced by the direct process, and they show that the object was not produced from fully liquid 'cast' or 'crucible' steel.

From the amount of carbon in the microstructure and other military-type artefacts recovered from the same or associated contexts, it seems possible that the fragment may be material from the steel tip of a projectile that may have been forged onto softer low carbon steel/iron. However, weld lines which could confirm this theory were not observed in the microstructure, although this may be down to the small size and fragmentary nature of the object.

Specimen 4030.1 was removed from a piece of slag that has the morphology suggestive of a smithing hearth bottom, although the microstructure of the slag is similar to some bloomery smelting slags. McDonnell (1983 & 1991) describes the crossover between some types of smelting and smithing slags 'Under the microscope, we can distinguish between microstructures of tap slags and [smithing] hearth bottoms, but [smelting] furnace bottoms sometimes look like tap slag and sometimes like hearth bottoms' (McDonnell 1983). However given that no tap slag which is diagnostic of smelting was found within the assemblage, it seems most likely that Specimen 4030.1 is a smithing slag.

4030.2 is probably also an iron smithing slag, possibly raked out from a hearth; the elevated levels of silica and alumina suggest some input from clay hearth or furnace lining.

4030.3 is possibly clay hearth lining material, it does not appear to be from a crucible (although it should be noted that the material appears to have undergone some post-depositional decomposition). Metallic inclusions within the material suggest that it relates to production of brass. Without analysis of some of the copper-based alloy artefacts recovered, it is not possible to say whether the composition of the metal blebs in the specimen is the same, or similar to, the metal in the artefacts. If the composition of the metallic blebs is representative of the 'finished' metal, it would be a comparatively soft brass.

Conclusion

The assemblage is pre-dominated by material from the Roman period, and most diagnostic residues appear to relate to iron smithing. Approximately 15.8kg of diagnostic residues were recovered from contexts dating from the Roman period, and of these, at least 11.4kg appear to relate to iron smithing. Only 122g of residues appear to relate to the production or working of non-ferrous metals. The post-medieval context produced approximately 2kg of diagnostic residues, around fifty percent of this material is fuel ash slag and around 950g of residues appear to relate to iron smithing.

Given the contexts that the Roman period material was recovered from, it is not possible to state whether it relates to domestic or military production. However the presence of military armour fittings and possible chain mail rings from the same or related contexts suggests that some of the residues could relate to the production and repair of military equipment. The period of deposition of the material coincided with a period of rebuilding at Chester, so some of the ferrous metal residues in particular, could relate to construction activities e.g production of nails.

Analysis has confirmed the presence of non-ferrous metal production or working, again given the context it is not possible to state whether this was the result of domestic or military activity. However, Bishop (2002) describes the use of brass in articulated Roman plate armour. Brass rivets made of soft brass 90-95% Cu and 10-5% Zn were used to connect hinges to plates, the soft alloy being chosen to make it easier to peen the rivets and secure the plates during manufacture. It would be interesting to investigate the composition of non-ferrous armour fittings recovered during the excavation to see how they compare to materials recovered from Corbridge discussed by Bishop. The relatively small quantity of non-ferrous residues within the assemblage suggests that it relates to small scale manufacture or possibly repair of artefacts; although, of course, it is possible that further deposits of metalliferous residues lie buried outside the areas excavated.

BONE ARTEFACTS AND CRAFT WASTE *by Iain L. Baxter*

Bone artefacts and craft waste include a cattle proximal metatarsal from pit [4041] (4030) Event 2 (Plate 2.20D) that has been sawn from the shaft in typical manner for the shaft to be converted to either a knife handle or split for the production of bone pins or needles (Armitage 1982b, Figure 47; MacGregor 1985, Figure 30). In the same context a partial sheep metacarpal was recovered with a circular hole drilled in the shaft (Plate 2.20E). Similar examples from Iron Age and Romano-British sites have been interpreted as artefacts used in the spinning of wool. A very problematic specimen is a cattle astragalus from Event 2 [4041] (4030) which, in addition to a chop or saw cut has an apparent drill cut on the lateral/plantar aspect of the distal articular surface and calcanial articular facet (Plate 2.20F). Two other bones exhibiting human modification comprise a cattle distal metatarsal from Event 3 [4002] (4009) with a flattened distal condyle which appears to be the result of rubbing (Plate 2.20C) and a horse metacarpal from Event 2 [3038] (3039) with the distal anterior surface smoothed resulting in obliterating of the anterior sagittal crest. These bones seem to have been utilized as impromptu rubbing tools. A sheep/goat metatarsal Layer (4001) has holes in the proximal articular and proximal posterior surfaces together with polishing of the shaft. A drilled and sawn bone fragment from Event 3 [2234] (2235) (Plate 2.20A) represents craft waste from the production of an indeterminate artefact. Worked red deer antler fragments include a cast base with saw cuts (cf. Ulbricht 1978, Figure 3 no.11; MacGregor 1985, Figure 32) discarded after preliminarily working found in Layer (4006), a hollowed antler beam fragment with a saw cut from Event 2 [5004] (5003) and a polished antler tine fragment from Event 3 [5007] (5005) (Plate 2.20 B, G and H).

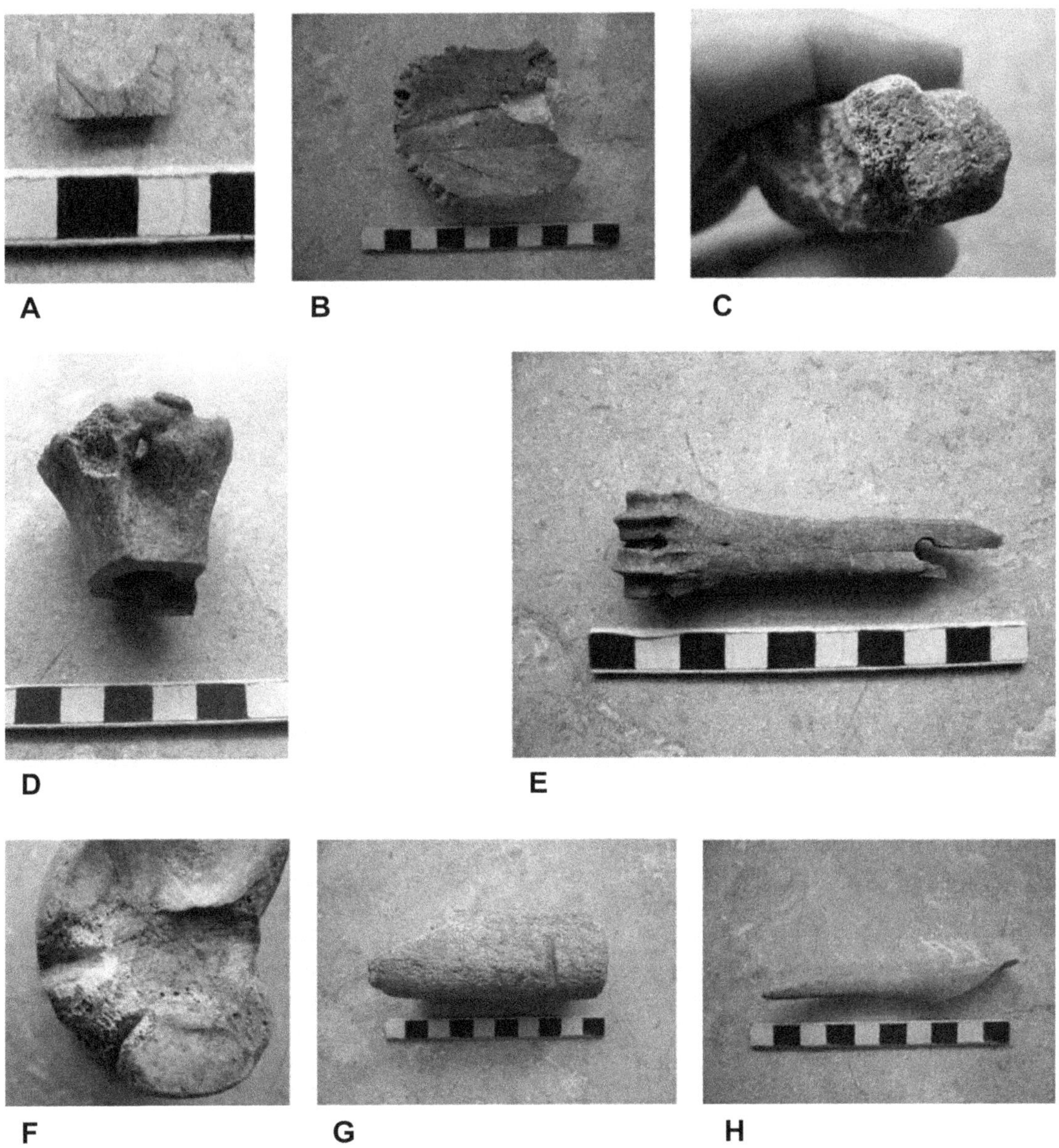

PLATE 2.20 ANIMAL BONE - WORKED A-H.

Catalogue

1. Craft waste. Drilled and sawn bone fragment. Sheep/ goat (*Ovis/ Capra* f. domestic). Length 15mm. 2235, Pit [2234], Group 112, Area A, Event 3. (Plate 2.20A).
2. Cast base with saw cuts. Red Deer (*Cervus elaphus*). 4006, Area C,. (Plate 2.20B).
3. Distal metatarsal with a flattened distal condyle as a result of rubbing. Cattle (*Bos* f. domestic). 4009, Pit [4002], Group 112, Area C, Event 3. (Plate 2.20C).
4. Sawn proximal metatarsal. Craft waste for knife or bone pins/ needles. Cattle (*Bos* f. domestic). 4030, Pit [4041], Group 112, Area C, Event 2. (Plate 2.20D).
5. Partial metacarpal with circular hole drilled in the shaft. Sheep/ goat (*Ovis/ Capra* f. domestic). 4030, Pit [4041], Group 112, Area C, Event 2. (Plate 2.20E).
6. Chop or saw cut astragalus with an apparent drill cut. Cattle (*Bos* f. domestic). 4030, Pit [4041], Group 112, Area C, Event 2. (Plate 2.20F).
7. Hollowed antler beam fragment with a saw cut. Red Deer (*Cervus elaphus*). 5003, Pit [5004], Group 112, Area E, Event 2. (Plate 2.20G).

4\. Polished antler tine fragment. Red Deer (*Cervus elaphus*). 5005, Pit [5007], Group 112, Area E, Event 3. (Plate 2.20H).

THE COLUMN BASE *by Malcolm Hislop*

A stone fragment, which represents slightly less than half of a Roman column base was recovered during the watching brief on the quarry (in 2008). It is 95mm high and has an approximate diameter of 340mm. The profile is formed from three principal tiers of continuous horizontal parallel grooving, the upper and lower grooves 6mm wide x 10mm deep, and the centre groove 8mm wide x 15mm deep. Two short lengths of a much less significant groove survive at the bottom of the piece. The alternating ribs are slightly tapered so that the flat outer edges are a little narrowed. The top lip is set back slightly from the sides, and on the upper surface there is a narrow groove approx 3cm from the edge and parallel with it. The stone was pierced by a central 4.5cm square hole extending from top to bottom, of which one face survives together with small fragments of the two returns. Such features accommodated the dowels that held the elements of a column together, and are also held to be one of the indications that a piece was turned on a lathe (Blagg 1976, 166–167). The character of this particular piece is consistent with having been lathe turned.

Catalogue

1. Roman column base; surviving height 95mm; diameter 340mm; Phase 5. (Fig 2.32)

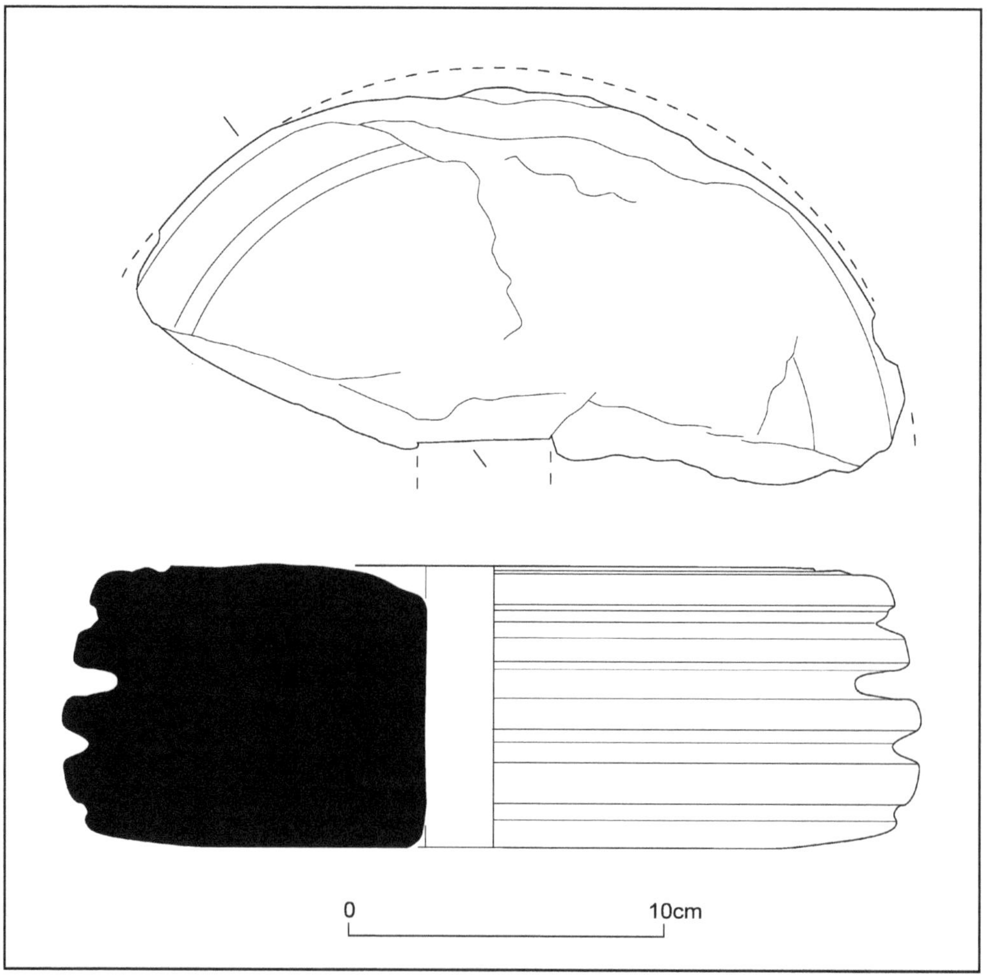

FIGURE 2.32
WORKED STONE.

THE ANIMAL BONES *by Ian L. Baxter*

Introduction

The animal bones were recovered from pits and linear features located in a series of excavation areas (A-C and E) running east to west across the site. Delamere Street is the first major excavation situated within the *canabae legionis* of Chester. Previous excavations at Chester have primarily focused on sites within the area occupied by the legionary fortress. An archive and unpublished draft report on five sites, located at Goss Street, Crook Street, Hunter's Street and Hunter's Walk, is held by Chester Archaeology (Cartledge 1991). These sites form the most useful comparator to Delamere Street. However, the assemblages from the time period (1st and 2nd centuries AD) represented by the Delamere Street deposits at the fortress sites is relatively small, 1326 fragments identified using a non-selective methodology, the number of measurable bones precludes detailed comparison, there are also outstanding concerns over the phasing of some sites and the draft report is in need of extensive revision. Nonetheless, it has been possible to compare in a general way the fortress assemblages with those from Delamere Street and with other military and urban assemblages from elsewhere.

Methods

The majority of the animal bones from Delamere Street were hand-collected. Consequently an under-representation of smaller bones from the main domestic species and bones from small wild mammals, birds, amphibians and fish is to be expected. The bones were recovered from features dating from the post-medieval period at the site do not appear qualitatively different to those dated to the Roman period although they are subject to greater leaching and recent fragmentation. They have been classified as residual Roman material following discussions with archaeologists at Chester Archaeology and have not been recorded.

The mammal bones were recorded following a modified version of the method described in Davis (1992) and Alberalla and Davis (1994). In brief, all teeth (lower and upper) and a restricted suite of parts of the skeleton was recorded and used in counts. These are: horncores with a complete transverse section, skull (zygomaticus), atlas, axis, scapula (glenoid articulation), distal humerus, distal radius, proximal ulna, carpal 2+3, distal metacarpal, pelvis (ischial part of acetabulum), distal femur, distal tibia, calcaneum (sustenaculum), astragalus (lateral side), centrotarsale, distal metatarsal, proximal parts of the 1st, 2nd and 3rd phalanges. At least 50% of a given part had to be present for it to be counted.

The presence of large (cattle/ horse size) and medium (sheep/ pig size) vertebrae and ribs was recorded for each context, although these were not used in counts. Where possible, cattle vertebrae have been identified as such. "Non-countable" elements of particular interest were recorded but not included in the counts. For birds the following were always recorded if present: scapula (articular end), proximal coracoid, distal humerus, proximal ulna, proximal carpometacarpus, distal femur, distal tibiotarsus, and distal tarsometatarsus.

The separation of sheep and goat was attempted on the following elements: horncores, dP_3, dP_4, distal humerus, distal metapodials (both fused and unfused), distal tibia, astragalus, and calcaneum using the criteria described in Boessneck (1969), Kratochvil (1969), Payne (1969 and 1985) and Schmid (1972). The shape of the enamel folds (Davis 1980; Eisenmann 1981) was used for identifying equid teeth to species. Equid postcrania were checked against criteria summarised in Baxter (1998).

Wear stages were recorded for all P_4s and dP_4s as well as for the lower molars of cattle, sheep/goat and pig, both isolated and in mandibles. Tooth wear stages follow Grant (1982).

Measurements in general follow von den Driesch (1976). All pig measurements follow Payne and Bull (1988). Humerus HTC and BT and tibia Bd measurements were taken for all species as suggested by Payne and Bull (1988) for pigs. Measurements taken on equid teeth follow Levine (1982). SD on dog long bones is measured as suggested by Harcourt (1974) and represents the midshaft diameter (msd).

Cattle horncores have been aged using the method of Armitage (1982a) and the morphology of the frontal bone recorded according to the typology published by Grigson (1976).

A full record of butchery marks for all species (following Lauwerier 1988) is recorded on the Access database.

Bone context, preservation and taphonomy

Most of the animal bones recovered from Delamere Street came from the infills of quarry pits originally dug to extract stone for the building of the legionary fortress that replaced the original timber building (Event 2). A much smaller assemblage was excavated from rubbish pits dug into these infills and from linear features (Event 3). Cattle bones and teeth are the most common elements recovered from the site. None of this material appears to be in primary context and generally consists of butchery refuse in secondary and tertiary deposition. It seems probable that these bones derive originally from midden deposits and were subsequently buried in pits during periods of "tidying up". There are many loose cattle maxillary and mandibular teeth. Some cattle long bones are highly fragmented, including many split long-bone shaft fragments comprising tertiary butchery waste, i.e. semi-industrial processing for marrow products etc. There is also a very small amount of household refuse, i.e. kitchen waste, within the pits. The animal bones are friable with many recent breaks but generally otherwise in reasonably good condition. The relative frequencies of skeletal elements for the three main domestic mammals are plotted in a sequence of survival suggested by Brain (1976) in Figure 2.33. In the cattle skeleton elements that bear little or no meat such as teeth and limb extremities are particularly common with bones that would have been included in the most important meat cuts, such as scapula, humerus and pelvis significantly less well represented. Cattle lumbar vertebrae are also much less frequent than cervical and thoracic vertebrae. This suggests that the cattle assemblage mostly consists of primary butchery waste with an admixture of secondary and tertiary butchery waste. The postcranial bones of sheep/goats and pigs by comparison are comparatively well represented from all parts of the skeleton indicating less intensive processing of the carcass and less selective disposal of refuse from butchery.

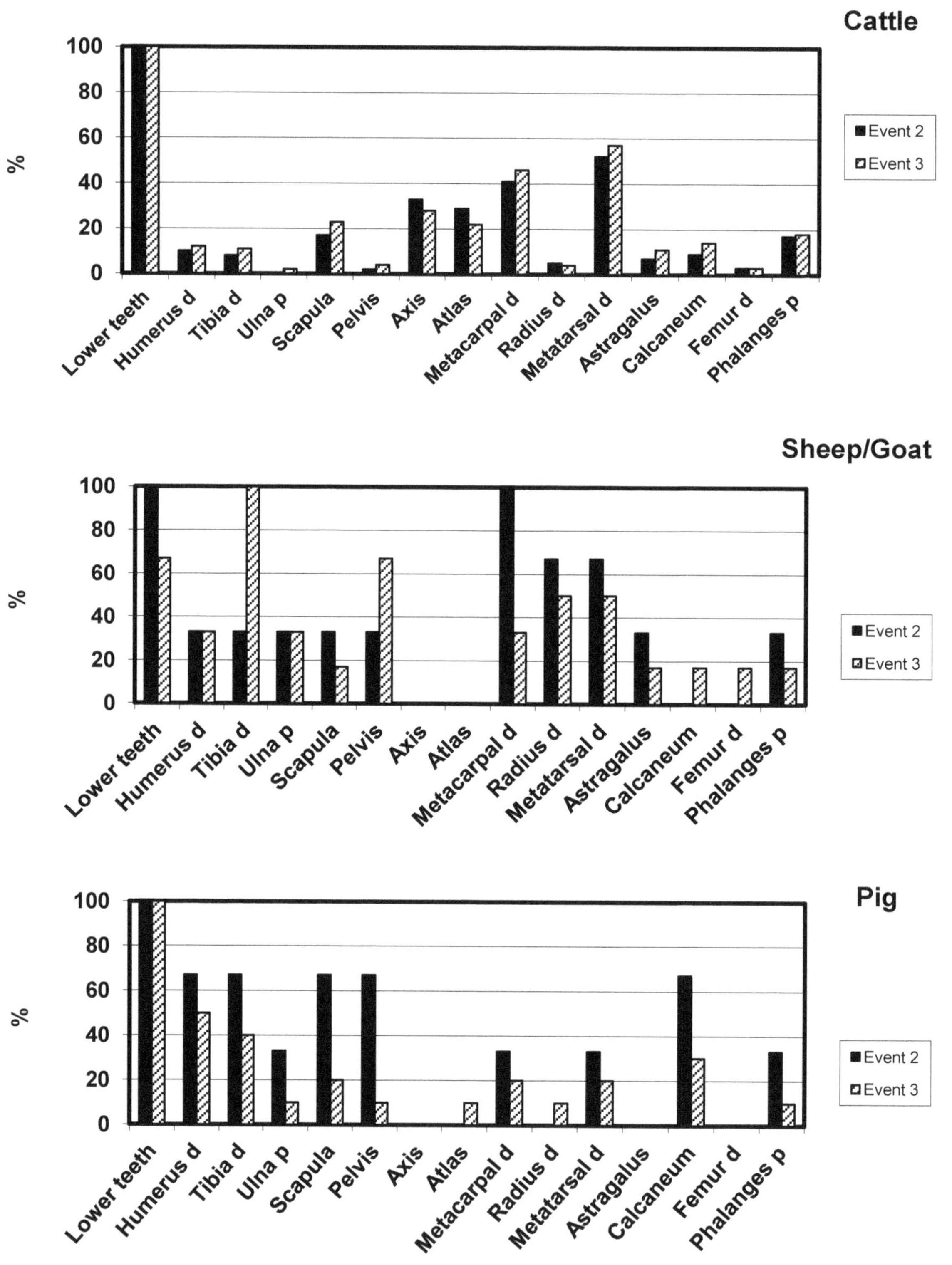

FIGURE 2.33 PERCENTAGE SURVIVAL OF THE BODY PARTS OF THE MAIN DOMESTIC MAMMALS. SEQUENCE BASED ON BRAIN 1976.

Frequency of species

A total of 4514 countable bone fragments were recovered by hand from the site (Table 2.25) and a further 105 fragments from the coarse fraction sample residues (Table 2.26). Cattle bones and teeth are by far the most common elements recovered from Delamere Street, accounting for 93% of the total assemblage, pig is the next most frequent taxon followed by sheep/goat (Table 2.25 and Fig 2.34). Horse, dog, wild mammals, domestic and wild birds are comparatively scarce. The assemblage from the large quarry pits in Area C is much larger than those from the other areas, accounting for 73% of identified fragments recovered from the site. However, there is little difference in the relative frequency of the main domestic mammals in each area. Minimum numbers of individuals (MNI) and numbers of identified fragments (NISP) are a very close match for all domestic species (Tables 2.27-2.30; Fig 2.34). The few bones recovered from environmental sample residues suggest that the preponderance of cattle remains compared with those of the smaller species does not necessarily result from collection bias (Table 2.26). The relative frequency of the major domesticates (NISP) has been illustrated for the Chester fortress sites, Colchester sites, the Heybridge small town in Essex, Lincoln, and the Nijmegen *castra* and *canabae* in the Netherlands for the 1st to 2nd century AD in Figure 2.35. The overall frequency of the main domestic mammals is very similar in the British urban sites and the Nijmegen *castra* with cattle most frequent and sheep/goat and pig variably next frequent. The only assemblage to display the massive preponderance of cattle fragments found at Delamere Street is that from the Nijmegen *canabae*.

TABLE 2.25: DELAMERE STREET, CHESTER. NUMBER OF HAND-COLLECTED MAMMAL AND BIRD BONES (NISP).

Taxon	Event		Total
	2	3	
Cattle (*Bos* f. domestic)	1705	2492	**4197**
Sheep/Goat (*Ovis/Capra* f. domestic)	38	66	**104**
Sheep (*Ovis* f. domestic)	(10)	(22)	**(32)**
Goat (*Capra* f. domestic)	(1)	(2)	**(3)**
Pig (*Sus scrofa*)	61	110	**171**
Horse (*Equus caballus*)	9	5	**14**
Dog (*Canis familiaris*)	2	7	**9**
Red Deer (*Cervus elaphus*)	1	+	**1**
Roe Deer (*Capreolus capreolus*)	-	1	**1**
Hare (*Lepus* sp.)	1	-	**1**
Domestic Fowl (*Gallus* f. domestic)	5	6	**11**
Goose (*Anser/Branta* sp.)	-	1	**1**
Red Kite (*Milvus milvus*)	-	2	**2**
Duck (*Anas platyrhynchos*)	+	1	**1**
Woodcock (*Scolopax rusticola*)	-	1	**1**
Total	**1822**	**2692**	**4514**

"Sheep/Goat" also includes the specimens identified to species. Numbers in parentheses are not included in the total of the period. "+" means that the taxon is present but no specimens could be "counted" (see text)

TABLE 2.26: DELAMERE STREET, CHESTER. NUMBER OF MAMMAL, BIRD AND FISH BONES (NISP) IN THE SIEVED ASSEMBLAGE.

Taxon	Event		Total
	2	3	
Cattle (*Bos* f. domestic)	7	81	**88**
Sheep/Goat (*Ovis/Capra* f. domestic)	-	3	**3**
Pig (*Sus scrofa*)	1	9	**10**
Mouse/Vole (Murid/Microtine)	1	-	**1**
Fish (*Pisces* sp.)	-	3	**3**
Flatfish (cf. *Pleuronectes* sp.)	(-)	(2)	**(2)**
Total	**9**	**96**	**105**

"Fish" also includes the specimens identified to species. Numbers in parentheses are not included in the total of the period.

TABLE 2.27. DELAMERE STREET, CHESTER. EVENT 2. BODY PARTS OF THE MAIN DOMESTIC MAMMALS BY NUMBER OF FRAGMENTS (NISP) AND MINIMUM NUMBER OF INDIVIDUALS (MNI).

Element	Taxon								
	Cattle			Sheep/Goat			Pig		
	NISP	MNI	%	NISP	MNI	%	NISP	MNI	%
Upper deciduous + permanent incisors							1	1	**33**
Upper deciduous + permanent canine							3	2	**67**
Upper deciduous + permanent premolars	101	17	**20**	0	0	**0**	5	1	**33**
$M^{1/2}$	344	86	**100**	2	1	**33**	3	1	**33**
M^3	146	73	**85**	2	1	**33**	0	0	**0**
Lower deciduous + permanent incisors	39	5	**2**	0	0	**0**	9	2	**67**
Lower deciduous + permanent canine							6	3	**100**
Lower deciduous + permanent premolars	275	46	**53**	6	1	**33**	13	3	**100**
$M_{1/2}$	313	79	**92**	11	3	**100**	4	1	**33**
M_3	163	82	**95**	5	3	**100**	2	1	**33**
Horncore	3	2	**2**	0	0	**0**			
Malar	56	28	**33**	0	0	**0**	1	1	**33**
Atlas	25	25	**29**	0	0	**0**	0	0	**0**
Axis	28	28	**33**	0	0	**0**	0	0	**0**
Scapula	30	15	**17**	1	1	**33**	4	2	**67**
Humerus dist	17	9	**10**	1	1	**33**	3	2	**67**
Radius dist	7	4	**5**	3	2	**67**	0	0	**0**
Ulna prox	0	0	**0**	1	1	**33**	1	1	**33**
Carpal C2+3	2	1	**1**	0	0	**0**	0	0	**0**
Metacarpal dist	69.5	35	**41**	5	3	**100**	2	1	**33**
Pelvis acetabulum	3	2	**2**	1	1	**33**	3	2	**67**
Femur dist	6	3	**3**	0	0	**0**	0	0	**0**
Tibia dist	13	7	**8**	2	1	**33**	3	2	**67**
Astragalus	12	6	**7**	1	1	**33**	0	0	**0**
Calcaneum	15	8	**9**	0	0	**0**	4	2	**67**
Centrotarsale	7	4	**5**	0	0	**0**	0	0	**0**
Metatarsal dist	88.5	45	**52**	3	2	**67**	2	1	**33**
Phalanx 1 prox	116	15	**17**	1	1	**33**	1	1	**33**
Phalanx 2 prox	27	4	**5**	0	0	**0**	0	0	**0**
Phalanx 3 prox	44	6	**7**	0	0	**0**	0	0	**0**

Unfused epiphyses are not counted. Only hand-collected material is included. The MNI has been calculated as follows: incisors have been divided by 8 for cattle and sheep/goat and by 6 for pig, deciduous + permanent premolars by 6, M1/2 by 4, phalanges by 8 and all other elements, except metapodials and vertebrae, by 2. Metacarpal = (MC1 + MC2/2 + MP1/2 + MP2/4) / 2; Metatarsal = (MT1 + MT2/2 + MP1/2 + MP2/4) /2, where: MC1 = complete distal metacarpal; MC2 = half distal metacarpal; MT1 = complete distal metatarsal; MT2 = half distal metatarsal; MP1 = complete distal metapodial; MP2 = half distal metapodial. Pig metapodials are considered the equivalent of cattle and sheep/goat half metapodials. % = frequency of an element in relation to the most common one (by MNI).

MT1 = complete distal metatarsal; MT2 = half distal metatarsal; MP1 = complete distal metapodial; MP2 = half distal metapodial. Pig metapodials are considered the equivalent of cattle and sheep/goat half metapodials. % = frequency of an element in relation to the most common one (by MNI).

TABLE 2.28. DELAMERE STREET, CHESTER. EVENT 3. BODY PARTS OF THE MAIN DOMESTIC MAMMALS BY NUMBER OF FRAGMENTS (NISP) AND MINIMUM NUMBER OF INDIVIDUALS (MNI).

Element	**Taxon**								
	Cattle			**Sheep/Goat**			**Pig**		
	NISP	**MNI**	**%**	**NISP**	**MNI**	**%**	**NISP**	**MNI**	**%**
Upper deciduous + permanent incisors							2	1	**10**
Upper deciduous + permanent canine							1	1	**10**
Upper deciduous + permanent premolars	169	29	**24**	3	1	**17**	9	2	**20**
$M^{1/2}$	343	86	**71**	6	2	**33**	6	2	**20**
M^3	153	77	**64**	4	2	**33**	2	1	**10**
Lower deciduous + permanent incisors	46	6	**5**	1	1	**17**	14	3	**30**
Lower deciduous + permanent canine							19	10	**100**
Lower deciduous + permanent premolars	367	62	**51**	10	2	**33**	9	2	**20**
$M_{1/2}$	489	122	**100**	16	4	**67**	10	3	**30**
M_3	242	121	**99**	3	2	**33**	12	3	**30**
Horncore	17	9	**8**	1	1	**17**	0	0	**0**
Malar	81	41	**34**	1	1	**17**	1	1	**10**
Atlas	26	26	**22**	0	0	**0**	1	1	**10**
Axis	34	34	**28**	0	0	**0**	0	0	**0**
Scapula	55	28	**23**	1	1	**17**	4	2	**20**
Humerus dist	28	14	**12**	3	2	**33**	9	5	**50**
Radius dist	9	5	**4**	6	3	**50**	2	1	**10**
Ulna prox	3	2	**2**	3	2	**33**	2	1	**10**
Carpal C2+3	6	3	**3**	0	0	**0**	0	0	**0**
Metacarpal dist	110.5	56	**46**	4	2	**33**	3.5	2	**20**
Pelvis acetabulum	10	5	**4**	7	4	**67**	1	1	**10**
Femur dist	5	3	**3**	2	1	**17**	0	0	**0**
Tibia dist	25	13	**11**	11	6	**100**	7	4	**40**
Astragalus	27	14	**11**	1	1	**17**	0	0	**0**
Calcaneum	34	17	**14**	1	1	**17**	5	3	**30**
Centrotarsale	8	4	**3**	0	0	**0**	0	0	**0**
Metatarsal dist	139	70	**57**	5	3	**50**	3	2	**20**
Phalanx 1 prox	176	22	**18**	3	1	**17**	5	1	**10**
Phalanx 2 prox	47	6	**5**	1	1	**17**	0	0	**0**
Phalanx 3 prox	57	8	**7**	0	0	**0**	0	0	**0**

Unfused epiphyses are not counted. Only hand-collected material is included. The MNI has been calculated as follows: incisors have been divided by 8 for cattle and sheep/goat and by 6 for pig, deciduous + permanent premolars by 6, M1/2 by 4, phalanges by 8 and all other elements, except metapodials and vertebrae, by 2. Metacarpal = (MC1 + MC2/2 + MP1/2 + MP2/4) / 2; Metatarsal = (MT1 + MT2/2 + MP1/2 + MP2/4) /2, where: MC1 = complete distal metacarpal; MC2 = half distal metacarpal; MT1 = complete distal metatarsal; MT2 = half distal metatarsal; MP1 = complete distal metapodial; MP2 = half distal metapodial. Pig metapodials are considered the equivalent of cattle and sheep/goat half metapodials. % = frequency of an element in relation to the most common one (by MNI).

TABLE 2.29. DELAMERE STREET, CHESTER. FREQUENCIES OF THE THREE MOST COMMON DOMESTIC MAMMALS BY NUMBER OF IDENTIFIED SPECIMENS (NISP) AND BY NUMBER OF INDIVIDUALS (MNI).

Taxon	Event							
	2				3			
	NISP	%	MNI	%	NISP	%	MNI	%
Cattle (*Bos Taurus*)	1705	95	86	94	2492	93	122	89
Sheep/Goat (*Ovis/Capra*)	38	2	3	3	66	3	6	4
Pig (*Sus scrofa*)	61	3	3	3	110	4	10	7
Total	1804		92		2668		138	

TABLE 2.30: DELAMERE STREET, CHESTER. EVENT 2. FREQUENCY BY MNI OF THE MAIN PARTS OF THE BODY OF THE MAIN DOMESTIC MAMMALS. DATA FROM TABLE 2.27.

Element	Cattle		Sheep/Goat		Pig	
	MNI	%	MNI	%	MNI	%
Teeth	86	**100**	3	**100**	3	**100**
Upper limbs	15	**17**	1	**33**	2	**67**
Lower limbs	8	**9**	2	**67**	2	**67**
Feet	45	**52**	3	**100**	1	**33**

"Upper limbs" includes scapula, humerus, pelvis and femur
"Lower limbs" includes radius, ulna, tibia, carpal, astragalus and calcaneum
"Feet" includes metapodials and phalanges

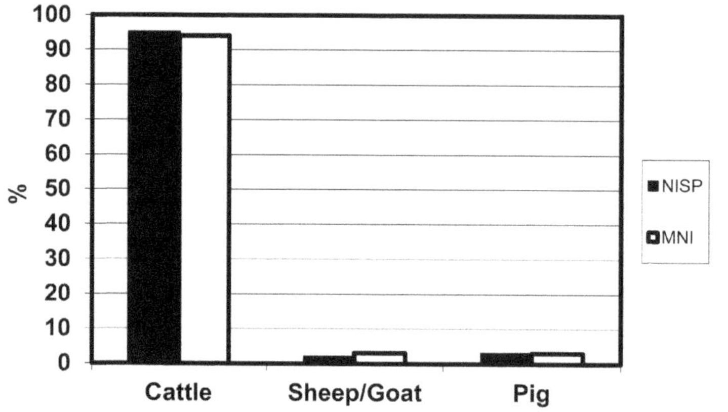

Event 2: Tot NISP=1804 Tot MNI=92

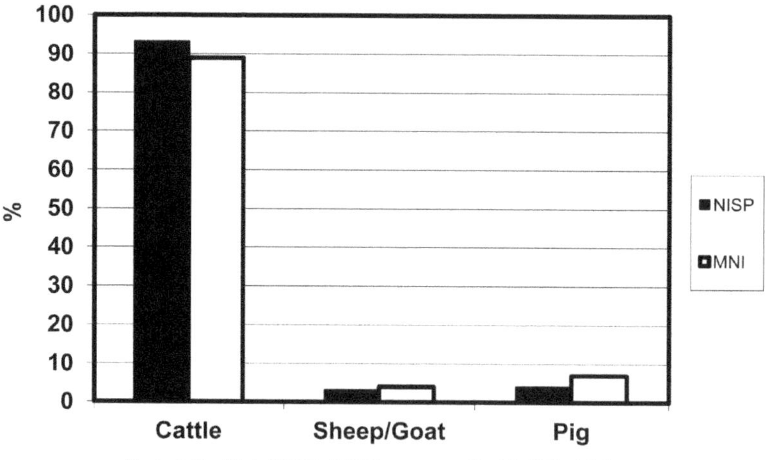

Event 3: Tot NISP=2668 Tot MNI=138

FIGURE 2.34 FREQUENCY OF THE MAIN DOMESTIC MAMMALS.

Chester sites based on Cartledge (1991); Colchester based on Luff (1993); Elms Farm based on Johnstone and Albarella (2002); Lincoln based on Dobney *et al.* (1996); Nijmegen based on Lauwerier (1988). (Chester sites include: Goss Street, Crook Street, Hunter's Street and Hunter's Walk. Colchester sites include: BKCJ1, COC1, CUL1, CUL2, CUL4, GBS1, GBS2, GBS3, BKC1, and BKC2)

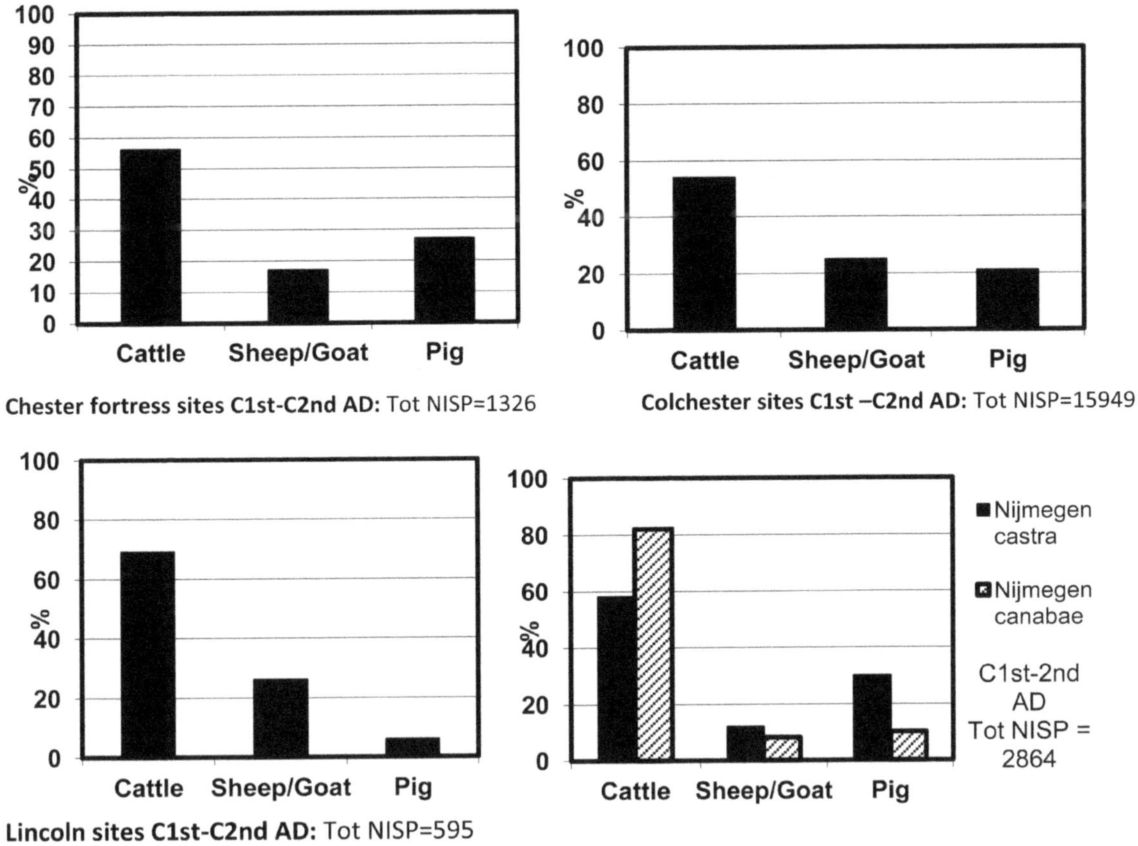

FIGURE 2.35 FREQUENCY OF THE MAIN DOMESTIC MAMMALS AT CHESTER FORTRESS SITE, COLCHESTER SITES, ELMS FARM, HEYBRIDGE, ESSEX, LINCOLN SITES, NIJIMEGEN *CASTRA* AND *CANABAE*.

Cattle

The relatively few cattle horncores recovered belong to small and short horned beasts in the typological scheme of Armitage and Clutton-Brock (1976) (Table 2.31). All of the crania seen have flat or convex frontal profiles (Table 2.32, Plate 2.21 A-E). Although most of the horncores from Delamere Street derive from adult cattle, juvenile and subadult cores (based on the criteria of Armitage 1982) are also represented (Table 2.31). In two cases the short horned cores are exceptionally flattened and these may represent a separate strain within the short horned population (Plate 2.21 G-H). Figure 2.36 shows scatter plots of the measurable cattle horncores in terms of size and shape. Mandibular wear stages, in many cases calculated from isolated lower third molars, are dominated by adult and elderly (i.e. cattle with M_3 in wear or worn at stage j+) individuals in the classification of O'Connor (1988) (Fig 2.37). However, it has recently been demonstrated (Jones and Sadler 2006) that lower third molars at wear stage j do not necessarily derive from animals older those with teeth at an earlier wear stage and that the position of the cement-enamel-junction needs to be ascertained before mandibles can be confidently ascribed to the "elderly" category. A cattle mandible from Delamere Street with CEJ at "y", i.e. above the alveolar border, is illustrated in Plate 2.22 G. Unfortunately, a significant proportion of the Delamere Street assemblage is composed of loose teeth and the author only became aware of this study half way through recording the material. The proportion of fused as opposed to un-fused bone epiphyses is never less than 76% (Table 2.33; Fig 2.38), which confirms that the cattle assemblage is primarily composed of dentally and skeletally mature animals. Only a very few deciduous teeth or bones of perinatal animals were recovered from the site. Based on the multiplication factors of Matolcsi (1970) applied to complete long bones the Delamere Street cattle range between 99-123cm at the withers with a mean of 107cm (n = 33) (Table 2.34). The cattle from Delamere Street are generally similar in size to those from other urban sites during the late 1st to 2nd centuries. The shape indices for both metacarpals and metatarsals separate into two groups, most probably representing males (entire or castrate) and females (Fig 2.39). The shape indices of the

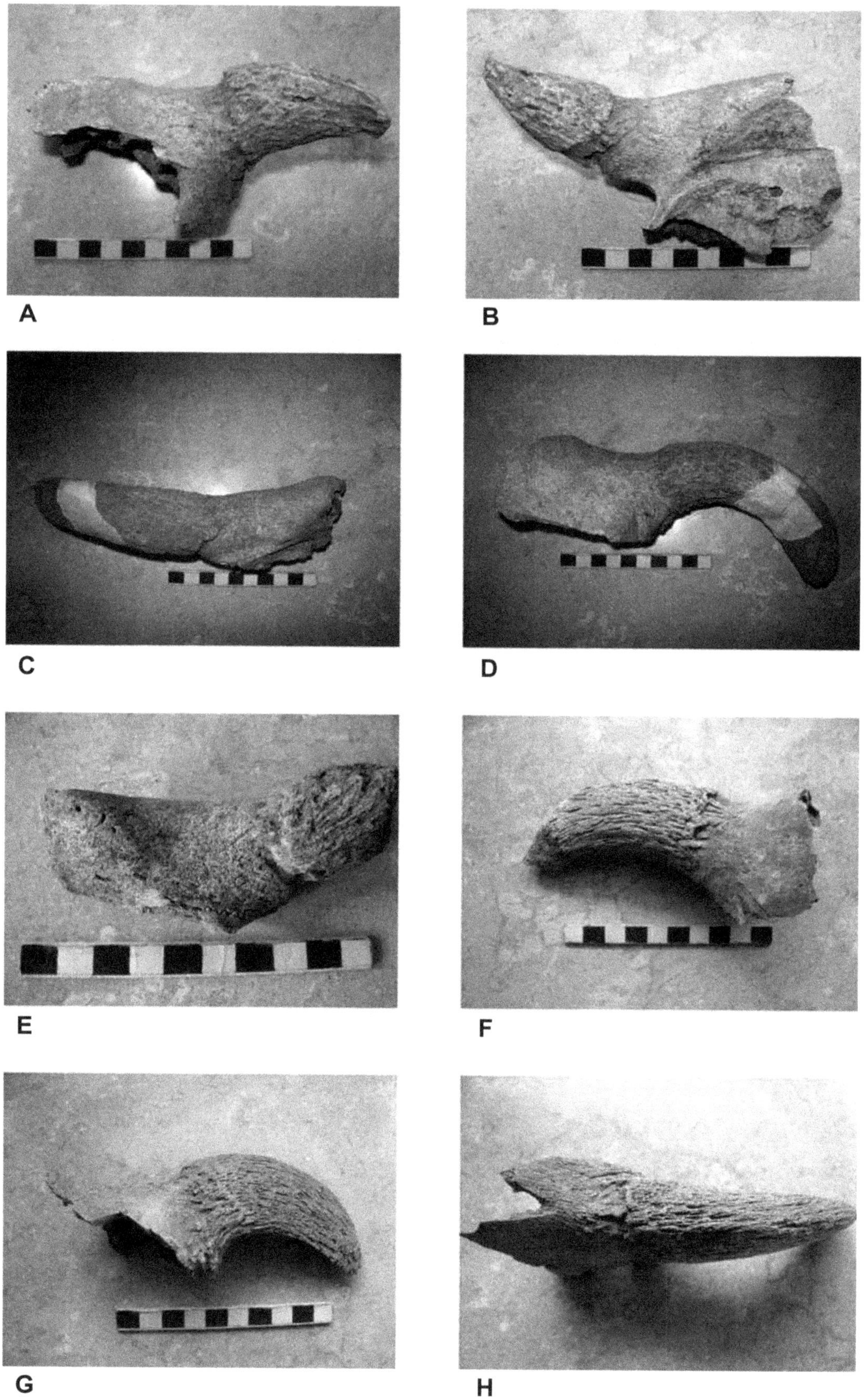

PLATE 2.21 ANIMAL BONE A-H.

A

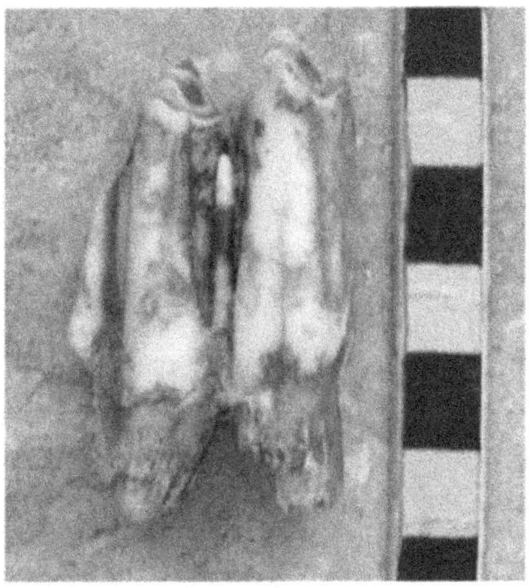
B

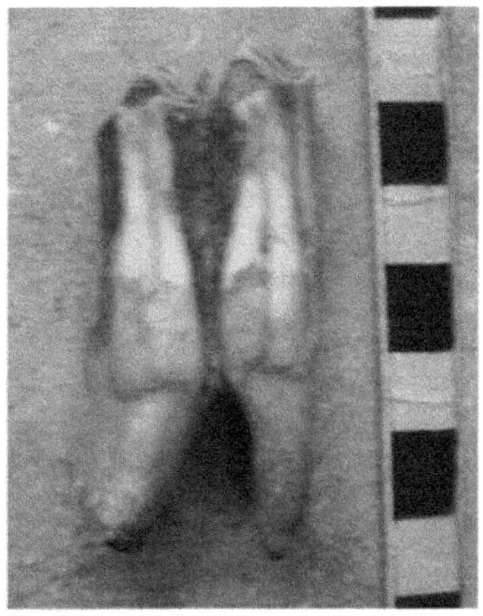
C

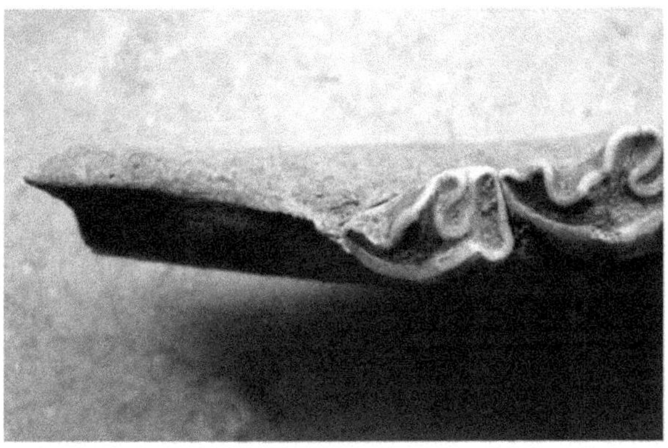
D

E

F

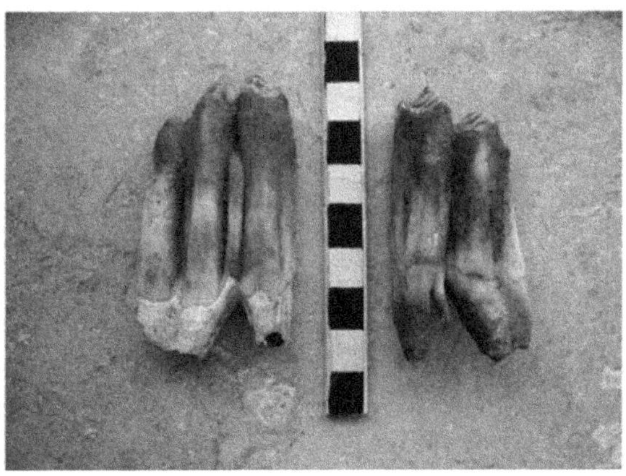
G

PLATE 2.22 ANIMAL BONE A-G.

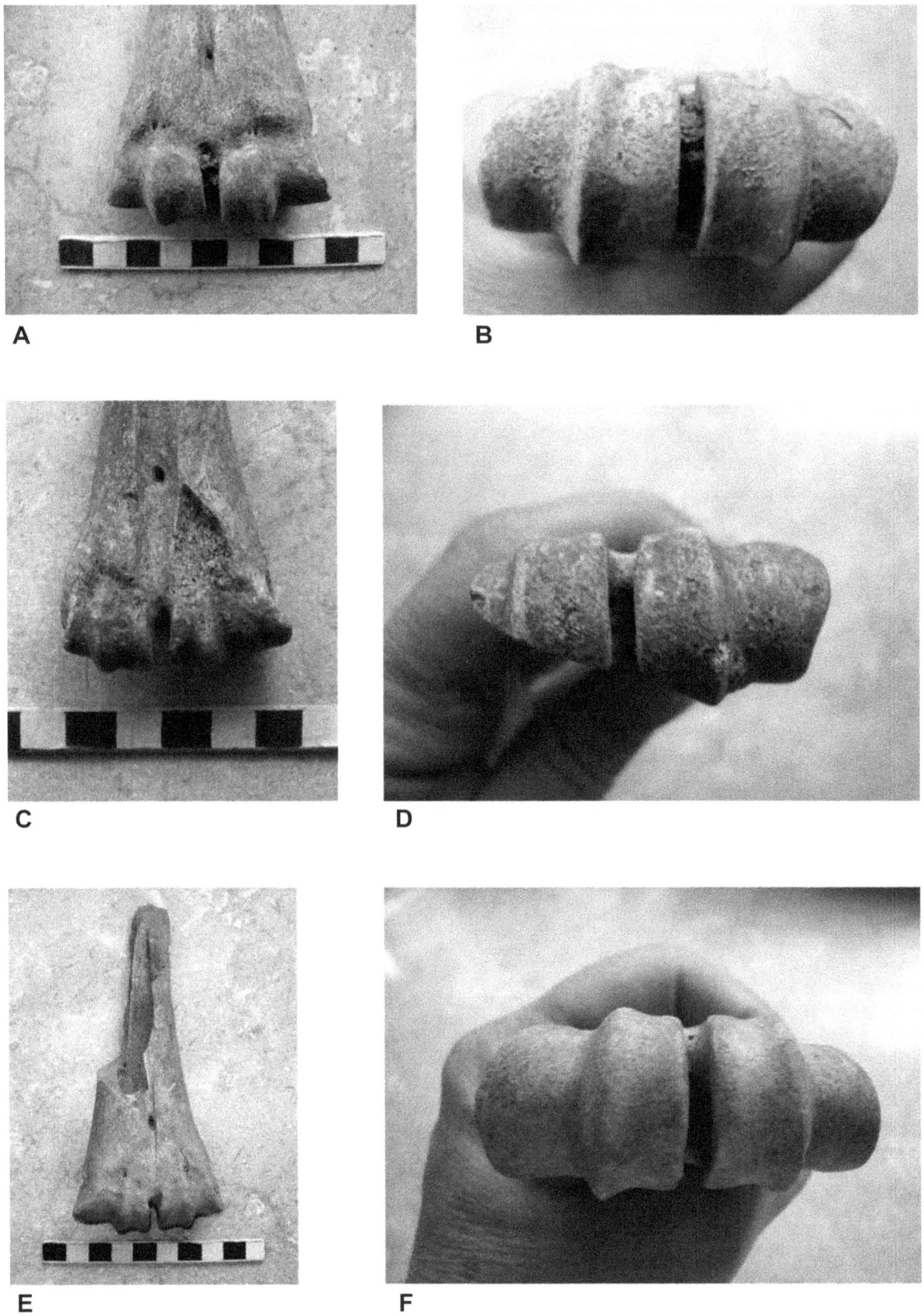

PLATE 2.23 ANIMAL BONE A-F.

A

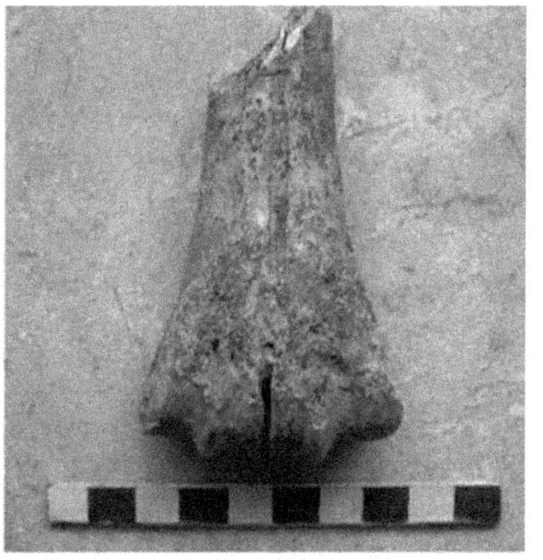

B

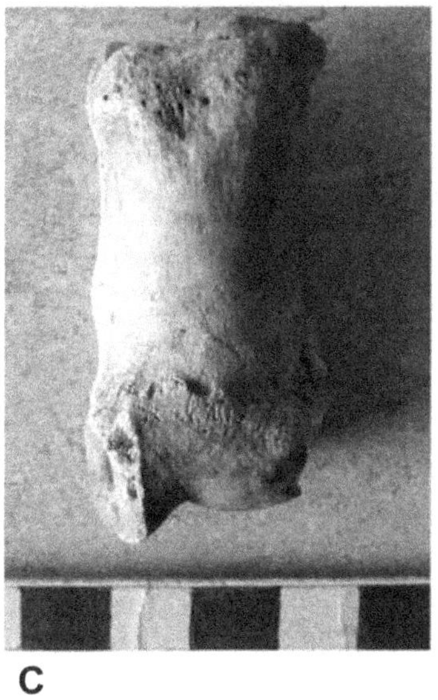

C

D

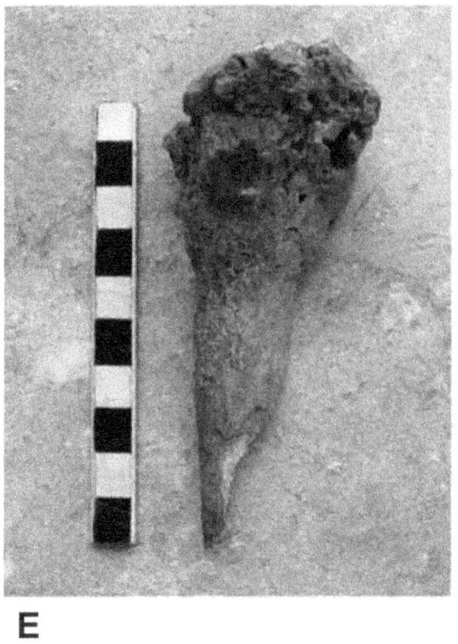

E

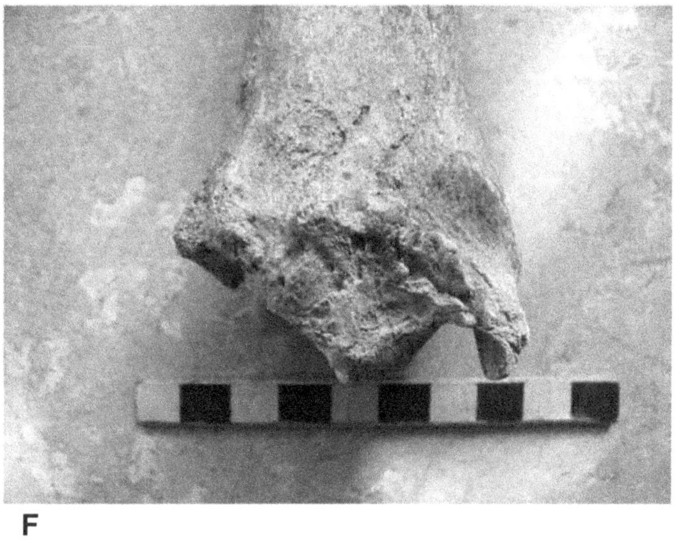

F

PLATE 2.24 ANIMAL BONE A-F.

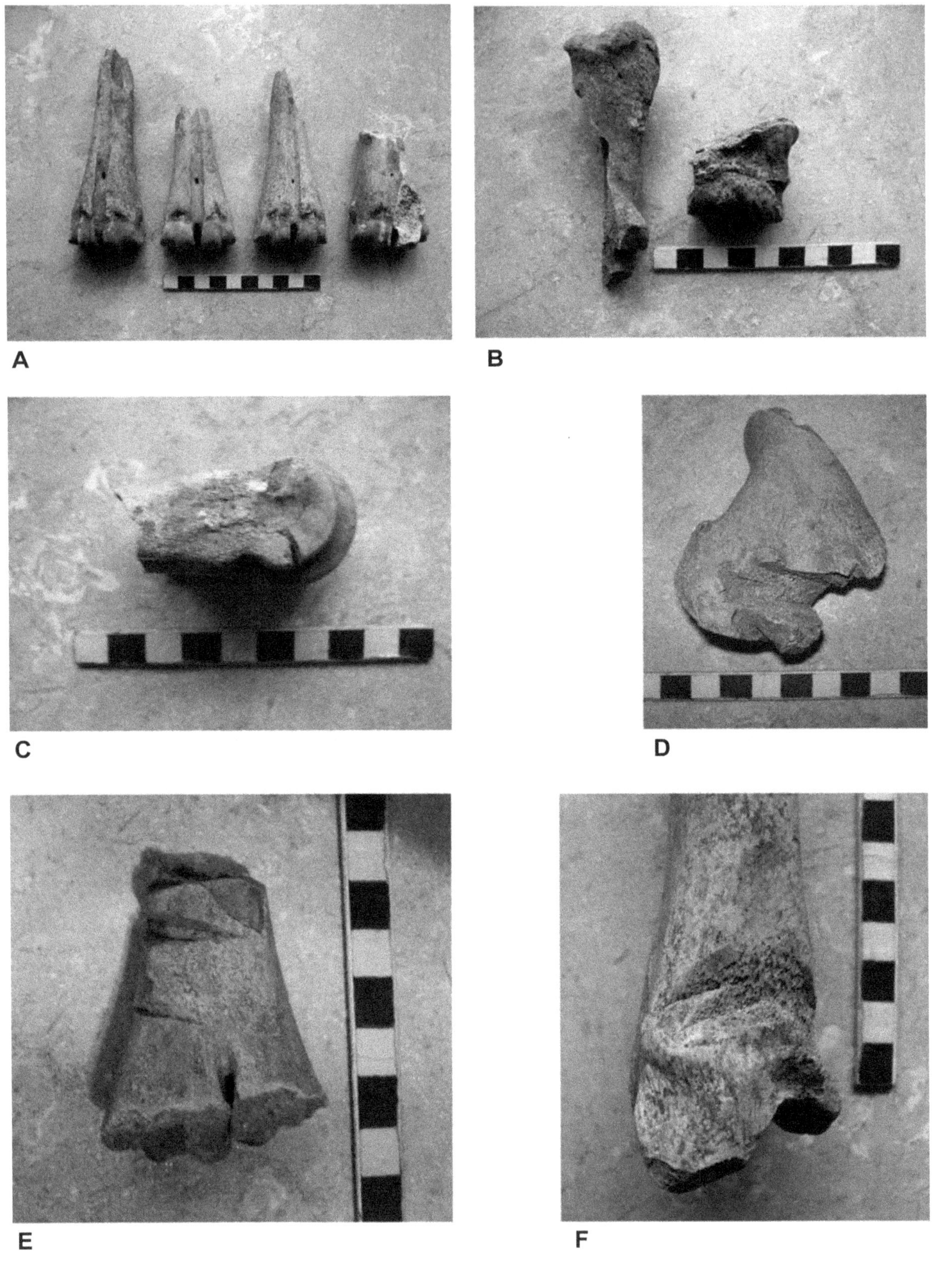

PLATE 2.25 ANIMAL BONE A-F.

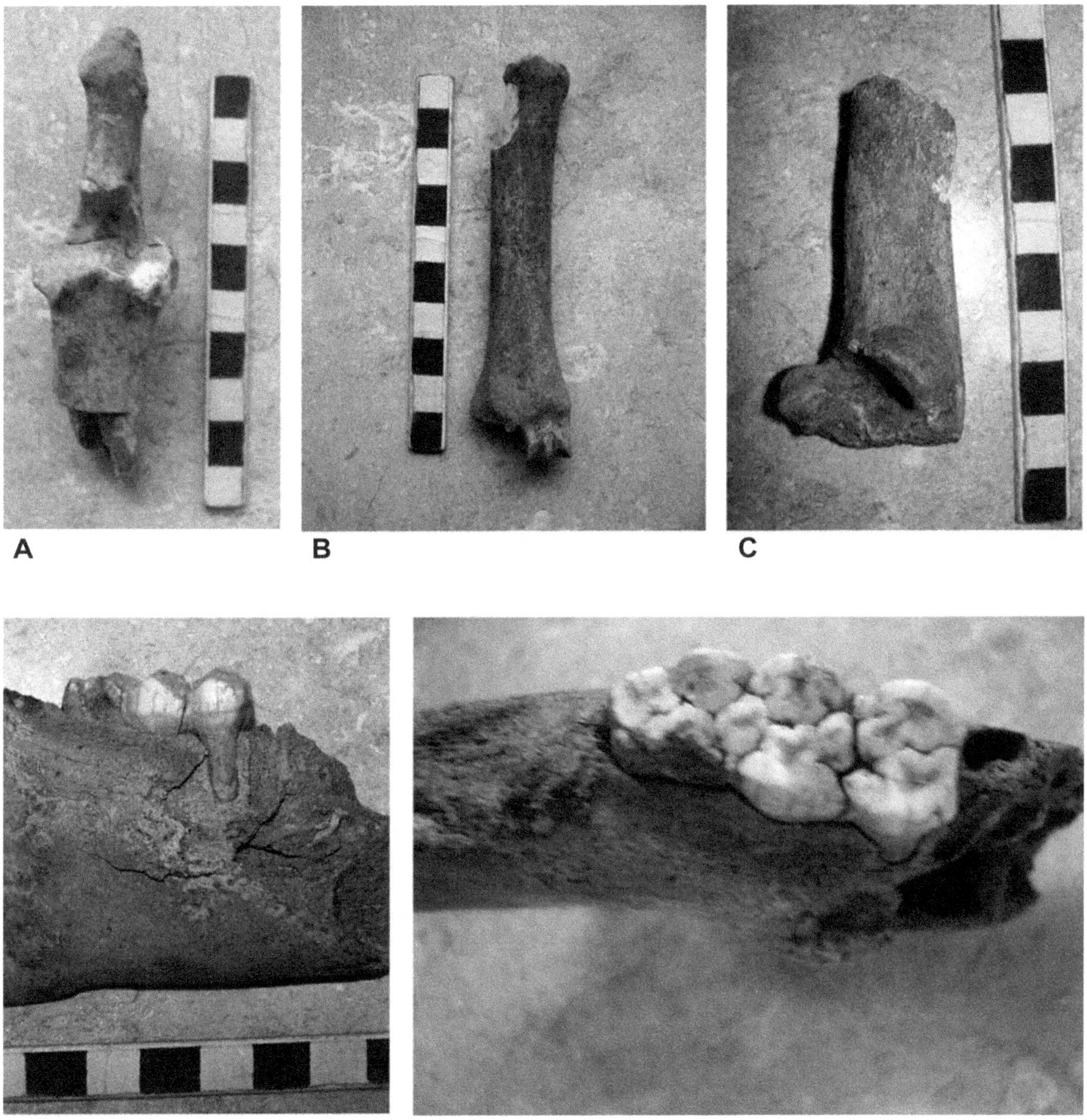

PLATE 2.26 ANIMAL BONE A-E.

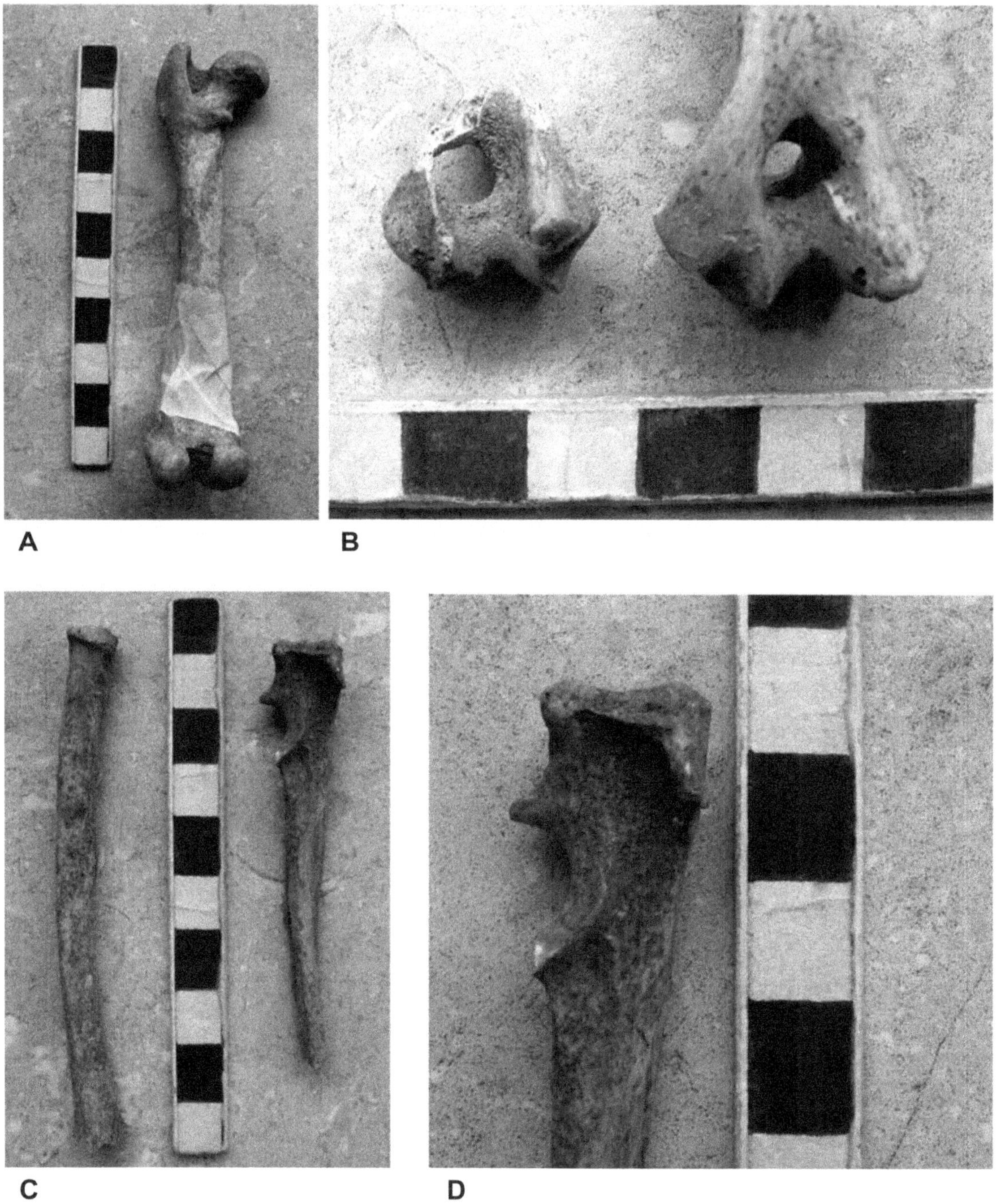

PLATE 2.27 ANIMAL BONE A-D.

TABLE 2.31. DELAMERE STREET, CHESTER. CATTLE HORNCORES. BASED ON ARMITAGE AND CLUTTON-BROCK (1976) AND ARMITAGE (1982)

Type	N	Age	N	Comments
small horn	1	adult	1	
short horned	17	juvenile	3	
		subadult	3	
		young adult	5	1 with flattened core
		adult	6	1 with flattened core

TABLE 2.32. DELAMERE STREET, CHESTER. CATTLE CRANIAL MORPHOLOGY. BASED ON GRIGSON (1976)

Frontal profile	Intercornual ridge
Flat	High single arch
Convex	Low double arch
Convex	Low single arch
Convex	Low single arch?

TABLE 2.33. DELAMERE STREET, CHESTER. NUMBER AND PERCENTAGE OF FUSED EPIPHYSES FOR THE MAIN DOMESTIC MAMMALS. FUSED AND FUSING EPIPHYSES ARE AMALGAMATED. ONLY UNFUSED DIAPHYSES, NOT EPIPHYSES, ARE COUNTED.

Element	Taxon								
	Cattle			Sheep/Goat			Pig		
	n	n_f	%	n	n_f	%	n	n_f	%
Scapula	85	85	**100**	2	0		5	4	
Humerus dist	45	45	**100**	4	3		12	12	**100**
Radius dist	16	15	**94**	9	6		3	1	
Ulna prox	3	1		4	2		3	-	
Metacarpal dist	181	170	**94**	9	8		9	6	
Pelvis acetabulum	13	13	**100**	8	8		4	4	
Femur dist	11	9	**82**	2	-		-	-	
Tibia dist	38	33	**87**	13	9	**69**	10	8	**80**
Calcaneum	21	16	**76**	1	1		8	3	
Metatarsal dist	228	208	**91**	9	4		9	2	
Phalanx 1	292	286	**98**	4	3		6	4	
Phalanx 2	74	74	**100**	1	1		-	-	

n = total number of fused/fusing epiphyses and unfused diaphyses; n_f = total number of fused/fusing epiphyses; % = percentage of fused/fusing epiphyses out of the total number of fused/fusing epiphyses and unfused diaphyses.
Percentages for total number of epiphyses smaller than 10 have been omitted.

astragalus (Fig 2.40 C) may possibly suggest the presence of two discrete populations although the sample sizes are very small.

Pathologies and anomalies

Small occipital perforations were observed in posterior frontal bones of cattle from [4002] (4009) and [3037] (3035) (Plate 2.21 C). These are most probably caused by a developmental disorder of genetic origin and its frequency within a population would appear to be indicative of common hereditary (Manaseryan *et al.* 1999; Baxter 2002; Fabis and Thomas 2011). Frequent dental anomalies affecting the Delamere Street cattle include absence of P_2 and reduction or absence of the M_3 hypoconulid (Andrews and Noddle 1975; Albarella and Davis 1994; 1996). The relative frequency of P_2 absence could not be quantified, but limited invasive analysis and external observation of the area of the mandible that should contain the tooth suggests that in many cases the tooth was lost in life with subsequent healing of the alveolar socket. Plate 2.22 F shows a sectioned mandibular fragment containing within it the roots of the broken P_2. In other cases there was no sign of alveolar resorption (Plate 2.22 D). A mandible from [1019] (1008) has P_2 absent and P_4 sitting in a dP_4 alveolus too big for it, perhaps a dP_4 fragment was retained *in situ* after the P_4 was in wear. Examples of lower third molars with reduced or absent third pillars are illustrated in Plate 2A-C and compared with a normal tooth in Plate 2.22G. The frequency of hypoconulid reduction overall is 8% and of absence 9% (Table 2.35). No anomalies of the mental foramen were observed in the cattle mandibles from Delamere Street. High incidences of dental abnormalities, together with the cranial perforations noted above, are considered to be due to population bottlenecks or result from inbreeding practices (Fabis and Thomas op. cit.). They are prevalent at Romano-British sites throughout Britain.

Most of the pathologies recorded on the cattle bones are common amongst animals used for draught and traction (Bartosiewicz *et al.* 1997). A wrist joint composed of ankylosed carpal 2+3 and proximal metacarpal found in [4002] (4013) (Plate 2.23 E) and a hock joint with centrotarsale and proximal metatarsal fused found in [4041] (4033) are spavins. Spavin is a degenerative chronic arthropathy (deforming and ankylosing) of the tarsal joint and less frequently of the carpal joint (Arthropathia deformans ankylopoetica tarsi/carpi) mainly affecting equids and cattle. An ankylosed C2+3 and metacarpal were recorded from the Oppidum at Manching (Boessneck *et al.* 1971, 58). A 1st phalanx from [4002] (4004) has Stage 3 high ring bone (Plate 4C-D). The most frequent traction related arthropathy observed is broadening of the distal epiphyses of the metacarpal and metatarsal (Plate 2.23 A and B). Only Stages 3 and 4 of Bartosiewicz *et al.* (1997) were recorded. In the metacarpal Stage 3 occurs in 8% of articular ends from Delamere Street and in the metatarsal 11% of bones are affected. Stage 4 is much less frequent (Table 2.36). A metacarpal with broadened distal epiphysis from [3012] (3011) also has grooving (eburnation) on the medial condyle and exostosis development on the anterior

TABLE 2.34. RANGE, MEAN AND COEFFICIENTS OF VARIATION (V) FOR THE WITHERS HEIGHTS OF THE DOMESTIC MAMMALS AT DELAMERE STREET, CHESTER.

Coefficients of variation only given for samples of at least 10 measurements.

Cattle (Matolcsi 1970)				
Skeletal Element		**GL (mm)**	**Withers height (cm)**	
Radius		244.5	100	
Radius		234.9	96	
Radius		280.3	115	
Metacarpal		171.1	106	
Metacarpal		172.4	107	
Metacarpal		170.2	105	
Metacarpal		170.8	106	
Metacarpal		198.1	123	
Metacarpal		173.4	107	
Metacarpal		179.9	111	
Metacarpal		161.4	100	
Metacarpal		172.0	106	
Metacarpal		174.9	108	
Metacarpal		173.4	107	
Metatarsal		190.9	104	
Metatarsal		192.9	105	
Metatarsal		203.1	111	
Metatarsal		182.4	99	
Metatarsal		182.3	99	
Metatarsal		189.1	103	
Metatarsal		190.9	104	
Metatarsal		192.5	105	
Metatarsal		215.0	117	
Metatarsal		209.0	114	
Metatarsal		194.6	106	
Metatarsal		196.6	107	
Metatarsal		192.0	105	
Metacarpal		175.8	109	
Metacarpal		176.7	109	
Metacarpal		162.2	100	
Metatarsal		195.2	106	
Metatarsal		193.7	106	
Metacarpal		176.9	110	
			Range: 99-123cm	
			N =	33
			Mean =	107cm
			S.D. =	24.22735
			V =	22.6

Horse (May 1985)					
Skeletal Element		**GL (mm)**	**Withers height (cm)**		**Hands**
Radius		368.1	151		15
Tibia		392.5	155		15

Sheep (Teichert 1975)				
Skeletal Element		**GL (mm)**	**Withers height (cm)**	
Metacarpal		110.5	54	
Astragalus		27.0	61	
Calcaneum		53.1	61	
Metacarpal		112.3	55	
Metacarpal		121.5	59	
Metacarpal		126.5	62	
Metacarpal		121.2	59	
Metacarpal		121.2	59	
			Range 54-62cm	
			N =	8
			Mean =	59cm

Goat (Schramm 1967)		
Skeletal Element	**GL (mm)**	**Withers height (cm)**
Metacarpal	104.3	60

Pig (Bone lengths only)		
Skeletal Element		**GL (mm)**
Mc.III		67.8
Mc.III		74.9
Mc.IV		68.0
Mc.IV		67.8
Calcaneum		62.5
Mc.IV		71.8
Mc.IV		81.3
Mc.III		76.9
Mc.IV		77.0
		N = 9

Dog (Harcourt 1974 and Clark 1995)				
Skeletal Element		**GL (mm)**	**Withers height (cm)**	
Femur		107.3	32	
Mt.III		54.1	39	
Mt.V		56.1	45	
Femur		101.7	31	
			N = 4	
			Range: 31-45cm	

TABLE 2.35. DELAMERE STREET, CHESTER. FREQUENCY OF REDUCED/ABSENT M_3 HYPOCONULIDS.

Total M_3s	Reduced		Absent	
	N	%	N	%
327	26	8	29	9

TABLE 2.36. DELAMERE STREET, CHESTER. FREQUENCY OF METAPODIALS WITH BROADENED DISTAL EPIPHYSES.

Total MCs	Stage 3		Stage 4	
	N	%	N	%
181	14	8	2	1

Total MTs	Stage 3		Stage 4	
	N	%	N	%
228	24	11	2	1

TABLE 2.37. DELAMERE STREET, CHESTER. EVENTS 2 AND 3 COMBINED (LATE C1ST-MID C2ND AD). BUTCHERY MARKS FOUND ON CATTLE BONES. BASED ON LAUWERIER (1988). NEW CODES WITH SUFFIX (I.E." B" ETC. NOT SHOWN SEPARATELY).

Mark No./ Elemt	1	2	3	4	5	6	7	8	9	10	11	12	13	14	15	16	17	18	19
mand	10	19	5	98	1		5				1			2					
scap	1	5	64	2		34		1	2		26					3	5		22
hum											36				1				
rad				1															
ulna			1			2							1						
Mc								1			2				140			1	8
pelvis		3	1	1	1								1			1			
femur																			
tibia	1					1						26	1		2				
ast	9	7	2	4			1	2					4	6	5	4	1	3	5
calc	5	3	19	3	4			1			1						9	1	
Mt	1				1		1						2	2	185	5			16
atlas		9		2			5						1	1	1				4
axis		34		1	5		1	7	1	5	4		1	3	4				

Mark No./ Elemt	20	21	22	23	24	25	26	27	28	29	30	31	32	33	34	35	36	37	38	tot
mand					1										1					**143**
scap	2		33			3	8			1		1								**214**
hum						1		1		1	4	3	19	19	15	3	3	11	2	**119**
rad						1	14	5				1		2	17					**41**
ulna																				**4**
Mc		1		1								2								**156**
pelvis							1			1										**10**
femur		9				1			2			3	2	1			7			**25**
tibia				1		1			14	1			4	22		1	4		2	**80**
ast	3																			**56**
calc																				**46**
Mt	1	5				2														**221**
atlas																				**23**
axis																				**66**

TABLE 2.38. DELAMERE STREET, CHESTER. MANDIBULAR WEAR STAGES (FOLLOWING CRABTREE 1989 AND O'CONNOR 1988).

ONLY MANDIBLES WITH TWO OR MORE TEETH (WITH RECORDABLE WEAR STAGES) IN THE $dP_4/P_4 - M_3$ ROW, ISOLATED UNWORN dP_4 OR WORN M_3 ARE CONSIDERED.

Taxon	Mandibular wear stages												
	A		B		C		D		E		F		Total
	n	%	n	%	n	%	n	%	n	%	n	%	n
Sheep/Goat	-	**0**	1	**7**	4	**29**	4	**29**	5	**36**	-	**0**	**14**

Taxon	Mandibular wear stages										
	Juvenile		Immature		Subadult		Adult		Elderly		Total
	n	%	n	%	n	%	n	%	n	%	n
Cattle	-	**0**	1	**0.2**	-	**0**	290	**68**	133	**31**	**424**
Pig	-	**0**	1	**7**	2	**14**	9	**64**	-	**0**	**14**

TABLE 2.39. COMPARISON BETWEEN THE FREQUENCY OF MEAT BEARING SKELETAL ELEMENTS AND METAPODIALS OF CATTLE AT DELAMERE STREET AND THE FORTRESS SITES AT CHESTER. FORTRESS SITES BASED ON CARTLEDGE (1991).

Skeletal Element	Fortress sites (all periods)		Delamere Street	
	n	%	n	%
Scapula	250	**16**	85	**14**
Humerus	176	**12**	45	**7**
Radius	126	**8**	16	**3**
Ulna	75	**5**	3	**0.5**
Pelvis	310	**20**	13	**2**
Femur	180	**12**	11	**2**
Tibia	143	**9**	38	**6**
Metapodials	263	**17**	409	**66**
Total	**1523**		**620**	

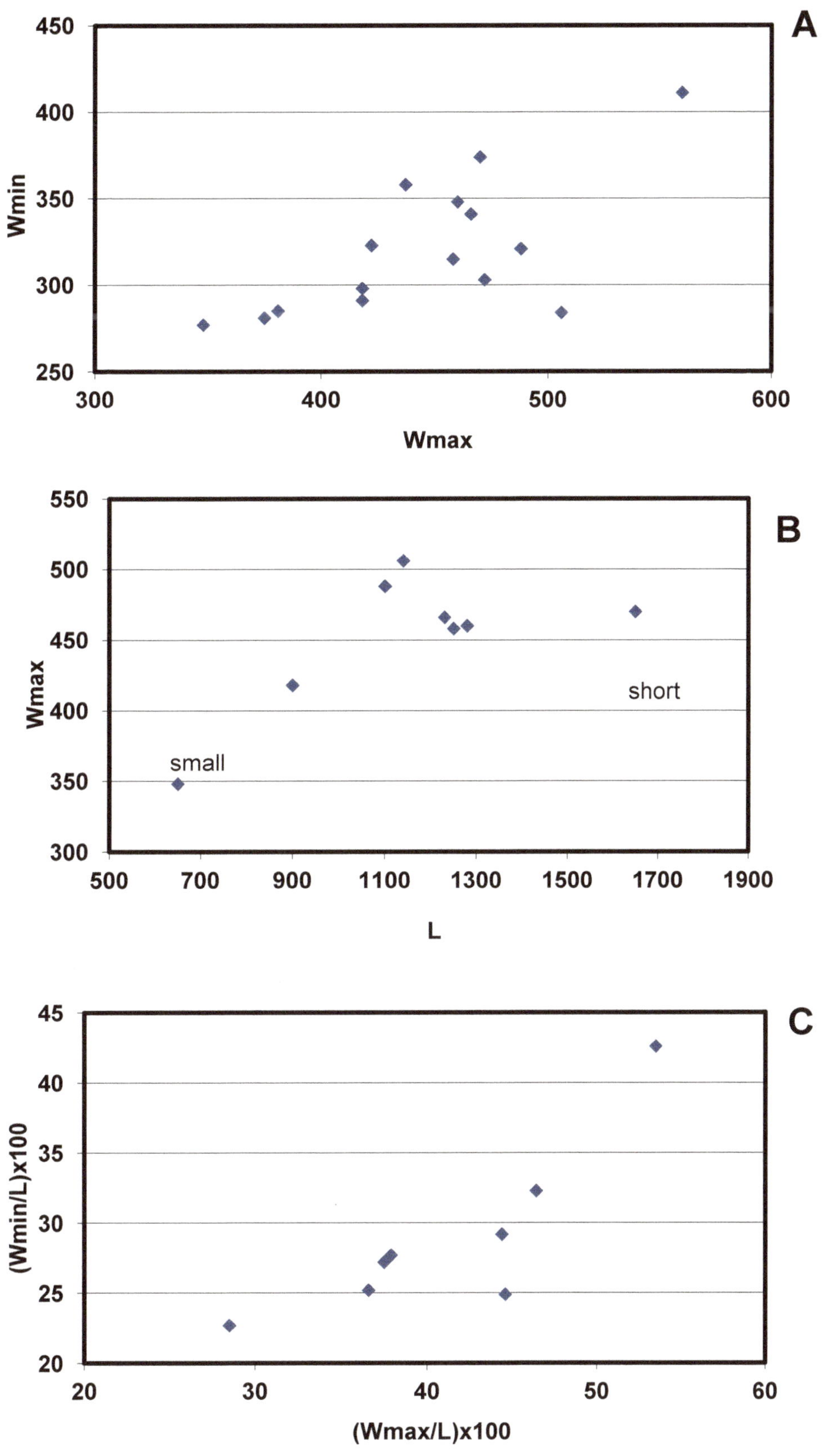

FIGURE 2.36 SIZE (A AND B) AND SHAPE (C) OF CATTLE HORNCORES. MEASUREMENTS IN TENTHS OF MM.

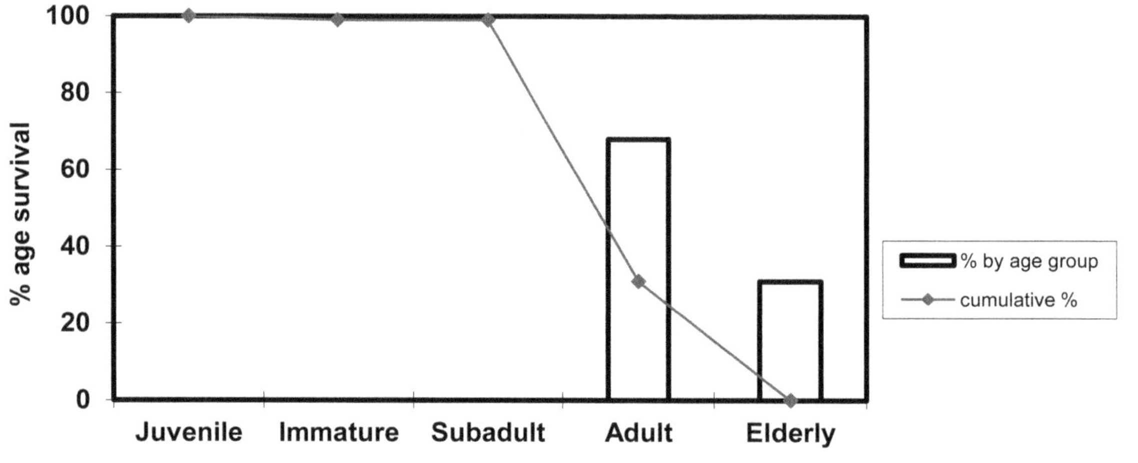

FIGURE 2.37 DISTRIBUTION OF CATTLE MANDIBLES BY AGE STAGE AT DELAMERE STREET. AGE STAGES AS DEFINED BY O'CONNOR 1988.

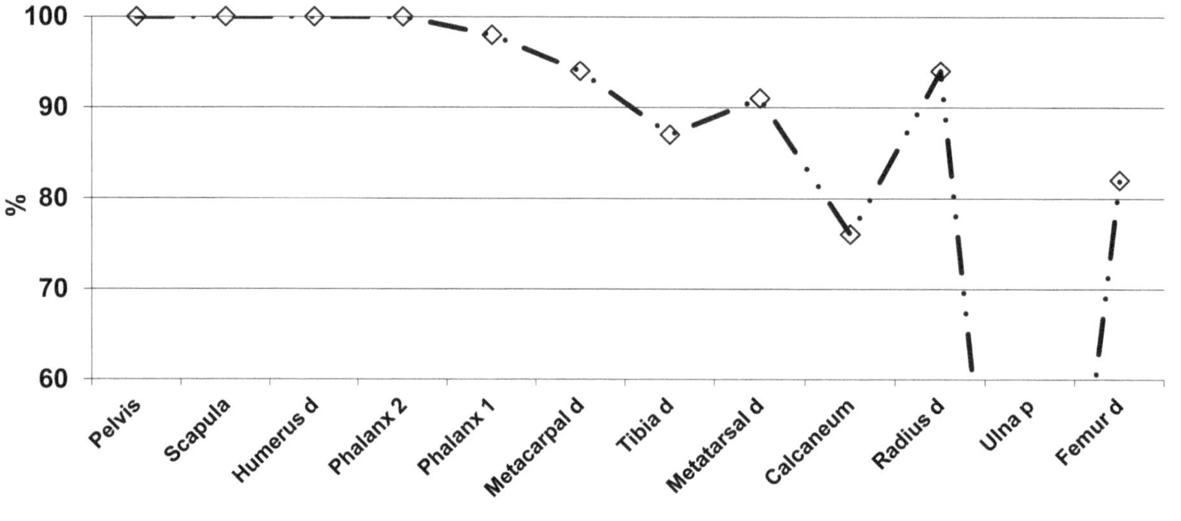

FIGURE 2.38 PERCENTAGE OF FUSED/FUSING CATTLE EPIPHYSES. FUSION SEQUENCE BASED ON SILVER 1969.

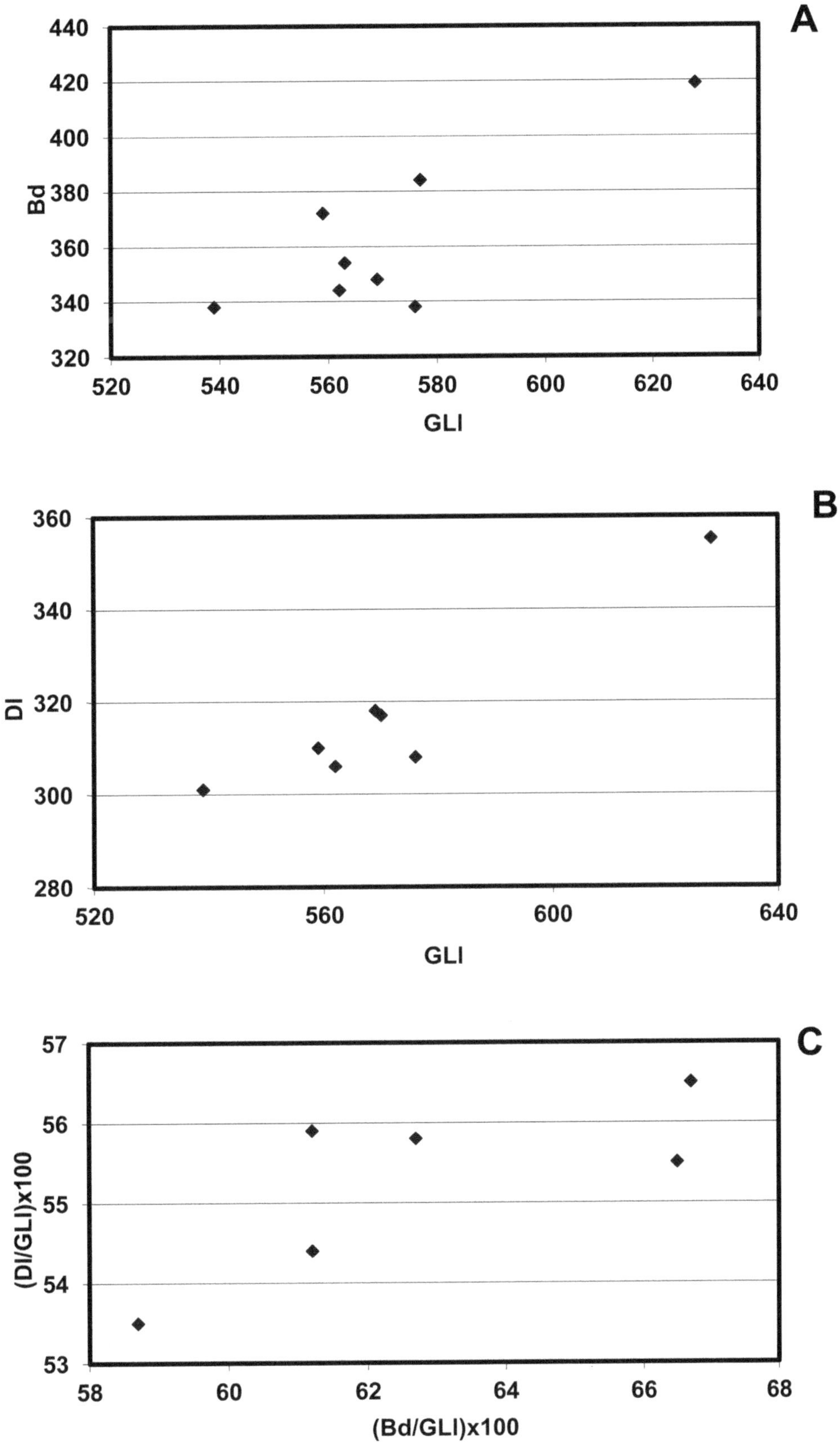

FIGURE 2.39 SHAPE OF CATTLE METAPODIALS.

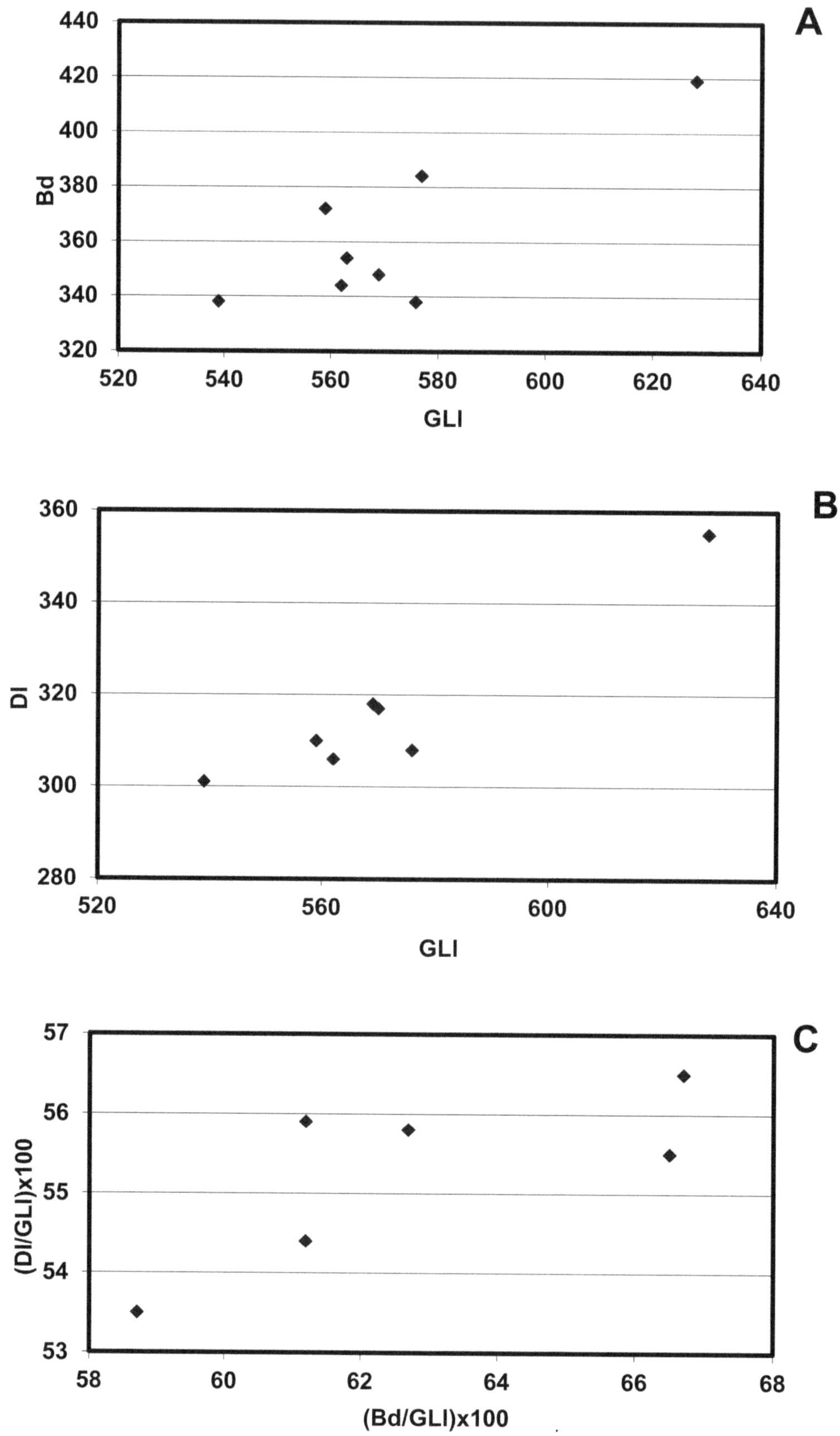

FIGURE 2.40 SIZE (A AND B) AND SHAPE (C) OF CATTLE HORNCORES. MEASUREMENTS IN TENTHS OF MM.

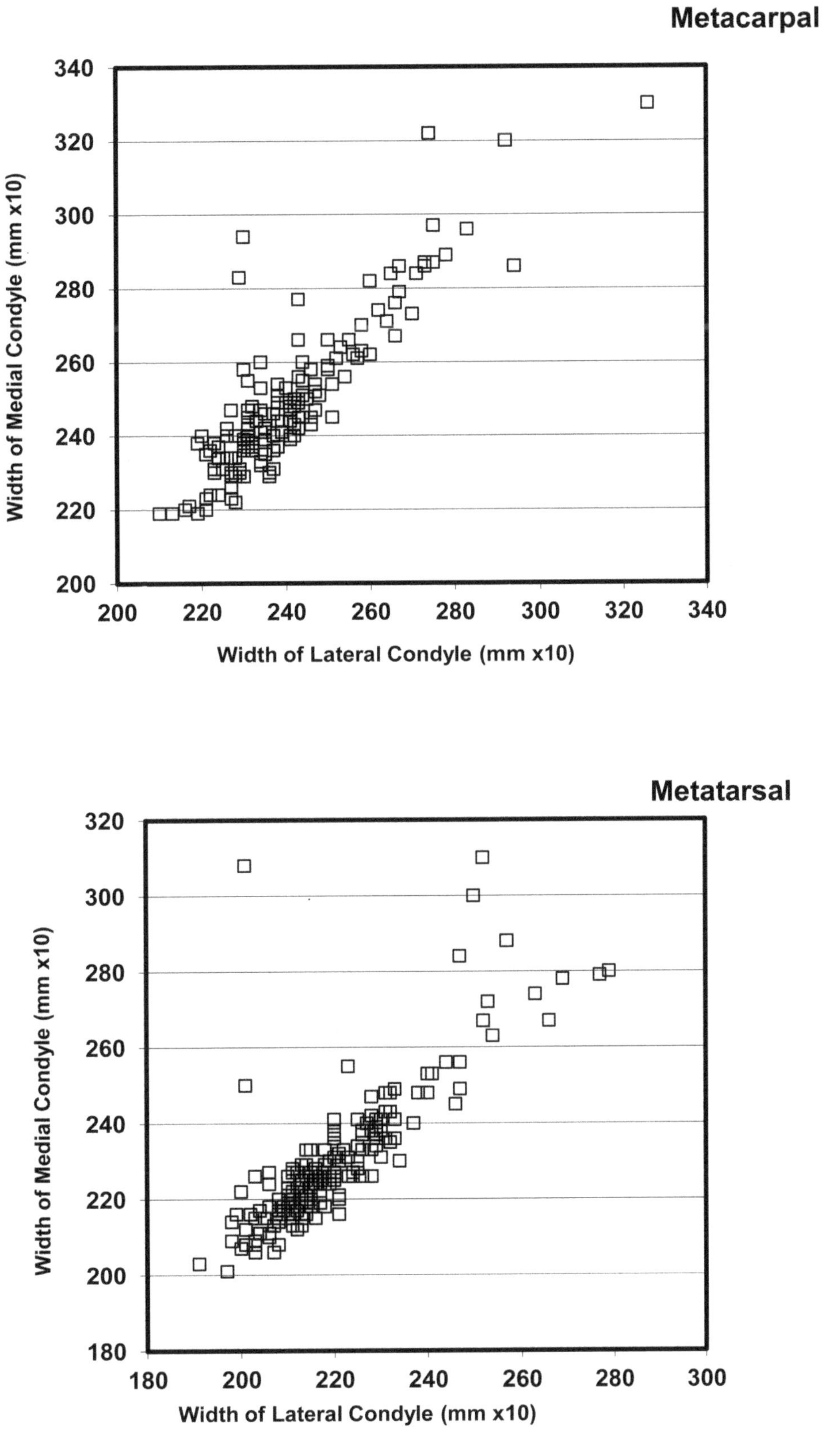

FIGURE 2.41 CATTLE METAPODIAL CONDYLE WIDTHS.

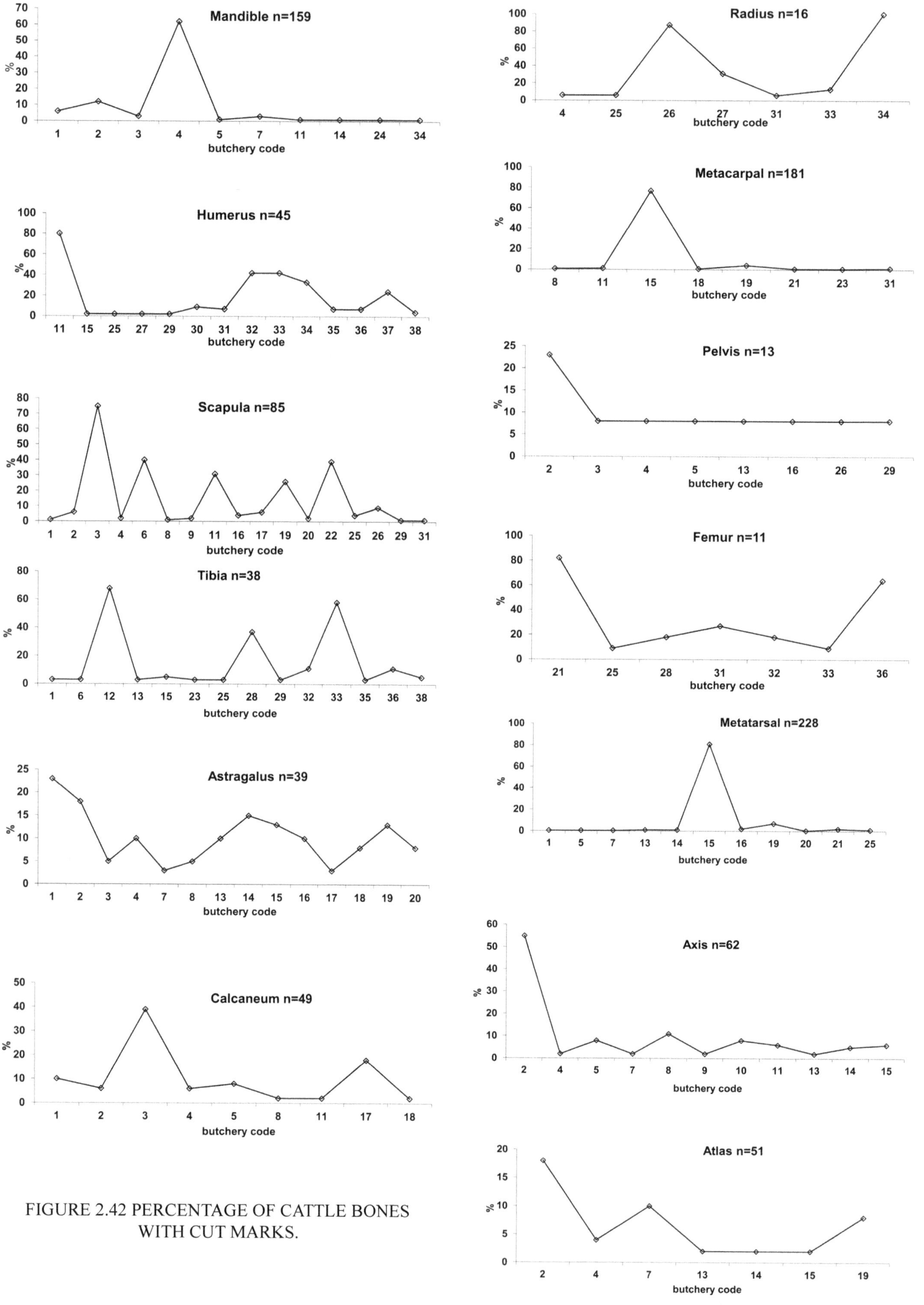

FIGURE 2.42 PERCENTAGE OF CATTLE BONES WITH CUT MARKS.

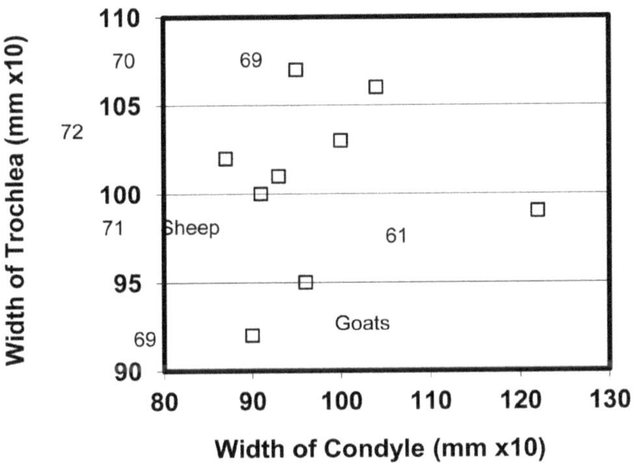

FIGURE 2.43 SHEEP/GOAT METACARPALS. BASED ON PAYNE 1969. INDEX WT/DVX100 (BOESSNECK 1969) INDICATED WHERE CALCULABLE.

surface above the articulation typical of osteoarthritis. In plots of metapodial condyle width bones with asymmetric broadening of the medial condyle group to the upper left of the main clusters in Figure 2.41.

Other pathologies of cattle bones include a metatarsal with distal epiphysis unfused from [1022] (1016) with a haematoma or tumour on the distal lateral shaft (Plate 2.23 A), a tibia from [5013] (5012) with exostoses above the distal articular surface (Plate 2.23 F) and an astragalus with slight polishing (eburnation) to the distal lateral articular surface found in (4001). A cattle rib found in [4002] (4009) has a swollen shaft posterior to the head probably resulting from traumatic injury.

Butchery

Butchery marks observed on bones have been recorded following the method of Lauwerier (1988). The results for cattle are presented in Table 2.37 and Figure 62.42. Very few butchery marks were noted on cattle horncores and none on cranial fragments. On the horncore these are most probably associated with the removal of the horn, perhaps for utilization of the sheath as a raw material for craft working. Few cut marks relating to skinning were observed on the cattle bones from Delamere Street. On the distal tibia marks 23 and 35 were recorded once each, on the astragalus mark 8 occurred twice and on the proximal metatarsal mark 1 was seen once. By comparison 90% of metacarpals and 92% of metatarsals had been chopped transversally through the shaft (15) although the exact position where this occurred varies considerably (Plate 2.23 E). This is assumed to be associated with marrow extraction (Lauwerier 1988, 155).

The most numerous mark recorded for mandibles (4) are a chop cutting through the diastema in front of the premolars, a mark that is produced when the almost meatless oral part of the lower jaw is struck off. This accounts for 69% of all butchery recorded on mandibulae. Next most frequent are chops resulting in cutting through the corpus behind the last molar (1) and cutting through the corpus between M_2 and M_3 (2) accounting for 7% and 13% of marks respectively. The only other marks observed at other than very low frequency are chops cutting through the mandibular corpus between P_4 and M_1 (3) and chop marks on the lateral side of the diastema (7) both at 3%. The dearth of marks produced when the lower jaw was removed from the skull (15, 16, 18, 19, 21-24) is at least partly a product of the recording system used for bones in this study, where only those fragments containing teeth are recorded, combined with the generally fragmented state of the mandibulae occurring before, during and after burial. Relatively few butchery marks were recorded on the atlas vertebra, to some extent because atlas fragments were frequently abraded. The most frequent was a chop through the centre of the bone (2) recorded in 39% of cases. On the axis vertebra the most common mark was a chop cutting off the lateral part of the facies articularis cranialis (2) on one or both sides, accounting for 52% of butchery recorded for this bone. Butchery marks on other vertebrae were not recorded but, in general, cervical vertebrae are transversally cut or longitudinally split and thoracic and lumbar vertebrae have been shorn of their lateral elements.

The scapulae from Delamere Street were extensively butchered with 214 marks recorded comprising 17 different types. The most numerous marks observed were chops cutting off the spina from the distal side (3) at 30%, chops cutting off the cranio-lateral side of the distal articulation (6) at 16%, chops cutting off the medial side of the articulation (22) at 15%, chops cutting through the caudo-lateral side of the distal articulation (11) at 12%, and chops cutting off the processus coracoideus (19) at 10%. Butchery marks 1-20 and possibly 25-29 are associated with the removal of the shoulder blade from the pectoral girdle. The humerus was also extensively butchered with 119 marks of 14 types. Disarticulation of the skeleton during primary butchery is represented by the following marks on the long bones: humerus 29-36 (67%); radius 4, 33-34 (49%); ulna 3 and 6 (three out of four); femur 28, 32 and 36 (44%); tibia 1, 28-29, 32-33, 35-36, 38 (61%). Butchery marks that can be associated with the boning of meat are exceptionally scare or non-existent. However, the chopping through of the diaphysis is very common: humerus 11 (30%); radius 26 (34%); tibia 12 (33%). The shafts were either cut through together with the meat attached in order to divide the meat into portions or, as with the metapodials, to facilitate extraction of the marrow (Lauwerier 1988, 157).

Sheep/Goat

Sheep form the majority of the caprid remains recovered from Delamere Street (Table 2.25). However, goats were also present in the assemblage and include a proximal radius and ulna fragment found in [4002] (4007) (Plate 2.24 A), a metacarpal found in [4041] (4033) (Plate 2.24 B) and an uncounted horncore fragment recovered from [2209] (2175). The width of the trochlea is plotted against the width of the condyle for complete sheep/goat metacarpals in Figure 2.43. The index width of trochlea/ diameter verticillus is displayed next to the specimens for which it could be calculated. Only one goat is indicated. The relatively few ageable sheep/goat mandibles show a generalized pattern of slaughter with a majority slaughtered at around one to two years and similar numbers slaughtered at two to four years and over four years (Table 2.38). The presence of younger animals is indicated a very few neonatal and immature remains. The few bones with epiphyseal ends preserved suggest that most sheep were slaughtered when skeletally mature (Table 2.33). Withers heights calculated from complete bones (primarily metapodials) using the factors of Teichert (1975) range between 54-62cm (n = 8, mean = 59cm) (Table 2.34). The goat metacarpal from [4041] (4033) came from an animal 60cm at the withers based on the multiplication factors of Schramm (1967) (Table 2.34). No pathologies were seen affecting the caprid assemblage.

Pig

Although far less frequent than the remains of cattle, pig teeth and bones are the second most common remains present at Delamere Street. This is typical of more "Romanized" sites (King 1978) and is generally the case with other excavated sites in Chester (Cartledge 1991). Upper canine teeth and empty canine alveoli were recovered from 4 males and 1 female; 18 male and 4 female lower canines and alveoli were recorded. Although female pig canines are smaller than those of males, it seems probable that the observed disparity is not due to a collection bias but to the fact that the assemblages recovered derive from what was primarily a consumer site rather than a producer site. That pigs were also being raised in close proximity to the points of deposition is illustrated by the very scarce inclusion of neonatal and very young remains. However, 64 % of the available pig mandibles (n = 14) came from young adult animals with M_3 in the early stages of wear (Table 2.38). The few pig bones with epiphyses preserved are also indicative of late slaughtering (Table 2.33). All of the measurable pig bones recovered certainly derive from domestic animals. The lengths of complete pig bones are listed in Table 2.34. However, a large and immature calcaneum fragment found in [4002] (4009) (Plate 2.26 C) may, possibly, derive from a wild pig. A mandible fragment from [4041] (4046) has exostoses developed buccal to the M_3 together with the site of a probable abscess (Plate 2.26 D-E).

Sheep/Goat and Pig butchery

The relatively few butchery marks recorded for sheep/goat and pigs (following Lauwerier 1988) are similar to those recorded for cattle, except that division at the midshaft of bones is much more frequent than any other evidence of butchery. Sheep and pig sized vertebrae and ribs form only a tiny part of the assemblage.

Other Domestic Mammals

Horse

Equid remains are relatively infrequent at Delamere Street. All bones and teeth recovered are consistent with belonging to horse (*Equus caballus*) and nothing was seen suggestive of donkey or mule. A radius found in [4041] (4030) and a tibia from (4015) in the same Pit came from animals of around 15 hands high based on the multiplication factors of May (1985) (Table 2.34). This is above average for Romano-British horses generally but more typical of military sites (Johnstone and Albarella 2002). Loose teeth recovered from several contexts came from horses aged between 7 and 15 years (Barone 1980; Levine 1982).

Dog

Isolated dog skeletal elements were recovered from several contexts. These all derive from small or medium sized animals (Table 2.34). The femur of a fairly large dwarf hound was found in [2293] (2307) (Plate 2.27 A). This came from an animal approximately 31cm at the shoulder. The bone has the typically expanded proximal and distal ends of chondrodystrophic types with a high midshaft diameter Index of 10.2 (Harcourt 1974). A very small distal humerus fragment found in Pit [4041] (4030) probably came from the same type of dog. In Plate 2.27 B it is compared with the complete bone from the York Road, Leicester dwarf hound (Baxter 2006). The femur of a dog with a withers height of 32cm found in (4030) may belong to a female dwarf type as the proximal and distal ends are not appreciably expanded but the msd Index is 9.0 (Baxter 2010a). The radius and ulna of another, non-chondrodystrophic, small dog were found in [2110] (2112) (Plate 2.27 C-D). In this animal there is a bony projection on the medial side below the ulna olecranon. The precise cause of this anomaly is obscure but might result from a combination of trauma and old age. The tallest dogs present in the assemblage are represented by metapodials: a 3rd metatarsal found in [5004] (5002) and a 5th metatarsal from [4041] (4006). Based on the multiplication factors of Clark (1995) these came from animals respectively 39cm and 45cm high at the shoulder. None of these animals are large compared with assemblages from other Roman sites (e.g. Thistleton; Baxter 2010b).

Wild mammals

Both bones and antler fragments of red deer (*Cervus elaphus*) were recovered at low frequency from Delamere Street. The antler fragments have evidence of working and are considered below. In addition a red deer mandible fragment containing P_4 and a butchered proximal radius were found in Pit [4041] (4046). The complete metatarsal of a roe deer (*Capreolus capreolus*) was recovered from [5007] (5005). Smaller wild mammals are represented by a hare (*Lepus* sp.) scapula found in [4041] (4030) and a mouse or vole tibiofibula recovered from a sample residue obtained from the same context.

Birds

Most of the few bird bones recovered derive from domestic fowl (*Gallus* f. domestic). Both males (cocks or capons) and hens are represented based on the presence or absence of a spur on the tarsometatarsus. Wild mallards or domestic ducks are represented by (uncounted) proximal radius fragments recovered from [4041] (4033) and [3038] (3039), and a femur with medullary bone indicative of a female in egg-laying condition (Driver 1982) found in [2291] (2290). The femur of a goose found in [5007] (5005) is very small for a domestic bird or its wild ancestor the greylag goose (*Anser anser*) and probably derives from a wild species. The humerus and ulna of a red kite (*Milvus milvus*) recovered from [2292] (2285) in Area A. This was a common scavenger of the period (O'Connor 1993). A woodcock (*Scolopax rusticola*) humerus found in [2292] (2283) may be evidence for wildfowling.

Fish

Only three fish fragments were recovered from the very few coarse fraction sample residues processed. Two of these are flatfish vertebrae consonant with plaice (*Pleuronectes* sp.) and were obtained from Area A [1020] (1021) and Area E [5007] (5005).

Discussion

Comparison of remains of the major domestic mammals, cattle, sheep/goats and pigs, at the sites within the area of the fortress and Delamere Street situated in the canabae has highlighted dramatic differences in the relative frequencies of these species in the two areas. This situation is closely paralleled by that at the broadly contemporary site at Nijmegen in the Netherlands (Figs 2.34 and 2.35). In both cases cattle remains are much more numerous in the canabae than the fortress. In his attempts to explain the disparity Roel Lauwerier (1988, 59-64) considered three possible hypotheses:

(1) In the castra relatively many small farm animals were eaten, in particular pigs.

(2) The pattern of meat consumption was actually the same in the castra and the canabae, but a proportion of the cattle were slaughtered in the canabae and consumed in the castra.

(3) The pattern of meat consumption was the same, but a proportion of the - larger - cattle bones were removed from the castra after slaughtering had been done.

The 2nd and 3rd hypotheses were rejected by Lauwerier because the frequency of cattle metapodials, a non meat bearing element that would be discarded during primary butchery of the carcass, was actually greater in the *castra* deposits than those of the *canabae*. However, if we consider the relationship between the cattle skeletal elements recovered at the Chester fortress sites and those from Delamere Street a significantly different picture emerges. It has been necessary to consider the total assemblage from the fortress as the sample size for the 1st to 2nd century is too small for useful comparison. Comparing the relative frequency of meat bearing skeletal elements to metapodials the latter account for only 17% of the total from the fortress but 66% from Delamere Street (Table 2.39). Even allowing for the probable exaggeration of pelvic fragments recorded for the fortress sites, these were heavily fragmented by butchery and at Delamere Street only the fragments including the acetabular part of the ischium were counted, the relative frequency of meat bearing limb bone fragments to those of metapodia is strikingly different. We have already noted that most of the metapodials from Delamere Street were chopped through the shaft, possibly to facilitate the removal of marrow. Another difference of potential significance is the intensive butchery of cattle scapulae at Delamere Street. This does not appear to be the case with scapulae recovered from the fortress, many of which were essentially complete (Cartledge 1991). The Delamere Street cattle assemblage, therefore, exhibits a greater selection of skeletal elements and more intensive butchery than the assemblages recovered from within the Chester fortress and a significantly different pattern of disposal to that documented for Nijmegen.

Summary and conclusion

The remains of cattle dominate the assemblage recovered from Delamere Street. These cattle were generally fairly small, though not excessively so for the period, and included both small horn and short horned types. The sheep and pigs were of a size and type typical for the period. The few horse remains came from animals of a size typical of Roman military sites. The dogs present in the assemblage are small to medium sized and include dwarf types. The presence of deer, hare and wildfowl is evidence that the diet was supplemented by hunting. A small number of bone artefacts, antler and bone craft waste were found among the animal bones. The composition of the cattle assemblage includes a significantly greater proportion of primary and tertiary butchery waste than sites excavated within the fortress area.

HUMAN BONE *by Sam Hepburn*

Introduction

Several human bone fragments were recovered from Roman quarry pits during the excavation. The remains represent a minimum number of three adults; this was achieved using guidelines set out in McKinley (2004) with the aid of skeletal recording sheets.

Preservation

A total of seven fragments were recovered (see Table 2.40). Two from Event 3 deposit (1008), the upper fill of pit [1019]: four from Event 2 deposit (4046) the lower fill of quarry pit [4041], and one unstratified femur retrieved from the building fill in the northeast corner of the site. All of the remains were disarticulated. Using the system of bone surface preservation recommended by McKinley (2004) the bone was recorded as Grade 1-2. The fragmentary nature of the bone has greatly reduced the amount of metrical information that could be derived from the remains.

Table 2.40: Summary of human bone remains

	1008	4046	U/S
Femur	1 (R)	4 (2L, 1R)	1 (R)
Humerus	1 (L)	-	-

Pathology/Trauma

The femoral head recovered from (1008) displays clean tool marks on the joint surface, which is consistent with new damage caused during the excavation process. The four femoral heads from (4046) all have similar tool marks present on them, although these are not as fresh as those on the bone from (1008) and may have occurred in antiquity.

The humerus from (1008) had a very pronounced deltoid tuberosity, which can be indicative that the individual in life was engaged in strenuous activity.

The unstratified femur had new cortical bone growth formation present on the margins of the femoral head, around the fovea capitis; and along the length of the linea aspera. This bone formation indicates that the individual had well developed muscular legs, and was probably engaged in activities that put great stress on the hip joint causing it to adapt and remodel itself.

Discussion

The bone from (1008) and (4046) was most likely from disturbed inhumations, which had been mixed with domestic waste (animal bone and pottery) and used to backfill a disused quarry pits (1019, 4002). The pottery from these features has dated it to the 2nd Century. The six pieces of bone, although they are fragmentary (head of a femur and the distal half of a humerus) they are both relatively large pieces of bone that show no signs of having been heat affected, and therefore are unlikely to have come from a cremation. Excavations of the early 2nd century Roman cemetery beneath Infirmary Field, Chester situated to the west of Delamere Street, revealed thirty-five inhumations of a comparable date (Newstead, 1914; 1921). This highlights that at this time there was a shift in burial practices from cremation to inhumation.

THE CHARRED PLANT REMAINS *by Pam Grinter*

The sampling strategy was set out in the written scheme with consultation with the City Archaeologist. The lack of well preserved organic remains in this part of the city was reflected in the large quantities of soil recovered from the site. Fifty litre archaeobotanical samples were collected from deposits in order to primarily recover charred remains (Table 2.41). The samples were taken from a range of features the majority of which were pits. These samples were assessed to determine if plant remains were present and of interpretable value. In addition samples were examined for the presence of other proxies (Beetles, pollen, plant macrofossils), but were found to be unsuitable for further analysis (pers. comm. Dr. Gearey).

In total 29 samples were selected for assessment – in most cases, selection was directly related to the significance of the archaeological context sampled. All material processed was recovered from securely datable contexts which were not subject to post medieval disturbance to ensure a secure result.

Laboratory Method

Sample volumes ranged in size according to the volume of the context. Environmental officers at Birmingham Archaeology used water flotation to process the samples, the flots and heavy residues were sieved to 500µm. Flots were then scanned by the author under a low-power microscope at a magnification of x15.

Results

Table 2.41 presents the results for the assessed samples. Charred plant remains were not present in any of the 29 samples examined. All the samples contained quantities of charcoal and most contained fragments of bone.

Conclusions

As charred plant remains were absent from the samples examined, no further archaeobotanical or palaeoenvironmental analysis was undertaken. This absence of preserved remains may be due to a total absence of archaeobotanical or palaeoenvironmental material. It would appear more likely it was due to preservation conditions present at the site. The soils were predominantly composed of sand with a small amount of clay, and these were deposited within relatively free-draining sandstone quarry pits. This may have resulted in poor preservation conditions for organic remains. This is further compounded by the absence of conditions suitable for the preservation of waterlogged deposits at the site. The absence of charred plant remains is somewhat surprising given the presence of other indicators of kitchen debris (*see Chapter 3, 4 and 5*).

TABLE 2.41: ASSESSMENT RESULTS FOR CHARRED PLANT REMAINS FROM DELAMERE STREET, CHESTER.

Sample No.	**Context**	**Context Type**	**Bone**	**Charcoal**	**fish bone**	**Waterlogged Plant Remains** Waterlogged Plant Remains	**Charred Plant remains observed (flot only)**		**Further Analysis**	**Comments on Flot**
							Grain	Chaff		
1	1008	FILL	-	+	-	-	-	-	NO	100% of the sample scanned. Charcoal and fragments of bone present. No Charred macrofossils. ASSESSED AS POOR.
5	2106	FILL	+	+	-	-	-	-	NO	100% of the sample scanned. Charcoal and fragments of bone present. No Charred macrofossils. ASSESSED AS POOR.
6	2123	FILL	+	+	-- -	-	-	-	NO	100% of the sample scanned. Charcoal and fragments of bone present. No Charred macrofossils. ASSESSED AS POOR.
7	2124	FILL	-	+	-	-	-	-	NO	100% of the sample scanned. Charcoal and fragments of bone present. No Charred macrofossils. ASSESSED AS POOR.
9	2111	FILL	+	+	+	-	-	-	NO	100% of the sample scanned. Charcoal, fish bones and fragments of bone present. No Charred macrofossils. ASSESSED AS POOR.
10	2126	FILL	-	+	-	-	-	-	NO	100% of the sample scanned. Charcoal and fragments of bone present. No Charred macrofossils. ASSESSED AS POOR.
13	2112	FILL	+	+	-	-	-	-	NO	100% of the sample scanned. Charcoal and fragments of bone present. No Charred macrofossils. ASSESSED AS POOR.
14	2113	FILL	+	+	-	-	-	-	NO	100% of the sample scanned. Charcoal and fragments of bone present. No Charred macrofossils. ASSESSED AS POOR.
15	2166	FILL	+	+	-	-	-	-	NO	100% of the sample scanned. Charcoal and fragments of bone present. No Charred macrofossils. ASSESSED AS POOR.

Sample No.	Context	Context Type	Bone	Charcoal	fish bone	Waterlogged Plant Remains	Charred Plant remains observed (flot only)		Further Analysis	Comments on Flot
							Grain	*Chaff*		
26	2248	FILL	+	+	-	-	-	-	NO	100% of the sample scanned. Charcoal and fragments of bone present. No Charred macrofossils. ASSESSED AS POOR.
35	2290	FILL	-	+	-	-	-	-	NO	100% of the sample scanned. Charcoal and fragments of bone present. No Charred macrofossils. ASSESSED AS POOR.
36	2283	FILL	+	+	-	-	-	-	NO	100% of the sample scanned. Charcoal and fragments of bone present. No Charred macrofossils. ASSESSED AS POOR.
37	2285	FILL	+	+	-	-	-	-	NO	100% of the sample scanned. Charcoal and fragments of bone present. No Charred macrofossils. ASSESSED AS POOR.
38	2250	FILL	-	+	-	-	-	-	NO	100% of the sample scanned. Charcoal and fragments of bone present. No Charred macrofossils. ASSESSED AS POOR.
41	2298	FILL	-	+	-	-	-	-	NO	50% of the sample scanned. Charcoal and fragments of bone present. No Charred macrofossils. ASSESSED AS POOR.
42	2316	FILL	-	+	-	-	-	-	NO	100% of the sample scanned. Charcoal and fragments of bone present. No Charred macrofossils. ASSESSED AS POOR.
43	2248	FILL	-	+	-	-	-	-	NO	100% of the sample scanned. Charcoal and fragments of bone present. No Charred macrofossils. ASSESSED AS POOR.
60	2219	FILL	+	+	+	-	-	-	NO	100% of the sample scanned. Charcoal , fish bones and fragments of bone present. No Charred macrofossils. ASSESSED AS POOR.
66	4004	FILL	+	+	-	-	-	-	NO	100% of the sample scanned. Charcoal and fragments of bone present. No Charred macrofossils. ASSESSED AS POOR.
67	4009	FILL	-	+	-	-	-	-	NO	100% of the sample scanned. Charcoal and fragments of bone present. No Charred macrofossils. ASSESSED AS POOR.
70	4003	FILL	+	+	-	-	-	-	NO	100% of the sample scanned. Charcoal and fragments of bone present. No Charred macrofossils. ASSESSED AS POOR.
71	4010	FILL	-	+	-	-	-	-	NO	100% of the sample scanned. Charcoal and fragments of bone present. No Charred macrofossils. ASSESSED AS POOR.
72	1411	-	-	+	-	-	-	-	NO	100% of the sample scanned. Charcoal and fragments of bone present. No Charred macrofossils. ASSESSED AS POOR.
75	4033	FILL	+	+	-	-	-	-	NO	100% of the sample scanned. Charcoal and fragments of bone present. No Charred macrofossils. ASSESSED AS POOR.
77	4038	FILL	+	+	-	-	-	-	NO	100% of the sample scanned. Charcoal and fragments of bone present. No Charred macrofossils. ASSESSED AS POOR.
80	3039	FILL	+	+	-	-	-	-	NO	100% of the sample scanned. Charcoal and fragments of bone present. No Charred macrofossils. ASSESSED AS POOR.
89	3050	FILL	-	+	-	-	-	-	NO	100% of the sample scanned. Charcoal and fragments of bone present. No Charred macrofossils. ASSESSED AS POOR.
90	4046	FILL	+	+	-	-	-	-	NO	100% of the sample scanned. Charcoal and fragments of bone present. No Charred macrofossils. ASSESSED AS POOR.

DISCUSSION *by Chris Hewitson and Kristina Krawiec*

The site of Gorse Stacks lies outside the defences of the Roman fortress of Deva and as such is part of the Chester extra-mural settlement or *canabae legionis.* In recent years Peter Carrington and David Mason have undertaken extensive reviews of the excavated evidence from the *canabae* and from within the fortress (Mason and Frost 2001; Mason 2007; Carrington 2006). Extensive archaeological excavation elsewhere within Chester in recent years has furthered our understanding of the development of the *canabae legionis*, with the eastern extra-mural settlement (Hayes 2005) and the amphitheatre in particular being subject to intensive investigation (Garner 2009). The area north of the city walls has only now been accessible for excavation and the following chapter seeks to place the results of these excavations within the context of the development of the fortress and *canabae.*

The Fortress

The fortress at Chester is situated on a natural sandstone outcrop, which sits above the bowl-like Cheshire Plain although the sense of this is now a little lost due to modern development. The plain is crossed by a series of major rivers including the Dee, the Gowy and the Weaver that are thought to have formed the tribal boundaries of the pre-Roman Iron Age communities, the Deceangli and the Cornovii (Mason 2007). Although large-scale nucleated settlement has yet to be found in the region there is the indication of the intensification in landscape clearance from the Iron Age onwards through the analysis of the palaeoenvironmental record (Birks *et al* 1997, 152). It is assumed that some form of local agricultural infrastructure was present at the time of the Roman conquest that could have been exploited by an advancing army. How formal this structure was and how useful it would have been to the advancing army is unknown. The supposed dispersed

A BRIEF TABLE OF EVENTS SHOWS THE ACTIVITY IN AND AROUND THE FORTRESS DURING THE TIME PERIOD OF THE EVENTS AT DELAMERE STREET.

Date	Activity in fort	Delamere street	Event	Emperor
AD50/60s	Paulinus' campaigns Cheshire firmly in Roman territory			
AD70s	Cerialis's campaigns and probable 1st fort at Chester by *Legio II Adiutrix* Expansion of the province			Vespasian 69-79AD
AD 80s	Campaigns in Scotland takes some of legion away Elliptical building started	Stone quarrying possibly begins, Ditch excavated	1	Titus 79-81AD Domitian 81-96AD
AD90s	Change in Legions, *Legio XX* take over Some buildings replaced in stone Elliptical building unfinished and used to dump rubbish	Stone quarrying in western area, Site cleared of stone quarrying waste Rubbish from Event 3 is accumulating in the fort	1,2,3	Domitian 81-96AD Nerva 96-98AD Trajan 98-117AD
AD100	Curtain wall completed	Stone quarrying, Ditch excavated	3	Trajan 98-117AD
AD100-120	Barracks rebuilt in stone Curtain wall completed in stone Building work in *canabae* Amphitheatre built in stone	Stone extraction	3	Hadrian 117-138AD
AD120-160s	*Legio XX* moved to Hadrians's wall 120AD Fort neglected, some buildings derelict and burnt down Civilian pottery production takes over	Disturbed burials? Clearing of burials? refuse pitting small amount of quarrying areas B and E Rubbish deposition in the quarry pits Less stone extraction refuse pitting	3	Hadrian 117-138AD Antonius Pius 138-161AD Lucius Verus 161-169AD Marcus Aurelius 161-180AD
AD196-210	196AD battle between Albinus and Severus, after *Legio XX* sent back to Chester 210 AD Elliptical building cleared rebuilding at fort recommences including baths, barracks, *canabae* and officers quarters	Disturbed burials? Clearing of burials The rubbish deposited at site could occur prior to rebuilding as clearance from within places like the Elliptical building	3	Commodus 177-192AD Severus 193-211AD

settlement of the Cheshire plain would have made the job of controlling the local population relatively difficult but the level of clearance suggested by the palaeoenvironmental evidence is indicative of a potentially much larger local population than the excavated record has so far revealed (Carrington 2007, 19). What evidence there is for settlement at the site of the fortress is limited to a small number of features identified beneath the foundations of the Elliptical Building, which was constructed in AD79 (Mason 2001, 32) and ploughmarks recorded beneath the Amphitheatre (Garner 2009, Garner pers. comm. in Hayes 2006). As Chester itself has been extensively developed, accessing these early deposits is problematic but the discovery of pre-Roman occupation of the site should not be discounted.

The arrival of the Roman army in the region during Paulinus' campaigns of the AD50s and 60s and Agricola with his campaigns into Scotland in the AD80s emphasises the strategic importance of a fortress at Chester. Its position at the upper reaches of the tidal flow of the Dee, provided a natural port and stationing point for the movement of troops along the coast for campaigns into Scotland and Wales. It is thought that the AD 70s saw the installation of the first fortress at Chester. Although the precise date for this construction is still a matter for debate, a date of AD 72-75 has been postulated (Mason 2001, 45--6). The primary phase of fort construction was under the auspices of the *Legio II Adiutrix Pia Fidelis* (AD 74-90) which was recruited predominantly from Gallia Nabonensis, Spain, Italy and the early colonies on the Rhine and Danube frontiers (Mann 1983).

The structures and defences of the first fort would have been built from timber, wattle-and-daub and turf. The initial defences would have consisted of a double-turf revetted rampart and palisade with wooden guard towers stationed at regular intervals along its length. Although the barracks and other administrative buildings were initially constructed in timber the majority of the high-status buildings such as the first *principia, horrea* and the *thermae* were constructed from stone from the outset in order to minimise the risk of fire (Pearson 2006, 17; Mason 2001, 63, 46).

Although the construction of the fortress was begun by *Legio II* in the AD80s, when the order came for the legion to be redeployed to the Rhine and Danube frontiers this resulted in a decline in the numbers of soldiers available for completing the building work (Strickland and Ward 1978, 27). This event marked a shift in emphasis from campaigning to administration at Chester at this time. The emperor Domitian ordered *Legio XX* (*Valeria Victrix*) to be recalled from Scotland, where the fortress of Inchtuthil was still under construction, to garrison the fort at Chester in AD88. The arrival of the Twentieth signalled a renewed programme of building works which were to see the timber fortress, now in a state of disrepair, transformed into the stone-built headquarters of the Legion. The construction of the curtain wall in stone was certainly underway prior to the arrival of the Twentieth and its completion must have occurred in the opening decade of the second century (Mason 2001, 90). The rebuilding of the barrack blocks was also begun at this time.

An hiatus in activity is recorded around 120AD which is ascribed to the movement of the Twentieth north to assist in the building of Hadrian's Wall and the Antonine Wall. This would have left only a small garrison at Chester and this is supported by the evidence from other excavations within the fortress which show an episode of decline, indicated by a build-up of refuse, which continued until the first third of the third century (Carrington 2007, 20). The Twentieth were to return to Chester, eventually, as a much depleted force having suffered heavy losses, this would explain why the refurbishment of the dilapidated buildings did not occur immediately after their return (Mason 2001, 156). After this period of inactivity large parts of the fortress that had been left open were cleared of refuse and buildings that were derelict were also cleared. This episode of rubbish clearance can be seen at the Elliptical Building where the original foundations were not re-excavated but instead the surface of the site was cleared of at least a century's worth of refuse and a new set of foundations was laid out (Mason 2001, 173).

The complex history of the fortress is mirrored in the extra-mural settlement and the evolution of the two can be seen as closely connected. The fortunes of the inhabitants of the extra-mural settlement would have been intrinsically linked to the movement of troops to and from the fortress. The next section will detail events from in the *canabae.*

The Canabae Legionis

The development of the *canabae* almost immediately followed the installation of the fortress (Mason 1987, 143), and was located predominantly along the arterial routes to the south, west and east of the fortress (see Mason 1988, 175, Fig. 2). Dwellings were centred around the crossing of the River Dee along the road as it passed out of the fortress to the civilian settlement at Heronbridge. The construction of the early phases of the *canabae,* like the fortress, were mainly in timber and remained relatively unaltered during the opening years of the second century with only large communal structures like the amphitheatre rebuilt in stone. The change to more widespread stone construction in the *canabae* appears to have occurred in the early 2nd century following the arrival of the Twentieth legion (Mason 2001, 141).

Occupation east of the fortress was focused on the road leading to Northwich, Manchester and ultimately York. This included the parade ground identified in 1966 which dominated the area, and structural remains along the line of Foregate Street. It is thought that the eastern area was the main commercial sector with evidence for industrial processes and other manufacturing activities with some associated settlement (Mason 2001, 103). Excavations in 2005 on the City Road revealed structural evidence dating to the 1st and 2nd centuries AD at a distance of over 400m from the fortress suggesting that the eastern part of the *canabae* was larger than previously thought (Hayes 2005). Possible agricultural land was also revealed by a series of smaller evaluations beyond this (Scruby 2001; Constable & Parsons 2001; Poole 2004) and Hayes (2005, 129) has suggested that this may represent the rural fringes of the settlement.

The western arm of the *canabae* occupied a narrow strip of land between the fortress and the river Dee harbour. Excavations, carried out mainly in the 1960s and 1970s, suggested the area was densely occupied before the 2nd century with buildings of good-quality stone construction, hypocausts and its own *thermae*, the Watergate baths (Mason 2001, 109). It has been postulated that this represents the upper-class area of the *canabae* (Mason 1999:59). Finally, the southern arm of the *canabae,* which lay close to the access route to a stone quarry on the southern banks of the Dee, contained the *mansio*[1] and the amphitheatre as well as several other buildings of unknown function.

The Chester *canabae* appear to have been abandoned in the mid fourth century around 350AD which is marked by a decline in coinage and the abandonment of the *mansio* and amphitheatre (Mason 2007, 20). At other fortresses it is postulated that the population of the *canabae* moved inside the fortress after the legion had been moved, although there is no evidence as yet to suggest that this was the case at Chester.

Occupation of the Northern Canabae AD 70--140

The area excavated along Delamere Street lies in an area of Chester's *canabae* that has seen limited previous research as it was occupied by the bus station and as such our understanding of its form, function and purpose is poorly understood. The area investigated lies to the east of the *Via Decumana*, the principal north road which enters the fort through the *Porta Decumana*. In the past there has been the suggestion that the *Via Decumana* was lined with burials and funerary structures as was common practice throughout the Roman Empire and is further implied by the presence of tombstones used to repair the North Wall (Esmonde-Cleary 1987, 175). As has been discussed the south and east have long been established as areas of intensive settlement. The site at Delamere Street lies between these two extremes and the evidence uncovered during the excavations, of stone extraction and waste deposition provides valuable information for the evolution of Chester's Roman settlement.

The previous absence of evidence in this northern area is perhaps best reflected by the writing of David Mason, the leading expert on Roman Chester, who noted the dearth of evidence for occupation in excavations first carried out in the 1970s (Mason 1978, 30).Despite numerous further investigations undertaken by the Chester Archaeological Service and other commercially funded projects the situation had changed little by the early part of the present millennium (Mason 2001, 111). The Delamere Street excavations, as well as work to the west at Tower Wharf (Towle 2007) have begun to challenge the view the area was devoid of activity. However, the density of activity seen in other areas of the *canabae* is largely absent here.

The presence of a defined 'clear zone' that operated around the perimeter of the fort where buildings were not permitted in order to prevent enemies using them as cover has been a matter of speculation for a number of years (eg. Mason 1980). Outside the fortress walls lay a defensive ditch and patrol track, the latter examined at Linenhall Street between 1961 and 1964. Beyond, there are hints of a zone *c* 150 m wide with a distinctive pattern of development. To the south, west and east the 'clear zone' from the Watergate Street baths to Souter's Lane, was occupied by substantial official buildings (principally the parade ground, extramural baths and amphitheatre). It appears that high-status buildings were permitted to encroach upon the 'clear zone' whereas more domestic and utilitarian structures were kept further back from the walls to prevent them providing cover for approaching enemies and to provide a clear field of fire. The limited evidence for structures in the northern arm of the *canabae* is perhaps another reflection of this clear zone or it may simply be that this area was not considered suitable for human habitation (Mason 2001, 101). Carrington (2006) speculates that the area to the north was a less favourable area for settlement and manufacturing due to its distance from the river and port, and may have been the focus of less permanent activity. The proximity of the River Dee would have been essential for the water supply and may have precluded settlement on the highest parts of the escarpment.

But why would there be so little material associated with the later 2nd and 3rd centuries at this site? The spread of the northern *canabae* would surely have led to some deposition but there appears to be very little material dating to after *c* AD140 in the assemblage, suggesting the absence of occupation was deliberate. Truncation caused by later activity such as the presence of the quarry-pits and later, post-medieval stone extraction cannot be discounted. As we shall see (*Chapter 3 below*) the area was extensively cleared in the 17th century after the English Civil War campaigns. This may preclude identification of earlier structural features, particularly those of timber construction where remains would typically be ephemeral. This does not account for the absence of residual later 2nd and 3rd century artefacts from the assemblage, which although surviving are not found in comparable quantities to those from the earliest period of Roman occupation.

[1] The *mansio* is a coaching inn type structure which was used to accommodate visiting dignitaries.

The cemeteries

A third possibility for the lack of structures in the northern *canabae* is that the *Via Decumana* was lined with a Roman cemetery. The evidence for this has come in two forms. The first is the large number of tombstones and pieces of sculpture that had been incorporated as building material into the north wall of the later medieval city (Jones 1888, 1-10; Haverfield 1890, 32). It is assumed that the preference for these tombstones indicates the presence of a cemetery in close proximity, with the tombstones providing a convenient source of building stone. These predominantly dated to the 3rd century AD and are presumed to have been recovered from the cemeteries that lined the *Via Decumana* (this has been recently reassessed and a ritual function to their deposition suggested; see Clay 2004). Further evidence for burials in this area is a lead coffin dating to the late-3rd century which was discovered during building work in the 1984 just to the north-east of the Northgate roundabout (*c* 200 m north of the fortress wall; Mason 2001, 185). Also a group of cremation burials was recorded in the 19th century *c* 650m north of the fortress at Abbot's Grange, which are presumed to have lain alongside the road north from Chester down the Wirral (Margary 1955, 2, 35, route 670; Jermy 1961; Thompson 1965, 48, fig 11 suggests more precise locations). Again there were no associated tombstones although it is possible that memorials erected at Abbot's Grange are among those that found their way into the North Wall.

Elsewhere the discovery of an earlier cemetery at Infirmary Field, to the west of the fortress, revealed a total of forty-six graves, predominantly inhumation burials with evidence dating them to the second half of the 2nd century and the 3rd century AD (Newstead 1914, 4--60). Small numbers of burials occur further south, at the end of Black Friars, on the edge of the Roodee, at Nuns' Field, and on St Olave Street/49--51 Lower Bridge Street. Recent excavations adjacent to Tower Wharf (Towle 2007) have revealed a series of inhumations well away from the line of the road leading north from the *porta decumana.* This tends to suggest that the *Via Decumana* was not the only location chosen for the interment of the dead from the *canabae* and fortress. However it is probable that higher status individuals were selected for burial alongside this important approach road as is seen in other British towns in the Roman period (Esmonde-Cleary 1987, 175--6).

The presence of human bone within the fills of the quarry-pits at Delamere Street may indicate that some graves were disturbed when the quarry-pits were excavated as some remains were present in the lower rubble fills. The human bone appears in small amounts and suggests that it may have become incorporated into the waste fill of the quarry through the redepostion of disturbed material. Detailed examination of the bones highlights pathologies indicative of strenuous activity, with some indication that the individuals had muscular legs and trauma stress of the pelvis. The bones could be from legionaries but the evidence could equally suggest they are associated with enforced slave labour. The use of slave labour to quarry stone was not unusual in the Roman period (Pearson 2006, 64--5) and specific examples of the use of Jewish captives exist for the Roman quarries at *Mon Claudianus* in Egypt (Forbes 1963, 177).

The excavations at Delamere Street suggest that a cemetery was probably located elsewhere, outside the clear zone, directly adjacent to the *Via Decumana* or solely on the western and northern sides of the fortress in designated zones associated with the River Dee. The fact that much of the existing evidence dates to the 3rd century AD may suggest alternative burial practices were in existence during the earliest part of the occupation of the fortress.

The Quarries

The Roman quarries at Handbridge have long been recognised as a source of stone for Chester and are located where the Lower Keuper Sandstone outcrops in cliffs either side of the River Dee. Associated with these quarries was a statue of Minerva set into a recess in the quarry face. It is suggested that stone has been worked at Handbridge since at least *c* AD 100 as excavations surrounding the statue recovered pottery of this date (Thompson 1965, 62--3). The extent of the quarrying in the readily available cliff faces would suggest that it was more than a short-lived episode, and that the stone was quarried for export to elsewhere in the hinterland including the fortress at Caernarfon (*Segontium*), North Wales. However the geological provenance of the building material used at Caernarfon to Chester itself is refuted (Pearson 2006, 140--1) and it may have been quarried from the wider area. Nonetheless it certainly implies that the use of this outcrop of Lower Keuper Sandstone was recognised as important from an early date.

The discovery of the quarry-pits at Delamere Street opens up new possibilities – namely that the Romans adopted a more sporadic approach to stone extraction that included exploiting the shallow beds of stone along the northern ridge around the fort and almost certainly involved the re-use of stone quarried from within the fortress. The quarry pits excavated probably represent only a percentage of those that were exploited during the Roman period within the boundary of the Delamere Street site as the later post-medieval stone extraction had truncated these earlier remains. The further stone extraction in the area is known to exist to the east where further excavation work revealed similar activity to that presented in this publication (Dodd *pers comm*).

Two types of quarry-pit were recorded at Delamere Street – the smaller borrow pits less than *c* 2m in diameter and large quarry-pits upwards of *c* 5m in diameter. The size of the former would suggest that these were small-scale pits associated with the stone extraction being incidental to the primary function of the pit. It has not been proven through the artefacts recovered or the environmental analyses exactly what that primary function may have been. The larger pits are a clearer sign of deliberate and organised extraction of stone for building purposes.

Examination of the pit walls suggests that the material was extracted along natural beds, with surfaces created along the fault-lines. Distinct tool marks are characteristic of the Roman quarry at Delamare Street and are indicative of 'scabbling' (excavation of trenches through the rock) of blocks and insertion of wedges to remove the stone from natural fault lines (Pearson 2006, 59-61). In addition to this a single instance of an abandoned pick-cut slot in the bedrock (Area A; Plate 2.14) was encountered. The depth of the pits, at less than *c* 2-3m, strongly suggests that the workmen of the Roman period only encountered the headrock[2] and the stone was not used for decorative or monumental blocks. The stone removed probably served as rubble for foundation courses or the core of walls and the irregular shape and profile of the pits supports this.

The quality of the stone visible in the Roman quarry-pits varied though overall it was weathered and extremely weak and therefore unsuitable for decorative or facing blocks. The situation of the Delamere Street quarries can be compared to those at Handbridge which have the principal advantage of having exposed cliff faces, thus allowing access to the lower beds of higher quality stone. Pearson (2006) notes that there is no such thing as good or bad building material, as stone can be adapted to various aspects of construction. The majority of Roman stonework was constructed on the principal of ashlar facing with a rubble and mortar core (2006, 74). This means that poorer quality stone could be extensively used in construction. Indeed not all stone would have been quarried from a single large pit, many smaller workings (borrow pits) would also have been undertaken on an *ad hoc* basis (*ibid*, 52).

Although the stone extraction cannot be conclusively dated to the arrival of the Legion XX in AD 88 and the resultant reconstruction in stone (*see above*) it can be put forward as a tentative possible early date for the stone removal. This also closely corresponds with suggested dates for the initial construction of the stone curtain wall that was added to the turf and timber rampart (Mason 2001, 90).

The curtain wall was constructed in the style *opus quadratum*, large unmortared blocks laid in regular courses with smaller blocks (*opus vittatum*) laid with a rubble core (*opus caementicium*) being employed for corner and interval towers (Mason 2001, 89). This would appear to preclude the use of the poorer quality 'ragstone' available in the surface deposits excavated on the Delamere Street site for much of the construction work. The close proximity of the stone would perhaps have meant that it was adopted due to its accessibility. In this case the close location of the pits to the fortress defences would suggest that they could have been rapidly backfilled after extraction to prevent their use as strategic cover for attacking forces. This rapid clearance of waste stone is indicated by the absence of more slowly accumulating primary fills in the large quarry-pits.

Estimates of the building stone requirement for the Flavian baths suggest *c* 33,000m³ (42,000 tonnes) were required for the baths alone (Mason 1999, 91-2) with additional material required for the reconstruction of the other buildings at the end of the 1st century AD. Another estimate lies in the reconstruction of the Severan fortress (*c* AD 220-30) that suggest potential volumes of stone at *c* 125,000m³ (around 300,000 tonnes) for the buildings within the fortress alone with a further *c* 60,000 tonnes for the construction of the defences (*ibid.*, 94-5). These volumes would suggest multiple sources for the stone but the logistics of transport from the Handbridge quarry would support the theory that reconstruction of the buildings within the northern sectors of the fortress utilised locally available sources of stone – even if this was of a lower quality.

Evidence from various locations within the fortress has suggested that an hiatus in building activity occurred from the period AD 120-160 associated with the posting of Legion XX to Hadrian's Wall (eg. Northgate Brewery site; Ward & Strickland 1978, 27). This may suggest that after an initial period of stone extraction, quarrying was much reduced as demand for stone declined.

The material remains – fortress or canabae?

The excavation at Delamere Street clearly represents stone quarrying activity specifically at the site but the subsequent reuse of the quarry pits to deposit refuse provides a greater understanding of the development from within the fortress at Chester. The pits were infilled with deposits of domestic refuse which represent one of the largest artefactual assemblages from a single site in Chester. The narrow date range for the accumulation of refuse, from the late-1st century to the mid-2nd century AD (*c* 90 to 160), provides a valuable insight into the economy of Chester at this time. The significance of the site perhaps lies not in the activity taking place on the site (ie stone extraction) but in the information the finds assemblage provides about the day-to-day domestic activities in and around the fortress.

One of the key questions lies in the origin of the refuse material. The assemblage has both military and domestic characteristics and it is open to debate whether it is from the *canabae* or brought from within the fortress. The array of items present in the assemblage is indicative of a range of activities and has associations with the full spectrum of social classes. The inclusion of some imported material, in the form of pottery and brooches, along with possible local products as well as items that would have been produced in the fortress may indicate that this assemblage originated from several sources.

The inference from the pottery is that the samian (*see Wild*) is of sufficiently high quality that it would have

[2] The material encountered in the initial few metres below the surface that has been subject to freeze/ thaw action and subsequently fractured. It was usually only suitable for rubble construction.

graced an officer's table. Cool's analysis of the small finds notes the even distribution of household and military items and suggests that the assemblage may therefore be derived from within the fortress but not exclusively from the officer class. Certainly a number of the brooches are types common in the West Midlands or Severn Valley region and would therefore tie closely with the presence of Legion XX at the fortress. This contrasts with the beaded torc, a distinctly Romano-British item not normally associated with military contexts. In simple terms the finds suggest the complex blend of military and civilian populations associated with the fortress and *canabae* even at this early stage in Chester's development. Fragments of military equipment in particular *lorica segmentata* were distributed in a number of the pits and this shows that some elements of the Roman military uniform were becoming combined with the waste.

Baxter's analysis of the animal remains compares the high proportion of cattle bones, comparable with Nijmegen in the Netherlands, as suggesting that some of the waste derived from the *canabae* (*see Chapter 8*). Although this cannot be conclusively demonstrated, given that the northern area may have been part of a clear zone around the exterior of the fortress, military derived waste disposal would be more likely to have been permitted than domestic disposal from the *canabae*. Added to this, that the waste appeared to be consistent with secondary or tertiary deposition and which would imply the material was re-deposited from its original midden context.

Some of the refuse material could be associated with the clearance of buildings, which is supported by the presence of iron nails, during the rebuilding of the fort in stone after the arrival of Legion XX. The tile is also a fairly mixed assemblage originating from a range of different building types with no particular bias towards a specific structure (*see Heke*). This again suggests some areas of the fortress were demolished wholesale and the rubble was collected and deposited along with other types of waste. Some of the brick and tile still had mortar adhering to them along with some evidence of burning, pointing to the assemblage coming from demolished buildings rather than from production processes. Although small in comparison with other tile forms the evidence of *tubuli* (box tiles); *parietales* (facing tiles) and *tegulae hamatae* (half-box tiles) suggest centrally-heated rooms and a high-status origin.

The fact that the Elliptical Building became an open rubbish dump during the hiatus in building activity may help to elucidate the source of the material at Delamere Street. The types of waste purported to have been dumped there ranged from food waste, metalworking debris and broken pottery and tile. This diverse range of material suggests that the Elliptical Building was used to deposit refuse from a number of sources, which may be what is reflected in the assemblage recovered from Delamere Street. As the rebuilding of the fortress was an ongoing process it is possible that the empty quarry-pits were a convenient place to dump refuse from the areas in the fortress undergoing rebuilding. The excavations carried out at the site of the barrack block inside the north wall showed signs of several phases of rebuilding in both timber and stone and some of the material dumped at Delamere Street could also have originated from the clearance of these old structures (Ward and Strickland 1978, 27).

The final timber phase ended in *c* 120 AD and was replaced by several phases in stone between 120-160 AD, the Elliptical Building lay derelict through this time and was cleared sometime around 210 AD although the material that built up on the site would have predated this by at least a hundred years. It is therefore suggested that the material recovered from Delamare Street may in fact derive from the Elliptical Building rubbish dump and as such represents the by-products of a range of activities from the heart of the fortress. This includes domestic waste, food waste, the by-products of metalworking and repair, animal bone working etc. The material deposited suggests an intermingling of waste which is precisely what occurred in the middening at the derelict site of the Elliptical Building. It would make sense for the disused quarry pits to be back-filled with the waste from within the fortress when the rebuilding was required. It would seem unlikely that the people of the *canabae* would have been removing such large quantities of waste to an area at the top of the escarpment as it would have been easier and more practical to deposit their waste in close proximity to the settlement. But it is apparent that domestic waste from civilian and military contextw were present at Delamere Street, suggesting a more complex pattern of waste deposition. The way in which the deposits formed at Delamere Street does not suggest discrete phases where the rubbish was laid down in layers, instead the contexts appear to be single episodes of large volume deposition which fits the theory of clearance of middens.

It is tempting to suggest that the activity at Delamere Street could be directly linked to the clearance of the Elliptical Building. One can only imagine the immensely unpleasant task of clearing a hundred years worth of refuse by hand and transporting it to the pits outside the fortress walls.

The Populace of Deva

What, if anything, do the material remains recovered tell us of the people of the fortress and *canabae*? The short time span of deposition means that the material evidence can be tied closely to the historical evidence. For the period of the late-1st to early 2nd century AD a clear policy of recruitment to the army was in place (for an overview see Carrington 2006; Mann 1983). A body of soldiers in *Legio II* came from Aprus in Thrace as part of the Ravenna fleet, but otherwise in common with much of Britain they came from Gallia Narbonensis, Spain and Italy. Towards the later 2nd and 3rd centuries AD a transition occurred towards local recruitment, especially from families in the *canabae (ibid)*. The majority of the evidence for the cultural make-up of the population comes from epigraphic study and has been extensively reviewed recently (see Dobson & Mann

1973; Mann 1983; Le Bohec 2000, 75-82, Saller & Shaw 1984; Carrington 2006). The question to be asked of the material remains presented here is: do they reflect the patterns in the epigraphic evidence?

From the ceramic assemblage were four graffitti *sherds (see Tomlin)* which represent the only evidence for writing from the site and of these one represents an illiterate mark (*X*) whilst another appears to be a numeral (*ii)*. The other two sherds bear partial names and appear to be associated with legionaries, in particular sherd G4 having the probable name *G(aius) V[al(erius) G(ai) f(ilius)] Pud[ens] Maceia (tribu.* The nomen *Valeriu*s reflects origins in Italy or *Gallia Narbonensis*. The predominance of Gallic imports in the samian *(see Wild*), is typical of a British ceramic assemblage of this period.

Of the personal items recovered the brooches appear to suggest native British origins for some of the legion particularly from the West Midlands/ Severn Valley area which may reflect the movement of the *Legio XX* from Wroxeter in AD 88 (*see Cool)*. The presence of the single beaded torc is interesting as it opens the possibility that even at this early stage of the 2nd century AD that new recruits to the military were derived from the native population. However this should be interpreted cautiously as the torc may not have belonged to a legionary and as Cool has suggested may represent a curio or the spoils of war. Perhaps these items do suggest that affiliations between the legionaries and the native population had begun as early as the late 1st century AD, but it is not possible to establish whether these relationships were in terms of association with native members of the *canabae*, wives or actual native recruits. The possibility that this item represents relationships between legionaries and women is furthered by the presence of glass beads and an earring. However, the absence of bone hair-pins, a ubiquitous object that is often used to represent women in the archaeological record during the Roman period, implies that women were not a strong presence within the fortress whence this refuse material likely derives. During the 25 years of service for legionaries marriage was prevented from the time of Augustus and not revoked until Septimius Severus in AD 197 (Carrington 2006). These items within a military context may reflect tokens of affiliation or affection or as they are broken, symbols of loss.

Economy and Trade

The ceramic evidence is also useful in terms of defining patterns of pottery production and the nature of the economic relationship between the fortress/*canabae* and the *prata legionis*[3]. The mix of fabrics and styles does not point to any one local source but may reflect common styles from Holt and Wilderspool and as yet undiscovered kiln sites from within the *canabae* (*see Timby)*. The presence of waster sherds contributes to a growing body of evidence first recognised in the salvage recording associated with the inner ring road (Carrington and Ward 1982; Swan 1984) that suggests the western or northern *canabae* may have been associated with pottery production at an, as yet, undiscovered site.

The ceramic building material recovered from the site was undoubtedly locally produced and fabric types strongly reflect those of the Holt industry (*see Heke*). Given the close resemblance in fabric types and the potential economic cost of transport from Holt at *c* 10km, localised production of this industry would seem probable. Mason's estimates for the later Severan fortress construction suggest *c* 930,000 *tegulae* alone would have been required (Mason 1999, 96). This is a vast quantity of material which may help to support the case for local tile production within the western and northern *canabae*.

Further evidence for the production of items within and around the fortress includes industrial deposits associated with metalworking. Cool (*above*) suggests that much of this waste may be associated with a *fabrica,* or workshop within the fortress. The slag and metallurgical residues that were recovered are indicative of smithing which further supports this theory. Fragments of military equipment, in particular *lorica segmentata,* were recovered from a number of the pits and this suggests that some of the more consumable parts of Roman armour were becoming combined with the smithing waste. Of particular interest was the evidence for copper alloy working (*see Chapter 6*) which although slight, correlates with the small finds evidence in suggesting that armour fastenings were changing from the Corbridge style cuirass (*fe*) to the Newstead style cuirass (*cu* alloy) and reinforces the suggestion that at least some of the waste is derived from a *fabrica*. Mason places the workshops on the western side of the fortress close to the Elliptical Building (Mason 2001, 129, fig. 76). Elsewhere within the fortress metalworking debris was recovered from the excavations along Eastgate which included crucible fragments and smelting waste (Lloyd Morgan in Mathews *et al* 1995, 47). There were also crucible fragments recovered from Deanery Fields (Newstead 1928, 11) and Princess Street (Newstead & Droop 1939, 7) from the 1st century AD. This may provide potential locations for metalworking but it is important to realise the general prevalence of slag and metallurgical residues as well as small metal artefacts generally within the Roman material record. In comparison, the northern area of the fort at Wroxeter was primarily used for industrial processes rather than occupation (Ellis and White 2006, 160) and the presence of locations within the *canabae* for metalworking should not be discounted as recent excavations within the western *canabae* at City Road suggest (Hayes 2005, 126).

As well as the production of personal items and armour the nature of the subsistence for the fort is also well illustrated by the Delamere Street assemblage. The faunal evidence shows a bias towards cattle which is in keeping with other legionary sites in Britain as well as further afield in Gaul and Germany (Carrington 2006, 17). The assemblage

[3] The hinterland estimated to be around 500km^2 (after Mason 1988; Carrington 2006).

also showed that pork was being consumed at the fort and as it was considered a high status meat it probably came from the table of the senior officers (King 1978). This is consistent with previous excavations conducted within the fortress, however as Baxter (*above*) points out, there is a noticeable disparity in the selection of cattle bones at Delamere Street in comparison with the other assemblages recovered from the site. The butchery waste from Delamere Street is both primary (conducted during initial slaughter) and tertiary (conducted to extract marrow fat etc.) suggesting possible semi-industrial processing of carcasses. The evidence points to both slaughter waste and processed carcasses (perhaps from stewing or stocking bones to remove marrow and gelatine). This contradicts the glassware assemblage (*Cool above*) which is of a much more utilitarian nature and combined with the presence of portable ovens (*Clibani*) suggests that some of the waste is derived from kitchens. Heke (*above*) points to the close proximity of the cook houses on the Abbey Green site as a possible source (McPeake *et al* 1980, 19).

A small proportion of the animal bone assemblage contains wild animal remains indicating that the diet within the area was supplemented by hunting. Intriguingly pit [5007] from Phase 1c contained a wider range of species including domestic fowl, goose, roe deer and flat fish, whilst a decline was noticed in pig remains. The evidence is too small to draw broad conclusions but may suggest that by the mid-2nd century AD, the absence of the Legion on campaign led to a more varied diet drawn from both game and cattle.

It may be possible that some of the area to the north was used as pasture for cattle and pigs as the presence of pig neonate remains within the faunal assemblage suggests that they were kept close-by as opposed to imported for market/ slaughter (*see Baxter above*). This raises the intriguing possibility of a long-standing and continual use of the area around Delamere Street for stocking animals prior to slaughter which extended into the medieval, post-medieval and early modern periods.

Conclusion

The excavations at Delamere Street have presented a wealth of data concerning many facets of life within the fortress during the 2nd century. The small finds not only provide glimpses of the ethnic make-up of the *Legio XX* but of their possible relationships with the native and civilian populations of the extra-mural settlement (*see Cool above*). In terms of the ceramic evidence, Chester is comparable with other sites in western Britain, including Gloucester, Wroxeter and Carlisle with localised civilian enterprises closely associated with the military. The site also contrasts with the situation at similar fortresses in south-east Britain where there was a greater proportion of continental imports. The frontier position of Chester was probably partly responsible for this as supply routes would have been less reliable than those in the south and thus the reliance on trade and interaction with the local population was more keenly felt. The whetstones of tentative Devonian and Welsh origin which were recovered from the pit fills fit well with existing understandings of trade routes around the western fringes of Britain. Overall much of the assemblage reflects the isolation of the western frontier but also the communication links between the outposts that lined it during the later 1st and 2nd centuries AD.

The theory presented for the nature of the deposition of this material is one related directly to the clearance of the Elliptical building in the centre of the fortress. Its use as a rubbish dump through much of the second century, with waste from the cookhouses, *fabrica* and officer's dining table becoming incorporated into what was essentially an open landfill. The removal of this waste in a single episode fits well with the large volumes of material dumped into the pits outside the walls of the fortress and explains the varied and relatively unabraded nature of the remains. The excavations at Delamere Street have given new insights into the daily lives of the legionaries stationed at *Deva*. The assemblage recovered is the largest ever published from a single excavation in Chester and it is hoped will be used to understand sites excavated in the future.

SECTION 3 – THE POST-MEDIEVAL AND EARLY MODERN QUARRY

This section discusses the post-medieval and early modern phase of the quarry but includes evidence of the residual medieval material recovered during work. The evidence discussed includes the evaluation of 2002, the excavations in 2006 and the watching brief on the removal of the quarry deposits in 2008.

The Excavations

by Kristina Krawiec, Sam Hepburn and Chris Hewitson

The following phases have been identified and will be discussed during the course of this chapter:

- Phase 2: 16th century quarry
- Phase 3: 17th and 18th century deposition.
- Phase 4: 19th and 20th century buildings.
  - Phase 4a: mid-19th century sandstone buildings groups 107, 108,109 & 111.
  - Phase 4b: late-19th century Methodist chapel and Delamere Street dwellings.
- Phase 5: 20th century deposition of material.

At its greatest extent the sandstone quarry [5063] measured 95m (E-W) x 50m (N-S) x 12.70m (deep). In the last phase of the quarry's working life it was rectangular in shape measuring 62m (E-W) x 41m (N-S) (Figs. 3.1 and 3.2). The decision was taken to divide this area into Quadrants A-D for the purpose of the watching brief (Fig. 3.1), with two smaller extensions leading off it, one to the southwest, Quadrant E (Plate 3.1) 10.80m (E-W) x 20m (N-S), the other to the northeast Quadrant F 22.80m (E-W) x 24m (N-S) (Figs. 3.1 and 3.3)

Phase 2 - 16th and 17th Century Quarry

The presence of a quarry at the site can be traced back to the 16th century through cartographic and historic sources. However, the exact extent and development of the early quarry is unclear. It is difficult to see evidence for earlier work in the quarry faces and none of the deposits uncovered in the quarry date to this period. There was evidence of medieval and early post-medieval artefacts within the deposits (*see below*) but these probably represent residual material or reuse. The quarry dominated the middle of the site by the 19th century and it is probable that the longevity of the quarry has resulted in the loss of the earliest evidence for extraction.

Only one feature conclusively displayed evidence for the early post-medieval period. Pit [2204], a Phase 1 quarry/waste pit was reused in the 17th century [2321]. The fill (2205) of the reused pit produced a minimum of three yellow ware vessels, which bore comparisons with those from a Civil War deposits from Beeston Castle (*see Dodd*

PLATE 3.1
QUADRANT E
SOUTH-WEST
FACING

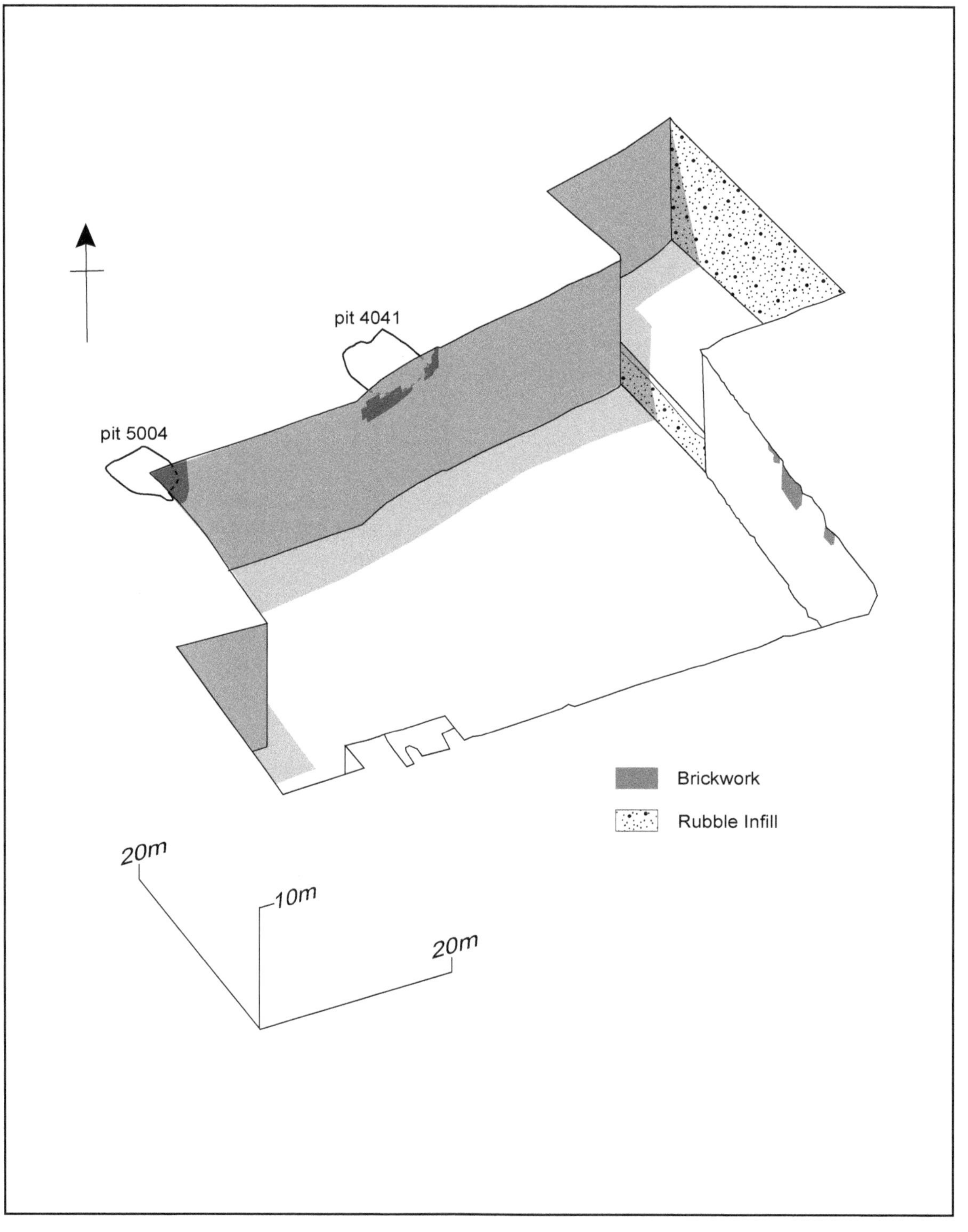

FIG.3.1 3-DIMENSIONAL IMAGE OF THE QUARRY

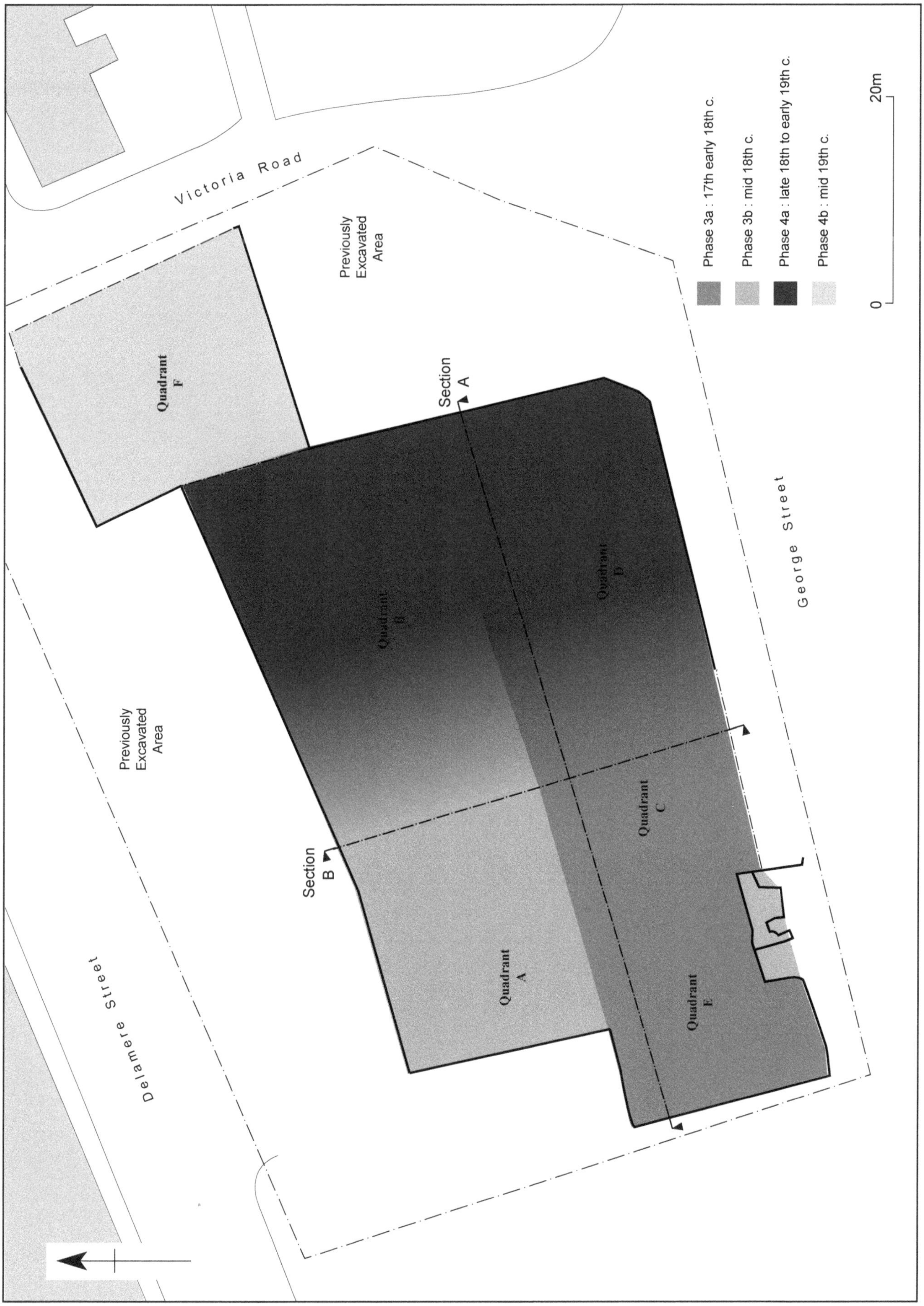

FIG 3.2 PHASING OF THE QUARRY

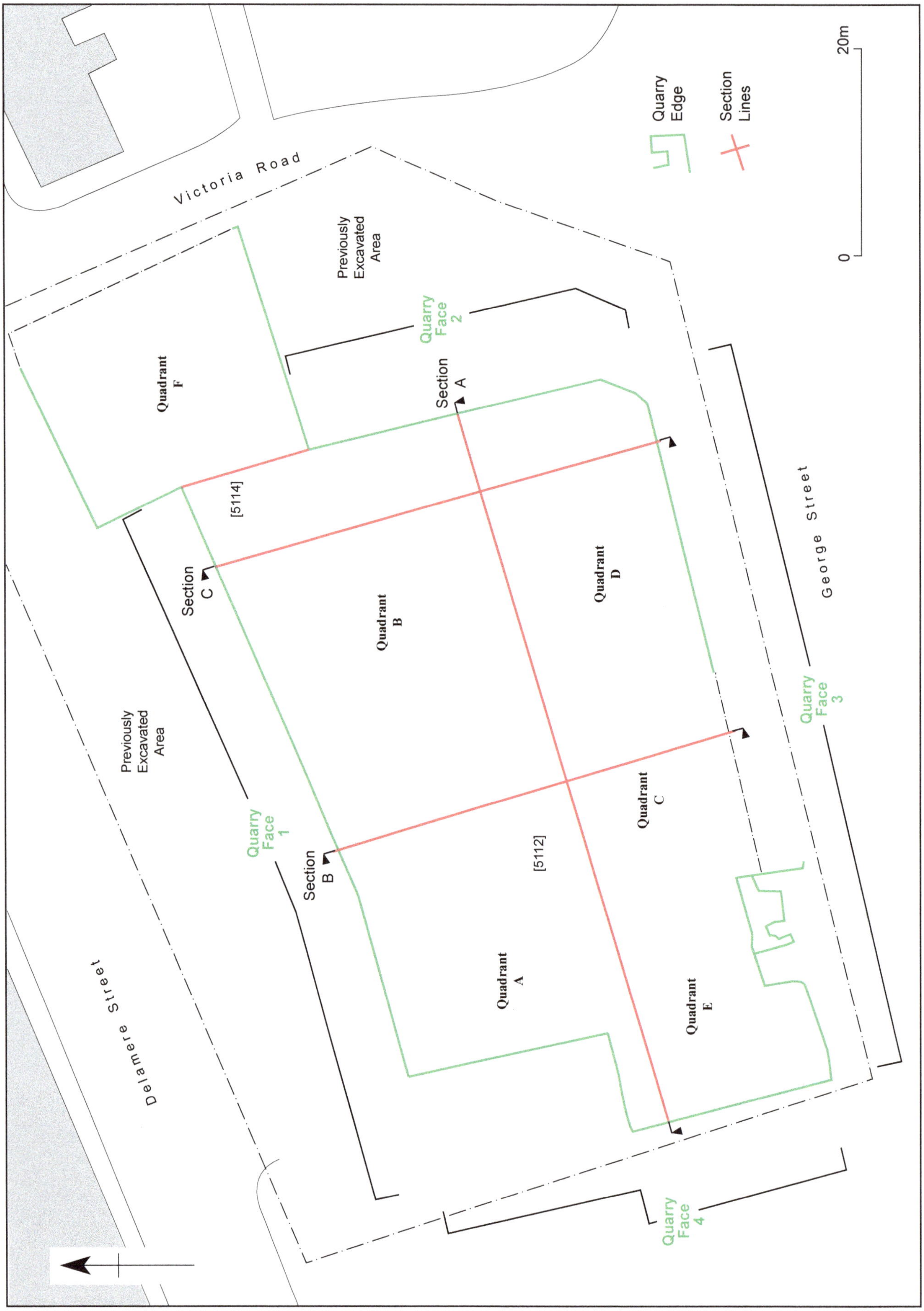

FIG 3.3 LOCATION OF SECTIONS AND PROFILES

in Pottery section). The Civil War association may point to the extreme nature of clearances north of the walls and may explain the absence of evidence for the earlier medieval periods.

Phase 3a: The Quarry by 1745

The extent of the quarry according to De Lavaeux's Plan of 1745 (Fig. 3.4), suggest it had expanded to dominate the centre of the site by this date. The exploitation of the red Lower Triassic sandstones, which is of the Sherwood Sandstone Group, at this site utilised the natural bedding planes. The beds of the rock sloped at an angle of 5-10° from south-west to north-east and were near horizontal in places. The beds varied in depth between 0.3 to *c.* 2m. The initial *c.* 1m below the ground surface was of poor quality stone and easily friable. It is likely this material was used for 'rag' stone (building rubble). Weathering appeared to be visible on some of the upper portions of the quarry face suggesting that exposure had occurred for some time, most notably in the northern face of the quarry in Quadrant A. This would appear to have remained open from the 18th into the 19th century and this may explain the friable quality of the upper 1-2m of the northern quarry face. The underlying deposits (c. 1-5m) were the finest quality, which lay in even beds of between 1 and 2m thick and were clearly the primary focus of extraction. South facing and west facing profiles are shown is Figures 3.5 and 3.6 and discussed below.

The exact depth of the quarry in the 18th century is unclear. The presence of sealed deposits securely dated to the period 1745 – *c.* 1760 (5214, see below) suggests that the quarry had been excavated to its maximum depth by this period. A small test trench was excavated below the level of layer 5214 to find the base of the quarry (not illustrated). The earliest quarry fill comprised a crushed sandstone (5231) deposit overlain by pale grey silt with sandstone inclusions (5230), and sealed by a further layer of crushed sandstone (5229), 3.9m deep in total. These deposits appear to represent a deep layer of quarry waste resultant from the extraction process. The nature of this deposition is unclear but it may have been the result of rapid infilling or periodic disuse of this area of the quarry allowing silt (5230) to accumulate before extraction resumed. No finds or other artefacts were recovered during the excavation of the test pit.

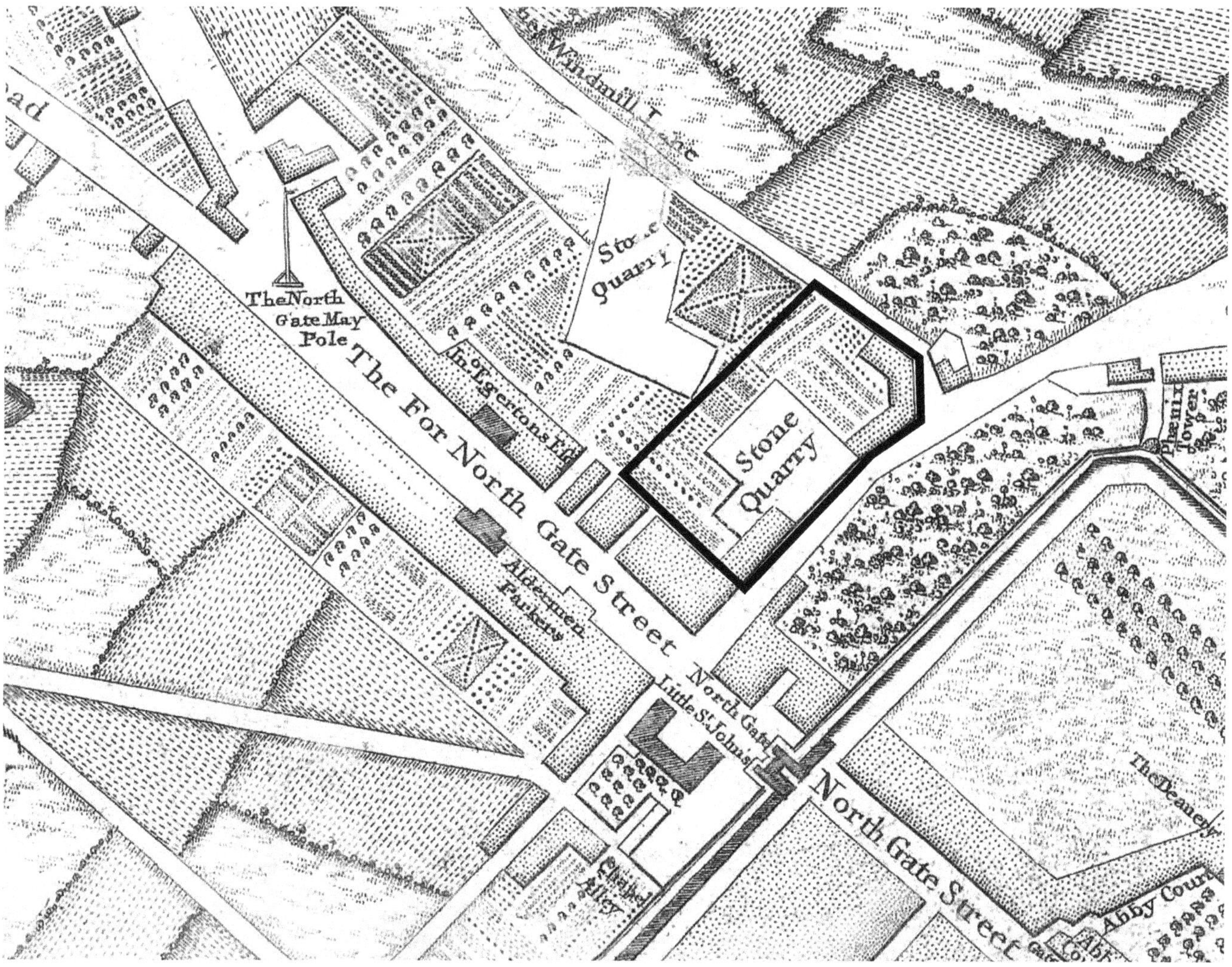

FIG. 3.4 DE LAVAEUX'S MAP 1745

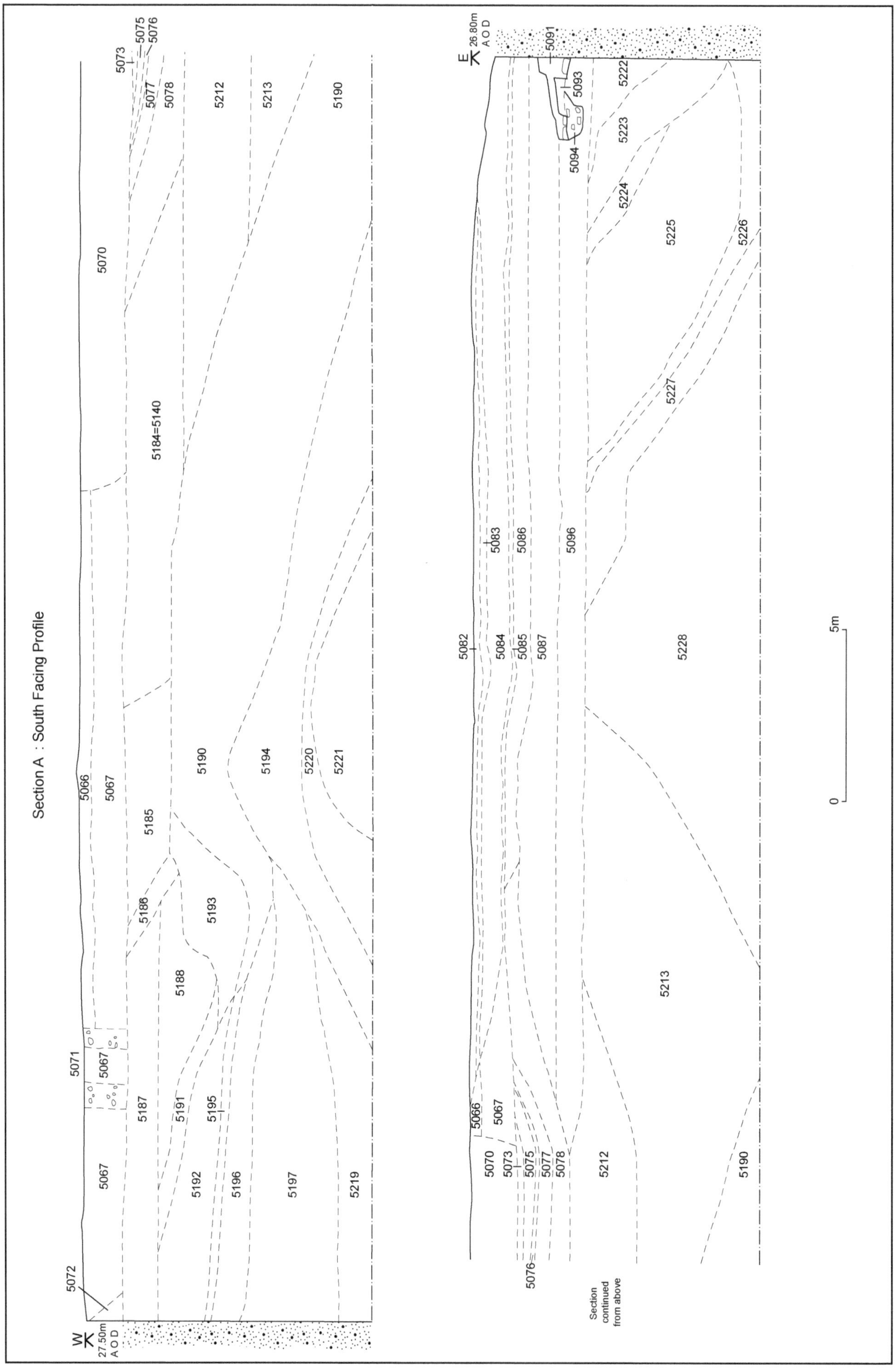

FIG. 3.5 SOUTH FACING PROFILE OF DEPOSITS

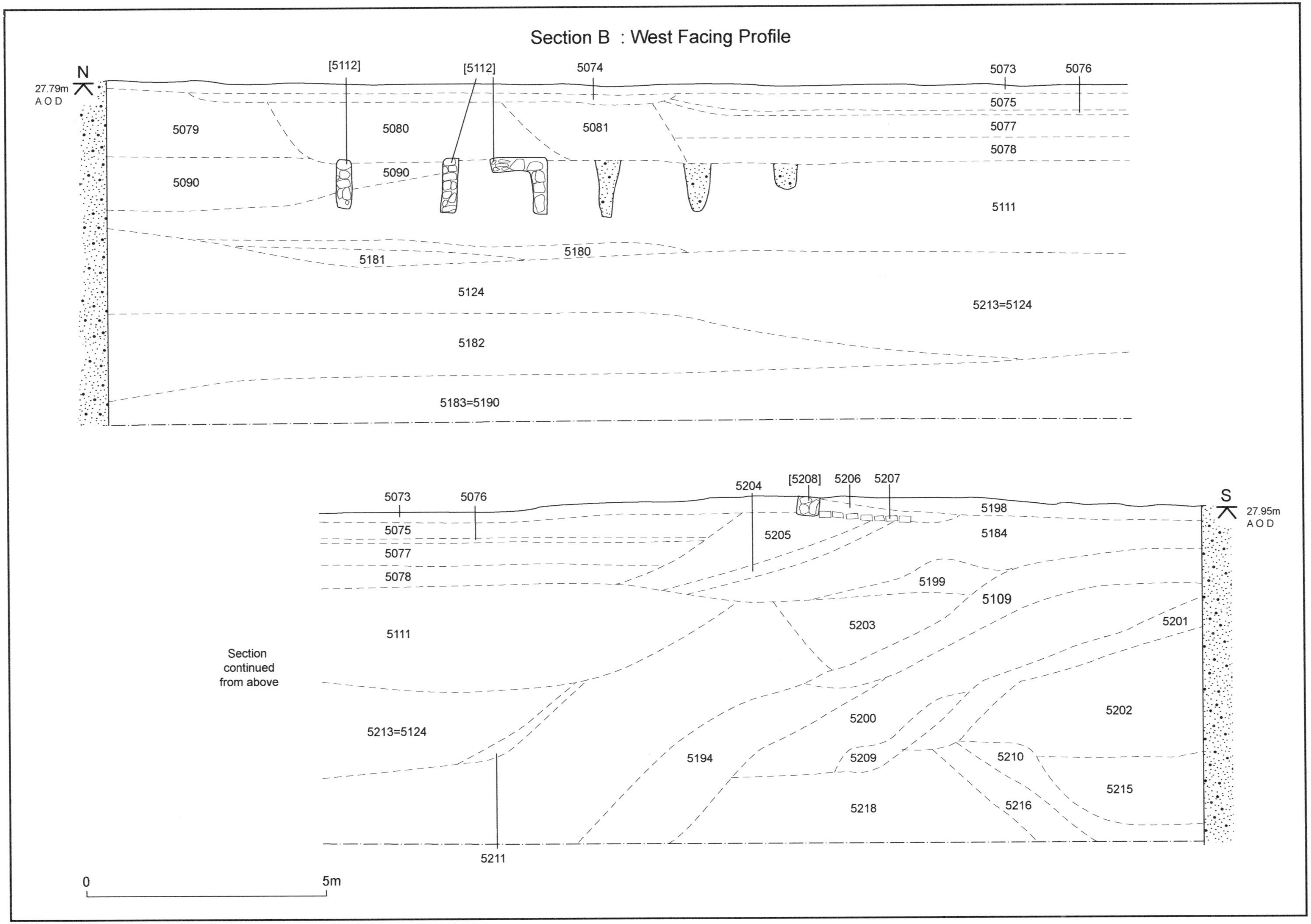

FIG. 3.6 WEST FACING PROFILE OF DEPOSITS

Phase 3b: The Quarry in the mid- to late-18th century

Deposition within the quarry appears to have occurred as a sequence of dumping episodes which resulted in rapid infilling followed by periods of levelling with quarry waste. Pottery from these dumping episodes appears to suggest a relatively confined period of deposition between the 1700s and 1760s. The earliest periods of deposition were found in Quadrants C and E, recorded from the formation level (18.60m AOD). Most of the deposits were demonstrated in section with the lines of deposition generally sloping from southwest to northeast suggesting a general process of dumping from the George Street. There were also discrete dumps of material against the quarry face which did not appear in section and are therefore not illustrated. The earliest deposit was mid-grey sand silt clay (5214; 19.62m AOD; Plate 3.1; not illustrated in section) which was situated up against the southeast corner of Quadrant E. This deposit contained 18th-century tin-glazed pottery, Staffordshire slipware including a thrown and trailed dish, clay tobacco pipe from the first quarter of the 18th century, glass onion bottle fragments, animal bone predominantly of horse (*Equus caballus*) and cattle (*Bos* f. domestic) and a quantity of disarticulated human remains in particular an ulna, fibia, tibia and a near complete frontal bone from the crania. Radiocarbon dates from the femur produced a range Cal AD 1160 to 1280. The context also contained residual floor tile with a *fleur de lys* pattern and with nail holes dated to the 14th or 15th century.

The next deposit in the stratigraphic sequence was a mid grey brown compact silt sand (5221, Fig 2.35) which in turn was overlain by (5220) a layer of red crushed sandstone. Overlying this was a deposit of mid-grey silt clay (5194; 20.82m AOD); medium to large sandstone blocks quarry waste mixed with mid-grey silt sand (5219; 21.82m AOD), and a demolition deposit of broken bricks and mortar (5197; 22.32m AOD), mixed with large quantities of 18th-century pottery, complete glass onion bottles, animal bone, 17th-century clay tobacco pipe, and a few residual pieces Roman glass and pottery. This was overlain by mid grey-brown sand silt clay (5196; 23.07m AOD), with brick and stone inclusions, 18th-century pottery, glass and animal bone; a narrow band of crushed mortar and brick (5195). This was then overlain by a 2.1m thick layer of crushed sandstone (5190). The This was then sealed by a thin layer of mid brown silt sand (5192) which in turn was overlain by a brick and stone dump, containing 18th-century pottery (5193; 24.02m AOD) equivalent to layer (5179).

The upper deposits of the western half of Quadrant E were a series of smaller dumps of material (5191-5185). The remains of a wall foundation (5071) were cut through these deposits and relate to the Phase 5 bus station structure.

The pattern of deposition in Quadrant C was similar to that in Quadrant E, sloping in a general orientation southwest to northeast away from George Street. Initial deposits appear to be crushed red sandstone quarry waste deposited in irregular dumps (5218; 22.45m AOD: Fig 3.6). On the southern side these deposits were overlain by a dark-brown silt-sand (5216) which in turn were overlain by a series of deposits containing red sandstone blocks mixed with animal bone, pottery and glass, suggestive of a dump of butchery waste (5210, 5215 and 5202; 20.95m AOD). These were then overlain by mixed fills of quarry waste, silt and brick (5201, 5209 and 5200). A series of tip dumps, sloping southwest to northeast overlay this of ash, charcoal and mixed silt deposits (5194, 5109 and 5203 24.92m AOD). These were then sealed by crushed sandstone quarry waste 2.1m thick (5190) that extended across much of the area and appears to represent a cessation in rubbish disposal.

The layers show distinct periods of dumping of waste and then similar tip-lines of quarry waste suggesting a continual process of deposition. The direction of slope recorded in the layers indicates that waste was being tipped from George Street and the quarry may have been used as a dump for waste from elsewhere in the city. Many of these deposits contain demolition rubble indicating a period of redevelopment in the surrounding area.

Phase 4a: The Quarry in the late-18th to early 19th century

The later phases of the quarry were located by the excavation, the watching brief and Test pit 4 of the evaluation. The main body of the quarry expanded to the west in the remainder of the 18th and 19th centuries. The physical evidence of sandstone removal suggests that this was the most productive period of the quarry and corresponds with the expansion of the City of Chester after the arrival of the canal. Exhaustion of the quarry quickly resulted in backfilling and a series of deposits which were laid down in the 19th century in the anticipation of redevelopment of the quarry area.

The western side of the quarry (Fig. 3.6) was levelled by a series of deposits including a dark-brown silt sand containing pottery and bone (5182; 22.60m AOD, Fig 3.6) that corresponded to a similar large dump in the quarry centre (5228, not illustrated). This was overlain by a thick layer of crushed red sandstone quarry waste (5124, 5213; 24.30m AOD, Fig. 3.6). Two thin layers of light brown silt sand (5181, 5181) overlay this deposit. This area was brought up to height by a 1.25m thick dump of mid-brown silt sand (5111, 5212; 26.20m AOD) that contained 19th-century pottery and animal bone.

In the eastern side of the main body of the quarry a series of layers dated to the 19th century suggesting rapid infilling of the eastern side of the quarry as it was discontinued and building up of the ground level. The lowest fill was crushed brick and sandstone rubble material (5121, 5136; 24.90m AOD, not illustrated) that contained 19th-century pottery. Overlying the rubble was a series of brown silt and sand deposits, and crushed sandstone layers (5135, 5138, 5148-5153). This was overlain by a deposit of mid-brown domestic rubbish dump containing glass, 19th-century

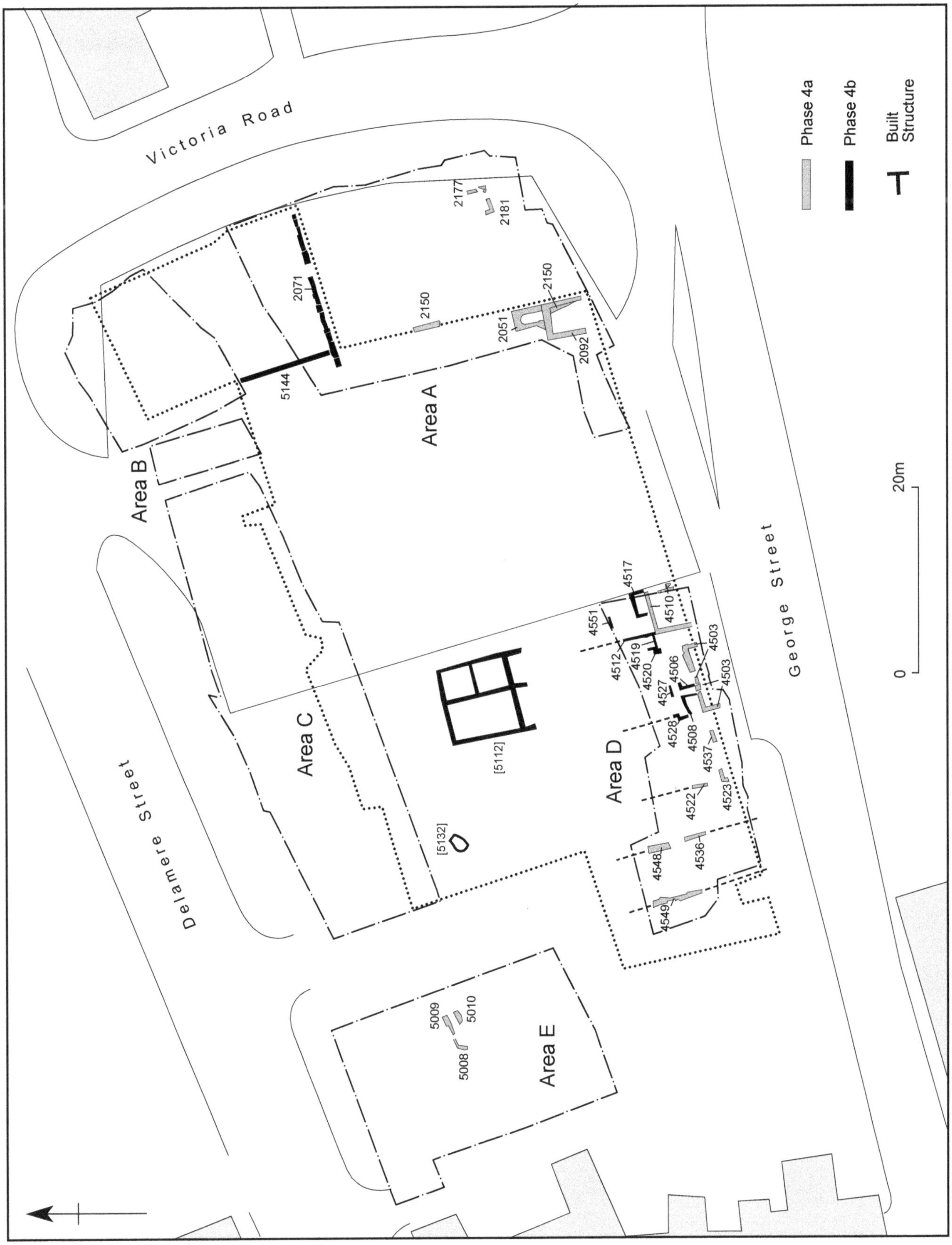

FIG. 3.7 PLAN OF PHASE 4A AND 4B BUILDINGS SURROUNDING THE SITE

pottery and animal bone (5137; 27.70m AOD). One of the latest of these layers of brown silt clay (5096) produced a series of domestic small finds including knife handles and brushes. Its location adjacent to the back-plots of the houses suggests domestic waste may have been discarded directly into the quarry.

Phase 4b: The Quarry in the mid-19th century

Sandstone buildings

This phase of activity is represented by a development of all the frontages (Fig.3.7). After the episodes of quarrying, the George Street frontage was redeveloped using sandstone and brick to form dwellings and ground floor shops. None of the earlier structural remains visible on De Laveaux's plan (1745) were located during the excavations (Fig. 3.4).

Groups 107 & 108

A series of wall foundations were aligned along George Street and uncovered during excavations in Area A. A cellar (2092) with adjoining apsidal room (2051, Area A) was recorded (Fig. 3.7). The walls of the cellar were of rough-coursed sandstone with rubble-core and lime-washed on their surfaces. Evidence of reuse of ashlar stone indicated that these may have been the waste products of stone masons working in the active quarry (wall 2092, Fig. 3.7). Due to the use of rubble in the majority of the construction (as can be seen in wall 2188) during this phase it is not really possible to date the buildings from the construction technique alone. Pottery was removed from the below the floor surfaces dating to the 19th century (S22, Fig.3.8). The Ordnance Survey map from 1869 shows a good correlation between the boundary walls and George Street frontage putting the earliest date at the mid-19th century (Fig. 3.9).

Group 111

In Area D, which was excavated parallel to the George Street frontage, a number of sandstone walls were recorded which relate to dwellings and ground floor shops. The most substantial of these remains was a large cellar (4510) constructed with sandstone rubble foundations. These were brought-up with brick walls laid on top that may be a slightly later development. An L-shaped wall of rough coursed sandstone rubble (4503) continued under the baulk southwards and probably represented the remains of another building. Fragmentary remains of sandstone walls (4522, 4523 and 4537) suggest further buildings may have existed to the west but were subsequently disturbed by later development. The chronology of the quarry means that the buildings cannot be earlier than the mid-18th century and it is probable that they developed in the later part of the 18th century as the quarry became in filled.

Another two boundary walls (4548, 4536, 4549) were identified in Area D which had been badly truncated by the bus station. These walls were dry-stone bonded and from the map evidence form boundaries between properties and the quarry edge (Figs. 3.7 and 3.9; Plate 3.2). Elsewhere

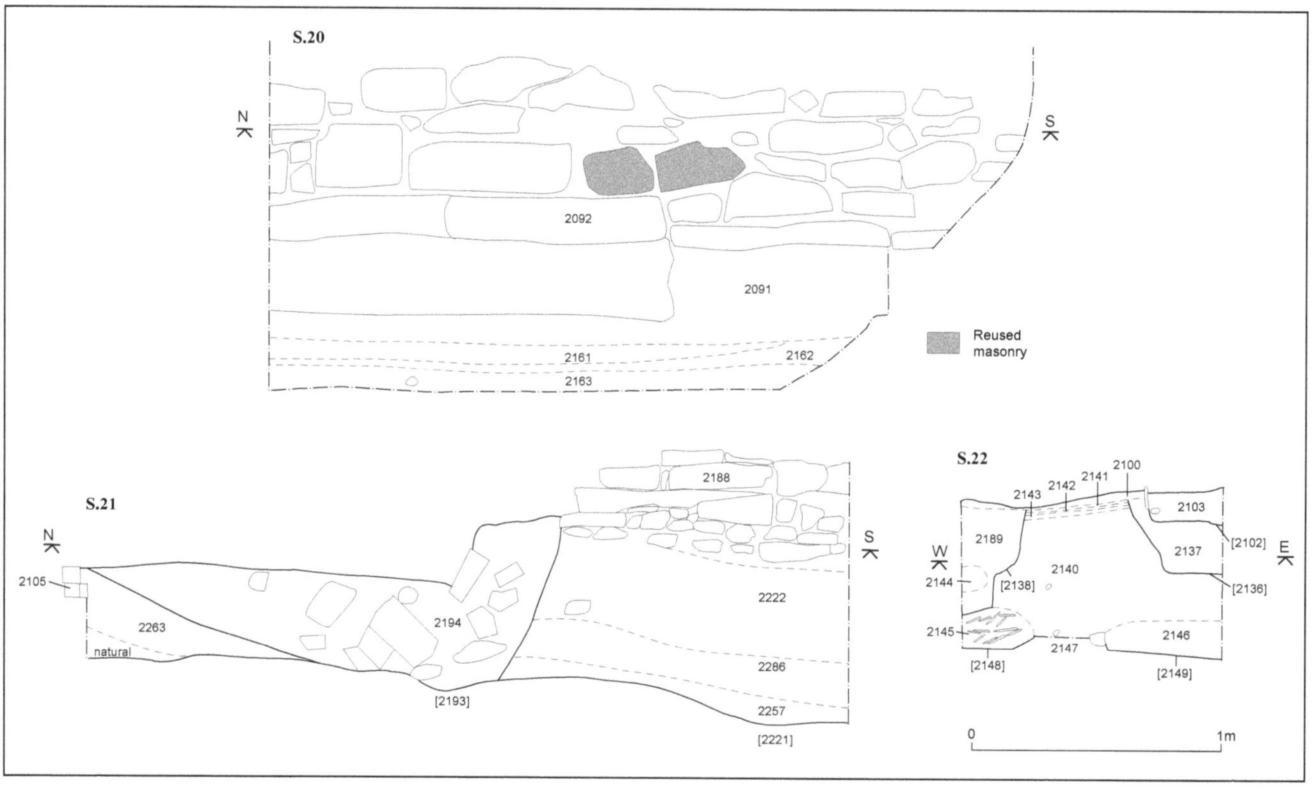

FIG. 3.8 SECTIONS 20-22

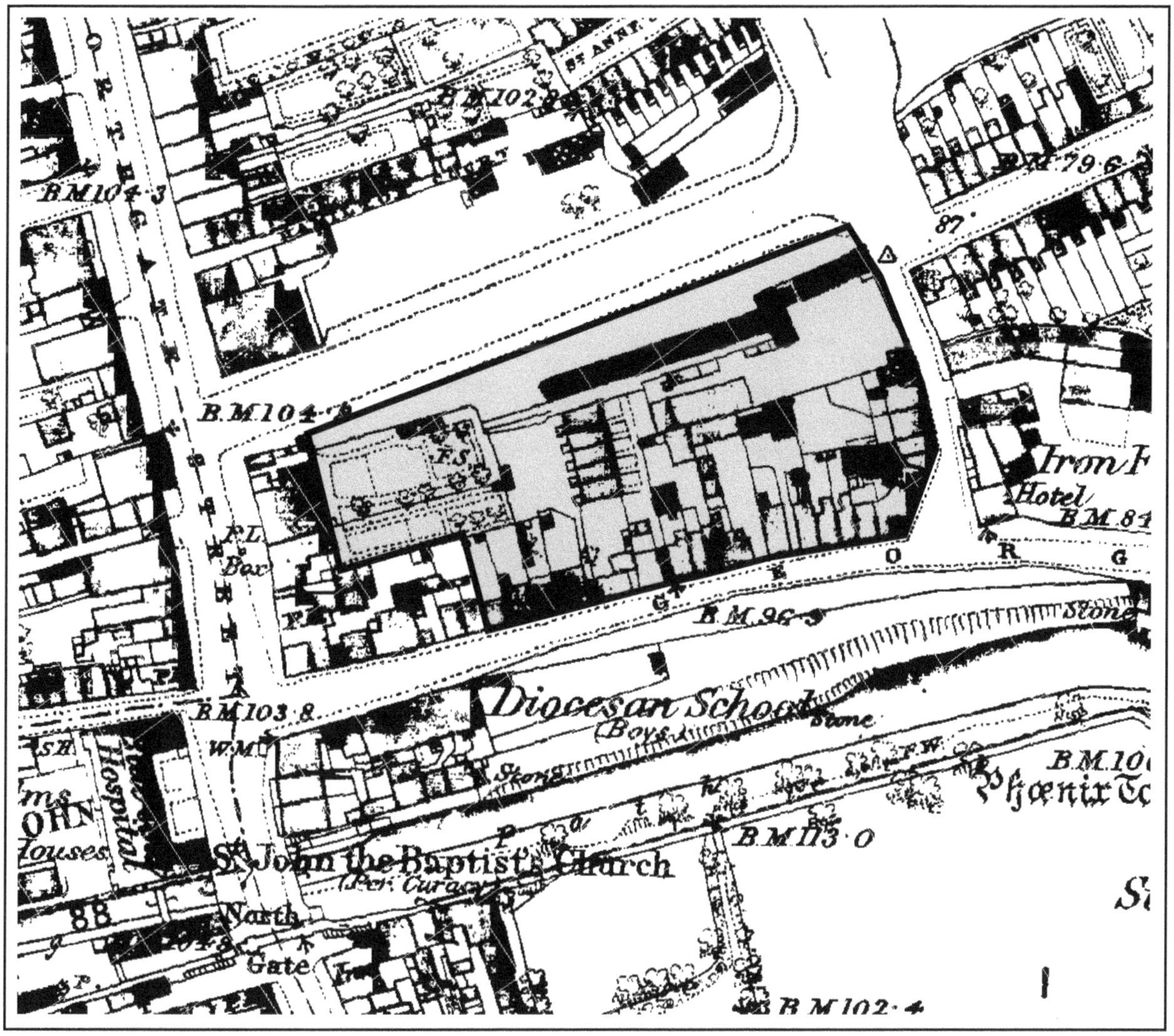

FIG 3.9 ORDNANCE SURVEY 1ST EDITION

boundary walls (5008; 5009; 5010) excavated in Area E also showed no signs of mortar and can only be assumed to be an 18th/ 19th century development.

Development of brick building - Groups 104 & 105

Further evidence for the development of the George Street frontage was recorded in Area D which included brick walls (4512, 4517, 4518, 4519, and 4520) and an external yard surface (4515). Pottery was recovered from what may have been a hearth rake out dump (4540) which had built up against wall (4512). The material is a mix of 18th and 19th century pottery which do not provide a secure stratigraphic sequence. This is not unsurprising given the level of disturbance in this area caused by the construction of the bus station.

Quadrant F and Group 109

Quadrant F was the only area of the post-medieval quarry that appears to have continued in use until the mid-19th century. The 1st edition map may show a large blank space that appears in the north-east corner, adjacent to George Street that suggests that the quarry pit was still visible although probably not in use (Fig. 3.9). This correlates with local resident recollections of the area where the quarry face was said to be exposed as late as the 1940s.

A mid-brown, silt clay (5143) with sandstone inclusions 1.6m thick, overlaid a short wall (5144), situated at 22.69m AOD (Plate 3.3). The wall, measuring 5m x 1m x 0.75m and aligned north-south, was construction consisting of large sandstone blocks bonded with a gritty grey mortar with a red brick section at its southern extent. The wall spanned the gap between the eastern end of Quarry Face 1, and the northern extent of Quarry Face 2, and affectively blocked Quadrant F off from the main body of the quarry. The wall overlaid deposit (5116) a mid-brown sand silt clay 1m thick, containing brick and stone inclusions, pottery, glass, tile and leather.

Walls had been placed along the edges of the quarry face, delineating Quadrant F of the post-medieval quarry. A large east-west orientated sandstone wall (2071, 2072) was recorded in Area A skirting the quarry edge. A similar wall was also noted in Area B, also orientated east-west (3013),

PLATE 3.2 AREA D WALL [4536]

which was constructed on the very edge of the post-medieval quarry (3024). These walls define the boundary of Quadrant F of the post-medieval quarry and probably are contemporary with its exploitation in the early to mid-19th century.

Phase 4b Buildings within the Quarry

The centre of the quarry does not appear to have been fully backfilled in the 19th century and was adopted for a row of five cottages visible on the 1st edition (1869) and the 2nd edition (1911) Ordnance Survey maps. These survived as foundations of a rectangular structure (5112; 27.83m AOD, Fig 3.6, Plate 3.4) measuring 8.50m x 9.60m x 1m. The walls were roughly constructed from large sandstone blocks bonded with lime mortar 0.30m wide, at least four courses were evident no particular bond was used. At least three rooms were evident in plan two measuring 3.50m x 3.50m and the third 4m x 7.40m. This was sealed by a series of thin demolition deposits (5079-5076, Fig 3.6).

Quadrant A contained several features of note, the earliest of which was a small sandstone lined pit 1.75m in diameter (5132; 26m AOD), cut into a deposit of crushed sandstone quarry waste (5126), with pottery and bone present. The primary pit fill (5133) was black organic silt, with leaf matter, animal hair, and leather, 0.5m thick. The secondary fill (5134) was black organic silt with lenses of sand 0.75m thick.

Phase 4c: late 19th century Methodist chapel and Delamere Street dwellings

Group 103

The quarry appears to be out of use and totally infilled by the early 19th century with buildings occupying the centre of the site. The documentary sources indicate that a new Methodist Chapel was built on the corner of George Street in 1888 (Bounds 1975:3). The chapel appears on the 1911 map of the city as 'Methodist chapel and Sunday school' (Fig. 3.10). The chapel served as an amalgamation of several smaller institutions which were no longer able to cope with the number of parishioners.

The excavation revealed that the chapel and Sunday school were very well preserved with parquet and minton tile floors intact (Plate 3.5). The distinction between the two buildings was very clear and is verified by the map overlay (Fig. 3.10). The toilet block also survived to a high level with the remains of toilet bowls and tiled walls. The external yard surfaces also survived with drains and piles of shattered roof tiles derived from the chapel demolition. Artefacts were also recovered from the demolition deposits overlying the chapel with the numbers from the prayer board, an Indian Rupee and various sherds of 20th century pottery.

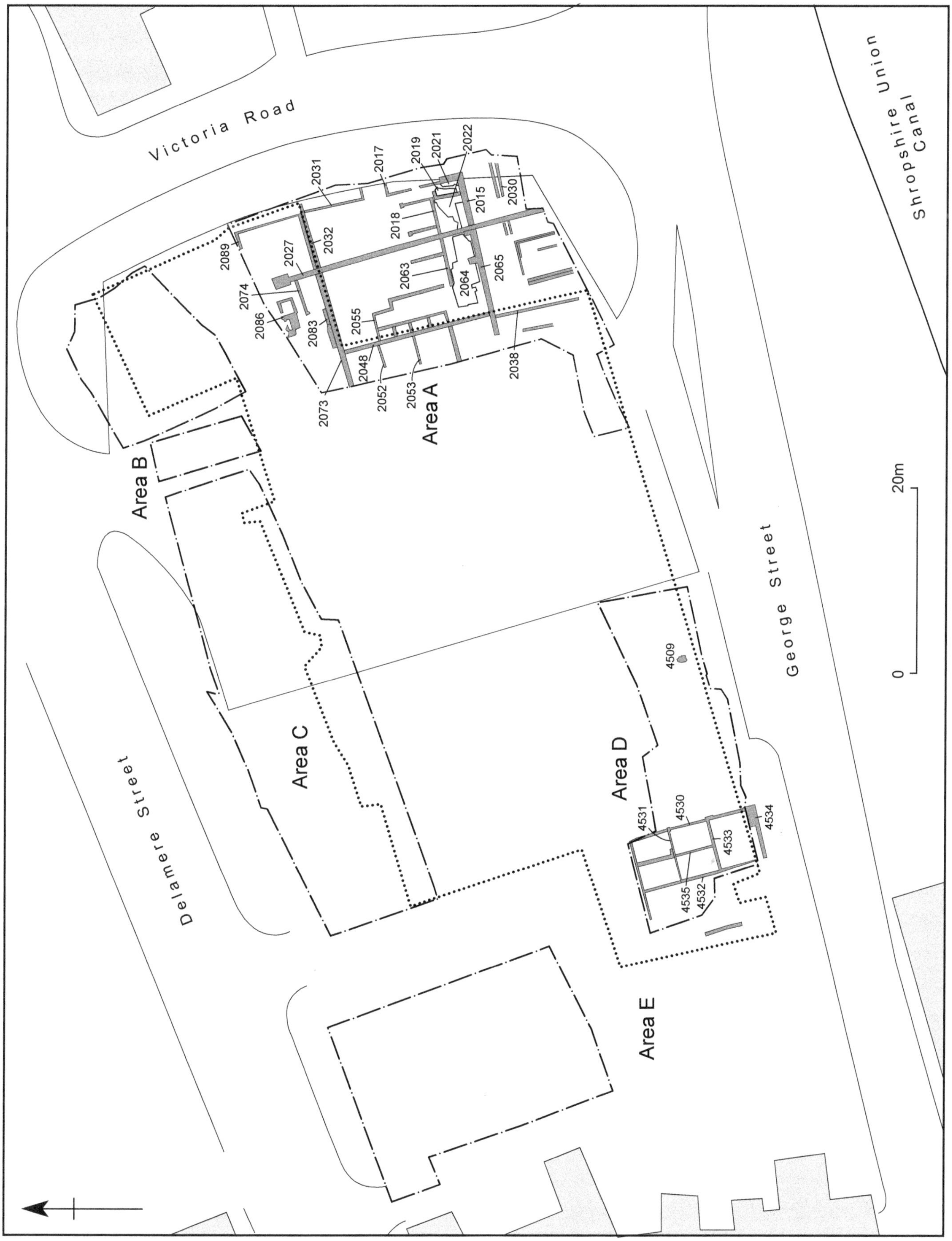

FIG. 3.10PLAN OF PHASE 4C BUILDINGS SURROUNDING THE SITE

The cellar was also identified below the rooms of the west wing of the chapel. This was of brick construction and was infilled by rubble derived from the demolition of the floors above. The last service was given on the 15th of February 1970 and demolition commenced immediately after this (Bounds 1975:20). It appears the building was razed to ground floor level with the rubble being used to level the ground for the next stage of building, the bus station.

PLATE 3.3 QUADRANT F WALL [5144] WEST FACING

PLATE 3.4 QUADRANT A STRUCTURE [5112] WEST FACING

PLATE 3.5 AREA A REMAINS OF THE CHAPEL

Phase 5: the bus station groups 100,101,102 & 114

Buildings within the site began to be demolished in the late 1960's culminating in the demolition of the Methodist chapel fronting onto George Street in 1970. The excavation of Area D revealed the remains of the bus station with walls left upstanding to 2m. The concrete floors were also preserved along with a cellar. The building was constructed using standard red brick in a Dearne bond. Throughout Area D and Area A large piles filled with hardcore were recorded, these relate to former bus shelters and other structures.

The upper layers of the quarry were a mixture of disturbed layers with pottery of 20th century and post-medieval date, glass and animal bone. The mixed nature of the layers resulted from disruption from the foundations of the old bus station.

FEATURES IN THE QUARRY

In several places along the top edge of the quarry face has been infilled with walling. In Quadrant A, the areas are infilled with large sandstone blocks (Fig. 3.11). In the northwest corner of Quadrant A, the infilling is with dry-stone wall rectangular sandstone. In the northern face in Quadrant A, on the northern quarry face, a further area of infilling is again in dry-stone sandstone blocks and likewise correlates with Roman, Phase 1 quarry pit 4041. A single episode of infilling occurred in Quadrant D on the eastern quarry face in sandstone blocks (Fig. 3.12). The infill was presumably undertaken to stabilise the quarry face during extraction and the use of sandstone blocks suggests that it occurred in the 18th century.

Elsewhere in Quadrant D, on the eastern quarry face these areas were filled with a coarsely bonded hand-made red-brick wall (2½" bricks) with a pale grey gritty mortar. These do not appear to correlate with earlier Roman quarry pits and the appearance of the mortar (unpointed) suggest that infilling had occurred from outside the quarry and had been done by the occupants of the properties along George Street. This may also explain why brick as opposed to readily available stone was adopted.

The majority of the quarry faces were marked with tool marks made during the process of sandstone extraction. Variations in the tool marks across the site did occur but the majority displayed distinctive hachure associated with extraction of stone by pick and adze. The tool marks in Quadrants A and E (Phase 3; Figs. 3.11, 3.12, 3.13) appeared marginally cruder than those in Quadrant F (Phase 4) in particular and may reflect chronological developments in quarry practice and tools. However, the techniques of quarry extraction remained unchanged from Roman times and refinement in tool marks may merely display variation in the teams operating in the quarry.

Rectangular pudlocks cut into the faces, varied in size from larger ones 0.30m x 0.40m x 0.35m, to smaller ones 0.10m x 0.20m x 0.05m. The distribution of pudlocks is unclear. The earliest areas of the quarry appear to have few. The largest pudlocks appear to be associated with the expansion of the quarry in Phase 4a (early-19th century), visible in Quarry Face 1 and 2 (Figs. 3.11 and 3.12; Plates 3.6 and 3.7). By the latest phase of the quarry (Phase 4b, mid-19th century) they are more frequent and smaller suggesting they perform a structural function such as supporting scaffold, platforms or ladders.

Examples at the Dorset cliffstone quarries were used for setting cranes, but these were usually set horizontally in the cliff top (Stanier 1996, 7). The variation in location of the pudlocks could suggest access was easier in the earliest and ultimate stages of the quarry and the restricted space may have resulted in cranes being fitted to the sides of the quarry to remove stone in the late-18th and 19th century.

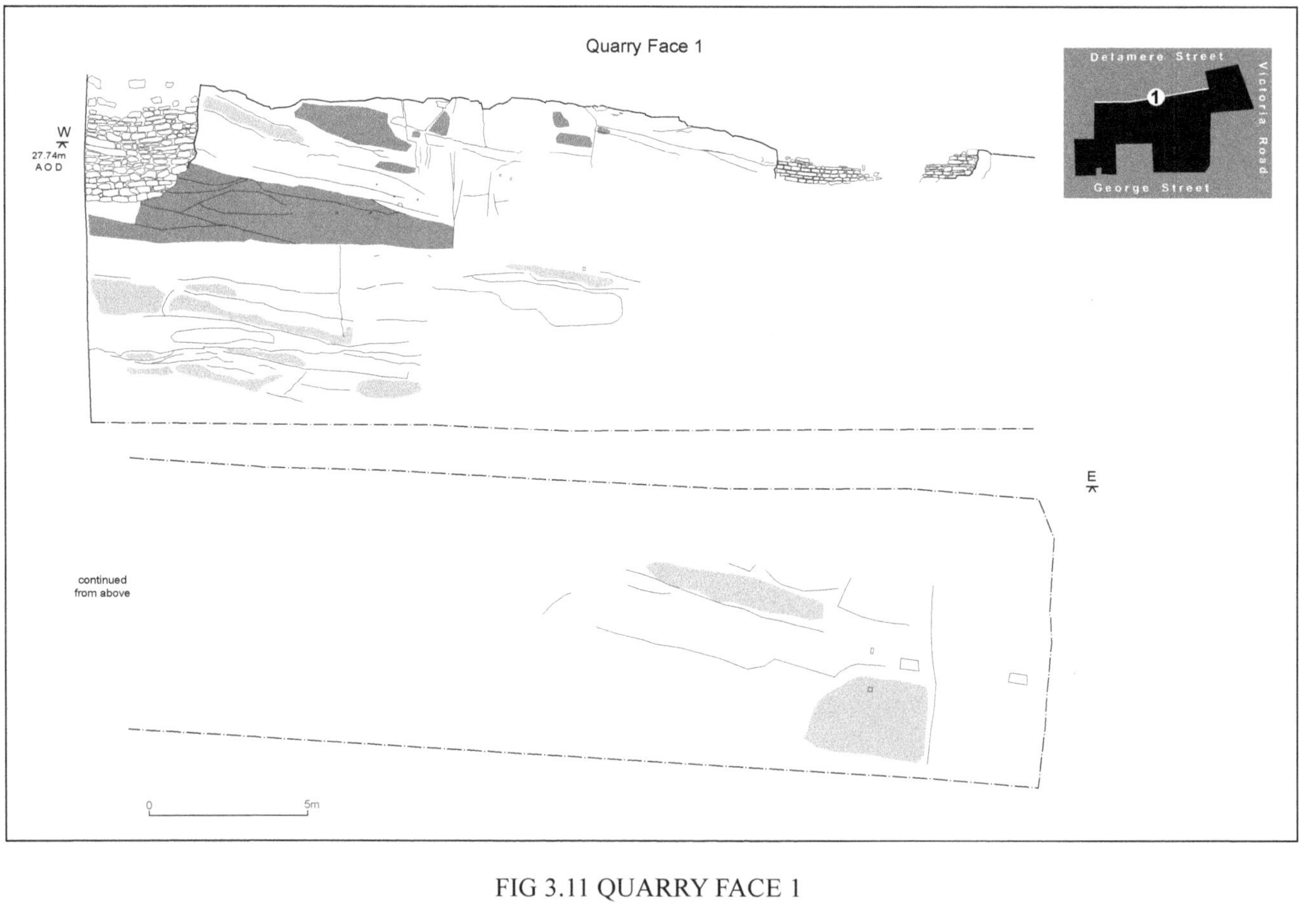

FIG 3.11 QUARRY FACE 1

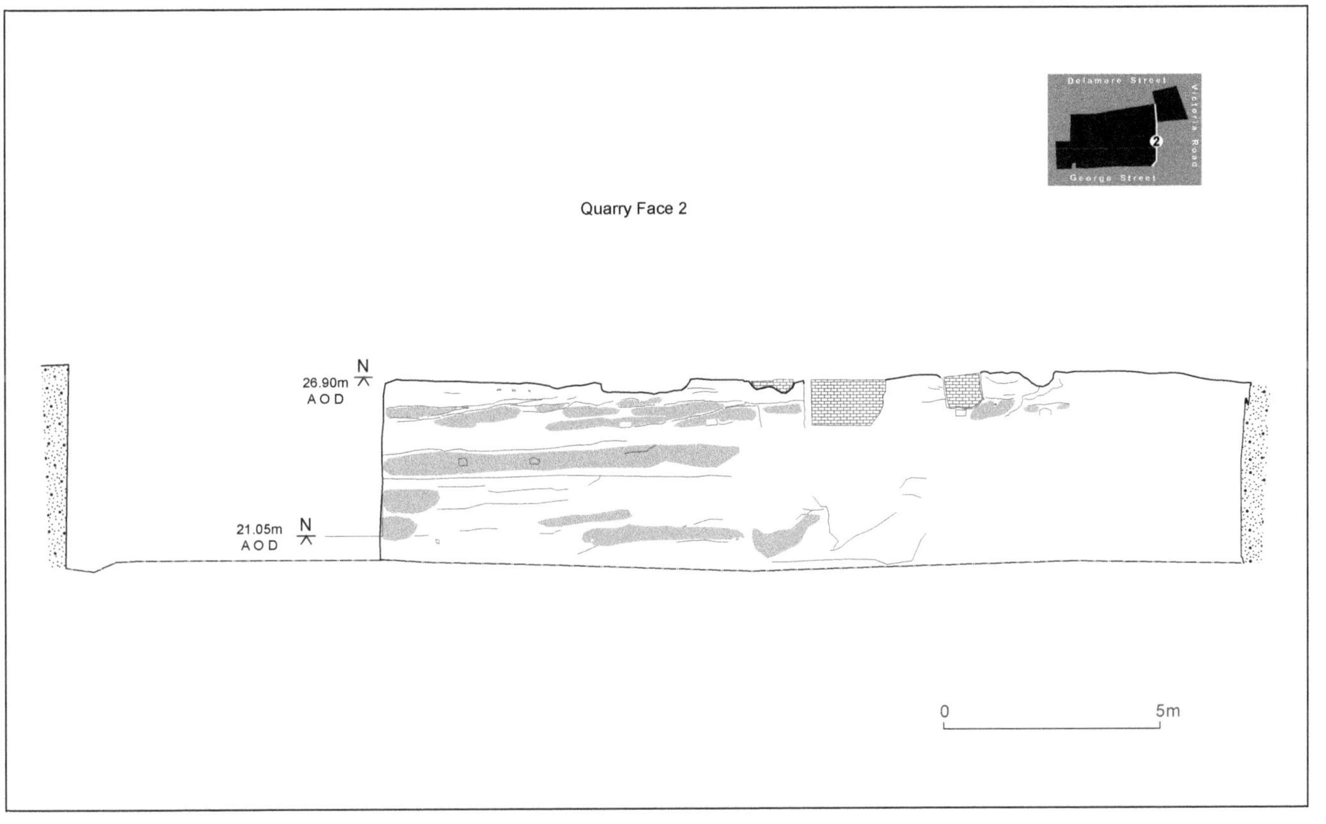

FIG 3.12 QUARRY FACE 2

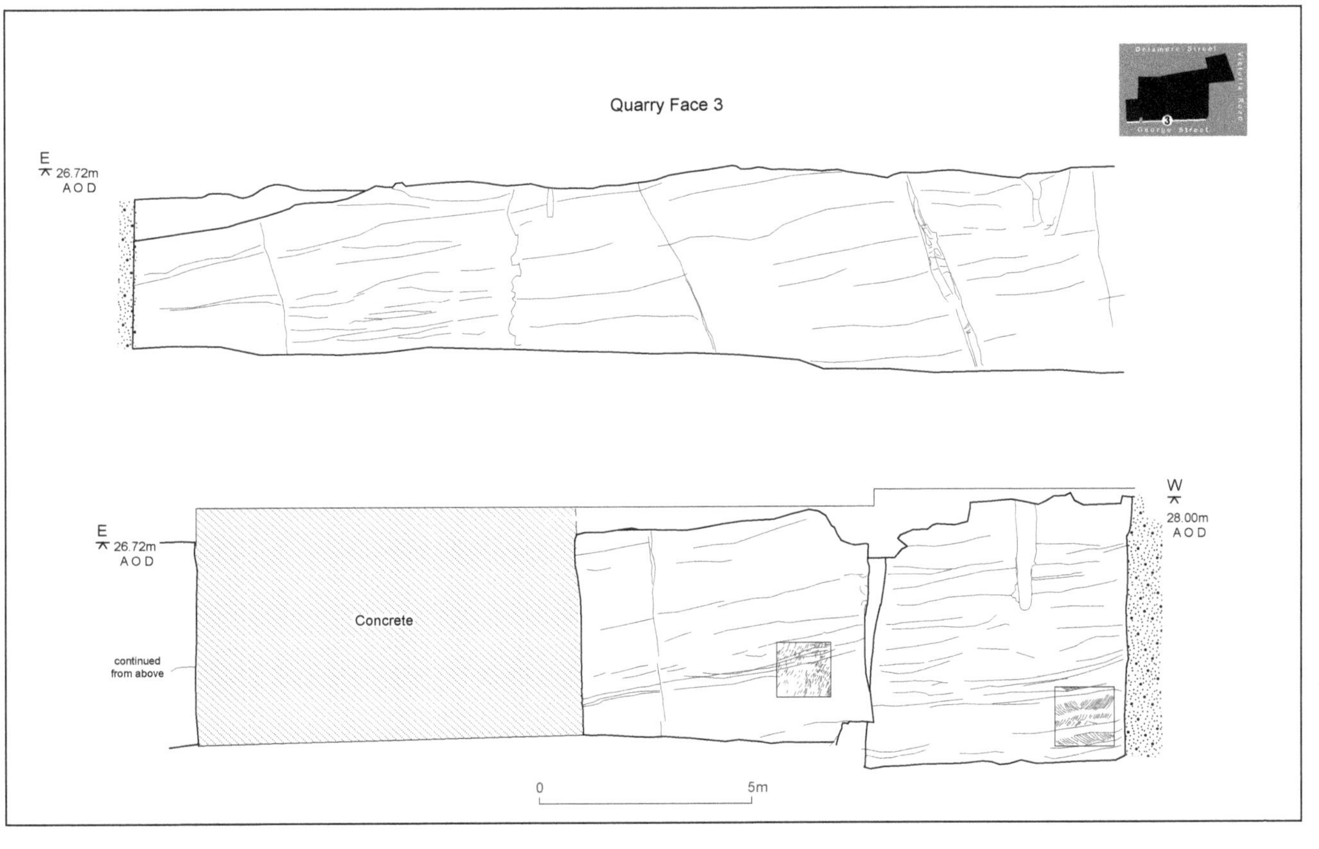

FIG 3.13 QUARRY FACE 3

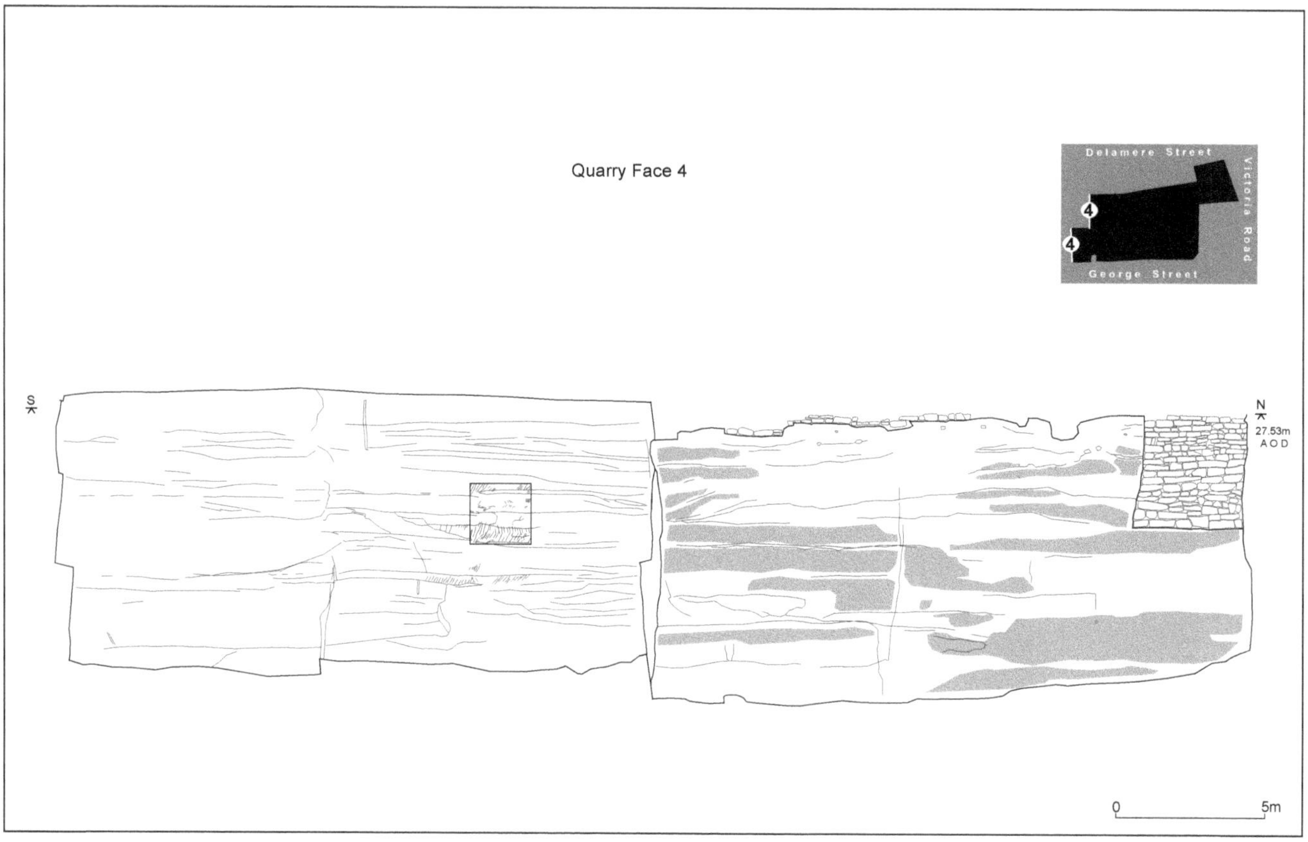

FIG 3.14 QUARRY FACE 4

PLATE 3.6 QUADRANT B QUARRY FACE 1

PLATE 3.7 QUADRANT C QUARRY FACE 2

PLATE 3.8 GRAFFITI MARK QUARRY FACE 3

PLATE 3.9 GRAFFITTI MARK QUARRY FACE 1

The faces were also marked with several instances of graffiti, workers initials carved into the stone. The instances of graffiti adjacent to a single pudlock may suggest an alternative function as a tool or dry store. No clear correlation can be established with the graffiti and it perhaps suggests that it is associated with stonemasons carving their mark.

Catalogue

1. Graffiti Mark. 'I' in Roman style font, no adjacent features, Quadrant E, Quarry Face 3, Strip Level 2, Phase 3. (Plate 3.8).
2. Graffiti Mark. Crude 'VE', no adjacent features, Quadrant B, Quarry Face 1, Strip Level 5, Phase 4a. (Plate 3.9).
3. Graffiti mark. 'JB' in Roman style font. Adjacent to pudlock. Quadrant D, Quarry Face 2, Strip level 4, Phase 4a.

THE ASSEMBLAGE FROM THE POST-MEDIEVAL QUARRY

Introduction

The assemblage from the post-medieval quarry must be viewed in light of the nature of the recovery. The assemblage was recovered during an extensive watching brief and as such was not necessarily recovered under the controlled conditions of excavation.

The quarry base was never reached, even at depths in excess of 6m below the current ground level and was instead piled at depth and there remains the possibility that material still survives at levels below the buildings constructed.

The recovery policy outlined above meant that often the selection policy was defined by an objective as opposed to a subjective viewpoint and this could lead to a bias in the results in a purely scientific sense. The visual nature of the recovery policy that was in place automatically leads to a bias to larger, lighter coloured objects, unlikely to fragment and easily discernable in dark soils. The nature of the watching brief meant a more scientific approach was not practical given the speed with which quarry deposits were excavated and removed from site.

The reality was the level of material recoverable varied vastly from the outlined expectancy. In contrary to the expected deep levels of well preserved material, stratigraphically secure, the assemblage was largely dominated by redeposited material. The quarry was dominated by continuous fills of disturbed 20th century material, the effects of the in-filling, demolition and reworking of the area as part of the construction of the Bus Station in the 1960s. A single area of the site in the southwest corner contained material that was stratigraphically secure and the material recovered dated to the 18th and 19th centuries. Material from these two sources have been treated differently during the course of the watching brief and during post-excavation. Those from securely stratified material have been referred to as key contexts and as such have been subject to full examination. Those from the mixed deposits dating to the demolition of the Bus Station have been termed non-key contexts and as such have been selectively examined.

It is important to realise that the quarry fills were not necessarily sealed contexts, like we would associate with pit deposition and were instead, large accumulation of waste material similar to middens or modern day open land fill sites. This suggests that deposits may have come from various sources, and not just be associated with a single episode of deposition. It is not unreasonable to see waste becoming mixed from various differing sources.

Given the obvious discrepancies in the assemblage an interesting collection of material has emerged. Some useful observations can be made about this area of Chester in the late post-medieval to early modern period. It is in this light that we should view the value of the material recovered from the post-medieval period.

POST-MEDIEVAL POTTERY *by Leigh Dodd*

Introduction

A little less than 3000 sherds of post-medieval pottery, weighing approximately 239800g, were recovered from the Delamere Street site during the excavation of quarry backfill deposits in 2008. The pottery spans the period from the early- to mid 18th through to the mid-19th, and comprises wares and forms that could be regarded as typical for urban assemblages of this broad period. An initial assessment of the pottery (Dodd 2008b) which included initial identification of wares present, 'spot' dating, and a basic minimum quantification comprising sherd count and weight – identified a number of key context groups that were considered to be worthy of further, more detailed analysis. The selection of these key context groups was determined by sherd size and condition, range and date of wares and vessel forms, and by the presence of either very little or no clearly intrusive later material. The pottery from these key contexts was subsequently quantified by sherd count and weight, with further additional quantification by Estimated Vessel Equivalents (EVEs), according to common ware names widely used and accepted within the North West/West Midlands region.

The report on the pottery from the quarry fills has been separated into two sections; the first details the pottery from contexts dated to the eighteenth century, and the second those from the nineteenth century. In each section and for each context two quantification tables are given; the first table provides details of the pottery according to common ware name quantified by rim EVE's, base EVE's, sherd count and weight, the second table quantifies the pottery in the same manner but shown as a percentage of the context assemblage overall. A discussion of the principal range of wares and forms recovered then follows at the end of each section together with any comments on dating. Wares that are clearly residual within the otherwise secure context groups have been quantified but are not included within the discussion. Appendices follow the main body of the report containing further tables for each context detailing the quantification of vessel forms identified according to ware.

A small number of vessels have been selected for illustration, in the form of photography, in order to visually illustrate the range and forms of vessels recovered, selection of sherds have been illustrated through photography, a method chosen over line drawings in order to demonstrate the range of glaze colours and patterns encountered across the assemblage.

Condition

Overall, the pottery from the quarry fill deposits is in a generally good condition with many contexts producing large sherds or several sherds from single, but smashed, vessels. The level of survival is probably due, in part at least, to the deposition of the pottery as rubbish into the excavated quarry, which in effect as been utilised as a large landfill site for the city of Chester during the later post-medieval period. A consequence of this context of deposition and recovery is, however, that none of the material from the key contexts or otherwise can be physically related to either a specific property or plot of land from which it originally derived prior to being deposited into the quarry. Naturally, this limits any useful comments regarding the origins and use of the pottery to the broad rather than the specific.

Range

A variety of 18th and 19th wares were present in forms that are predominantly of domestic consumption. However, a handful of industrial wares were identified including sugar cone moulds, in contexts dated to the 18th century, and saggar-like vessel fragments in contexts dated to the 19th century. In addition, a number of 19th stoneware bottles bearing the marks of a limited number of Chester wine merchants were also recorded.

In the earthenware category blackware, slipware and mottled ware generally dominate the 18th century assemblages. These wares, particularly the black wares and slipwares, were produced at countless potteries throughout Britain during this period although it is more than likely that the Staffordshire potteries along with the local Buckley, Flintshire, and Prescot, Merseyside, potteries would have dominated the material available to and purchased by the inhabitants of Chester during this period. Tin-glazed ware was present in most 18th century contexts but was far from abundant. Other earthenwares present included speckled ware, slip-coated buffware, Midlands Purple ware, unglazed coarsewares and small quantities of yellow ware. Early creamware was scarce implying that the 19th century contexts were deposited prior to the dominance of this ware during the mid- to late-18th century when this type of pottery was produced, sold and consumed on a vast scale. Small quantities of refined Staffordshire earthenwares were present in the form of Agate ware and refined redware. Brown stoneware was also present, although in quantities that fall far short of that for the white-bodied salt-glazed stonewares which dominates the finer end of the range of wares present.

Domestic imports from those areas of Britain outside of the Midlands and the North West, comprised a small number of sherds in North Devon gravel-tempered ware, a heavy, gritty ware, usually coated with an olive-green glaze, that is not uncommon on excavated post-medieval sites in Chester (Edwards 2008, 191). Any number of the tin-glazed sherds present may have been imported from the major manufacturing centres such as London or Bristol, but equally vessels manufactured in Liverpool are also almost certainly represented amongst these fragments. Continental imports are dominated by German stoneware, primarily the grey-bodied and blue decorated Westerwald wares, along with small quantities of brown salt-glazed Rhenish wares. Imports from more distant lands are restricted to a small number of sherds of Chinese export porcelain.

The 19th century contexts were dominated by mass-produced wares, many of which could have been manufactured in any number of potteries throughout Britain. Blue transfer-printed wares and other pearlwares dominate the tablewares whilst utilitarian vessels, predominantly large storage jars, are usually of the heavy-bodied blackware variety. Stoneware beverage bottles were recovered in large quantities in both light and dark glazes. Indeed, stoneware bottles dominate the pottery recovered from the 19th key contexts overall.

Dating

The 18th key context groups contained a range of wares and forms commonly in production during the early- to mid-18th century. The almost complete absence of creamware from the majority of the key context groups suggests that their deposition pre-dates the accepted period of dominance of this mass-produced ware as *the* fashionable ceramic during the later 1760s and 1770s (Barker 2007, 33). Furthermore, the white salt-glazed stoneware forms and decoration encountered on some tin-glazed vessels need not date any later than the mid to late 1750s; this date is supported further by the glass bottle forms also recovered from these contexts (*see* The Post-Medieval Glass *below*).

Backfilling of the former stone quarry with refuse of which ceramics clearly formed a large component, seemingly continued until the middle of the 19th century; the wares, forms and marks encountered within the 19th century key context groups imply that this material was being deposited up until the 1850s. The first edition *Ordnance Survey* map of Chester, surveyed in 1872 and published in 1875, indicates that the area formerly occupied by the stone quarry had become heavily developed beginning a new chapter in the urban development of this small area of Chester.

The 18th Century Pottery

All contexts and associated pottery discussed in this section are accompanied by Tables 3.1-3.14.

Agate ware

The characteristic streaky appearance of agate ware was achieved through loosely mixing red and cream coloured clays, producing a body reminiscent of the stone after which the ware is named. Produced in North Staffordshire from around *c* 1740, or quite possibly earlier, agate ware was popular throughout the mid 19th century and continued in production until the 1760s or early 1770s (Barker & Halfpenny 1990, 31). Sherds from agate ware vessels were scarce amongst the Delamere Street assemblage with only a handful of more common forms of cups, mugs and bowls being represented.

Blackwares

These black-glazed red-bodied earthenwares formed a major part of the overall assemblage of 18th century pottery. The predominant forms were large cylindrical storage jars, with club rims and horizontally applied handles, and deep dishes with flanged rims. A body sherd from a cistern complete with hole for a bung or tap was recovered from (5197; Tables 3.1 and 3.2). Smaller storage jars and bowls were also present, and chamber pots were not uncommon. However, sherds from smaller and finer bodied blackware vessel forms, such as cups and mugs, were particularly scarce. The more unusual forms represented were a dripping pan and a skillet, both recovered from (5179; Tables 3.1 and 3.2).

Brown Stonewares

Sherds from straight-sided mugs with turned banding were recovered in small quantities from contexts (5109; Tables 3.3 and 3.4) and (5210; Tables 3.5 and 3.6). Those sherds in a low-fired buff-orange fabric coated in a glossy orange-brown glaze have a thin white margin beneath the glaze, a feature distinctive of Nottingham or Derbyshire brown stoneware mugs (Jennings 1981, 221). Sherds in a similar fabric, attributed on this basis to Nottingham or Derbyshire, have been recovered from recent excavations in Chester (Edwards 2008, 233). The remaining brown stoneware vessels were manufactured in a harder, grey fabric, and may have been produced in London, Staffordshire or elsewhere. A single body sherd from a bottle coated with a light-brown salt-glaze was recovered from (5196; Tables 3.7 and 3.8), this vessel is also likely to be of English manufacture rather than continental.

Creamware

A single rim sherd from a moulded creamware plate with 'Queen's' shape rim was recovered from (5109).

Midlands Purple ware

Sherds from vessels in developed midlands purple ware of late 17th or early 18th century date were scarce. Those sherds that were recovered were all from jars with black glazing to the interior and exterior of the base.

Mottled ware

Mottled ware forms one of the dominant earthenware types encountered within the 18th century key context groups. Straight-sided mugs were particularly well represented and were of the cylindrical type decorated below the rim and above the base with distinctive bands of ribbing. These vessels compare closely with examples from an excavated pit-group encountered in Bridge Street, Chester, dated to the early 18th century (Edwards 2008, 239, nos. 95-99 & 101–102). Other mottled ware forms encountered at Delamere Street included round-bodied cups, shallow dishes, bowls, chamber pots and a jug or bottle.

North Devon Gravel-tempered ware

Although not uncommon on excavations in Chester, this heavily tempered gritty ware does not generally appear in large quantities. The Delamere Street site is no exception, with only a single base sherd from a jar or bowl, with internal green glaze, being recovered. North Devon gravel-tempered ware was traded by sea, often to America; its presence in Chester is thought to be the result of such sea-borne trade along the west coast of England, perhaps being sold as 'tradeable ballast' (Ratkai 1996, 24).

Porcelain

Two sherds of Chinese export porcelain were present amongst the overall assemblage. Both sherds are from cups with under-glaze blue floral decoration. One of these sherds, a rim from a coffee cup from (5179), has been further embellished with over-glaze red enamelling and gilt decoration; the rim has also been gilded. This particular range of colours is known as the *Imari* palette and dates this cup to approximately the second quarter of the 18th century. It was not, however, uncommon for further decoration of Chinese coffee cups to be undertaken in Europe, a practice begun early in the 18th century (Riley 2009, 38-41).

A single sherd of English porcelain was recovered from (5109). This sherd is from a blue decorated moulded sauceboat, complete with handle terminal scars, but of which only a tiny part of the floral design exists. The style of the moulded decoration and position of the handle terminals compares closely with an excavated sherd from a seemingly similar sauceboat recovered from the Longton Hall porcelain factory, Longton, Staffordshire (Barker & Cole 1998, 16, Fig. 16b). This sherd was from a particular design of moulded sauceboat commonly produced in Staffordshire white salt-glazed stoneware also during the mid 18th century. It is perhaps unwise to attribute the Delamere Street sherd to Longton Hall based on this comparison alone, however, other characteristics of the sherd, such as the wax-like glaze and an area of bubbled and slightly discoloured glaze close to the upper of the two handle terminals, are also characteristics found on extant pieces attributed to this factory (Godden 2004, 259). Longton Hall produced a significant amount of porcelain for a ten-year period between 1750 and 1760 (Barker & Cole 1998, 4).

Refined redware

Refined red earthenwares were the first of the fine-bodied earthenwares to be produced by the Staffordshire potters and are thought to have been in production at some date prior to the mid-1720s. These wares were also the first of the local Staffordshire wares to be produced using separate biscuit and glost firings (Barker & Halfpenny 1990, 23). A single rim sherd from a bowl, from context (5210), represents the only piece of this ware amongst the 18thcentury key context groups. The rim of this bowl has been highlighted with a cream-coloured slip; a practice encountered on similar but biscuit-fired waster bowls, cups and other vessels from excavations in Shelton, Stoke-on-Trent, and dated to *c* 1730–40 (Barker & Halfpenny 1990, 26–27. figs. 5–6).

Rhenish stoneware

Rhenish, or German brown stoneware, was limited to a single sherd comprising the rim and upper part of the neck to a bottle. This sherd, probably a Cologne or Frechen product of late 17th or early 18th century date, was in a grey fabric coated externally with a brown salt-glaze.

Slipware

Along with blackwares, slipwares dominated the overall 18th century pottery assemblage. Buff-bodied press-moulded dishes were by far the most common of the slipware vessel forms. These dishes, more often than not with typical piecrust rim, but some with scalloped rims pierced for hanging, were predominantly decorated with combed or combed-and-feathered cream and brown slip. Some embossed press-moulded dishes were also present, the embossed designs picked out with varying shades of mid and dark brown slip. Although incomplete, these designs appeared to include examples typical of north Staffordshire potters such as Samuel Malkin of Burslem, operating in the period *c* 1720–40; a fragment from a dish from (5197; Tables 3.9 and 3.10) decorated with what was probably a 'sun-face' design is particularly reminiscent of designs attributed to Malkin (*See* Cooper 1968, fig. 248; Orr 2003, 254, fig. 7).

Sherds from thrown slipware dishes decorated with trailed cream slip on a brown slip background typical of the second half of the 17th century were not present in any great number and it would seem that by the early eighteenth century the press-moulded dishes were perhaps the preferred choice of the citizens of Chester. A sherd from a thrown and trailed dish was recovered from (5214; Tables 3.11 and 3.12); this particular vessel, possibly a Buckley product, having been further decorated in the *sgrafitto* manner with a stylised tulip, a design popular during the 17th and early 18th centuries.

Slipware vessels other than dishes were scarce amongst the overall eighteenth-century assemblage. A sherd from a buff-bodied posset pot coated in cream slip and decorated with wavy lines in brown slip highlighted with cream slip jewelling was, however, recovered from (5109). This vessel probably dates to the late 17th or early 18th century.

Slip-coated buffware

This buff-bodied ware, coated in a brown slip that appears black under the lead glaze, was used for the manufacture of both fine and heavier utilitarian vessel forms during the period *c* 1710–60 (Barker 1997, 34). The Delamere Street sherds are from a variety of vessels including straight-sided mugs, press-moulded dishes, jars and chamber pots. The vessels represented were generally undecorated with the exception of a couple of dishes decorated internally with trailed cream slip lines.

Speckled ware

This type of earthenware is glazed in a similar streaky brown manner to that of mottled ware. However, it has been distinguished here apart from the fine-bodied buff-coloured mottled ware (*see above*), on the basis of its heavier, coarse pink-buff fabric that often contains clay lumps amongst other coarse inclusions. Both in this respect, and in respect of the heavier utilitarian forms recovered which included dishes, jars and chamber pots, as compared to the cups and mugs of the finer buff-bodied mottled ware, the ware compares closely with a type of earthenware termed *mottled ware* recovered from excavations in Prescot, Merseyside (*See* McNeil 1983, 49–94).

Contexts (5179), (5193; Tables 3.13, 3.14) and (5197) all produced large sherds from dishes, often forming complete profiles. A dish in a buff fabric coated with a speckled clear brown glaze recovered from Abbey Green, Chester (Rutter 1987, 27, fig. 6, no. 20) is not too dissimilar in profile and size to the dishes from both (5193) and (5197). The underside of one such dish from (5193) was heavily sooted suggesting use in an oven or hearth. A dish from (5197) had been crudely coated on the exterior surface with a thick cream-coloured slip; this slip coating is not, however, visible in section beneath the glaze on the interior surface. The profile of a rim from a chamber pot recovered from (5197) compares closely with an example from Prescot (McNeil 1983, 85, no. 74).

Tin-glazed ware

Tin-glazed ware, or delftware, vessels were present in small numbers but a variety of forms were identified. These included jars of various sizes, plates, dishes, bowls and chamber pots. The jars were predominantly undecorated being coated in white glazes, some pink tinged and others with a greenish-blue tinge. A fragment from a jar decorated with cobalt blue horizontal lines and vertical brush strokes recovered from (5210) is perhaps of Netherlands origins.

Fragments from small, plain white glazed, ointment or drug jars were present in a number of contexts, a near complete example was recovered from (5197) standing 6cm tall; these forms of jar can be broadly dated to the period *c* 1700–70.

Rims from plain white and bluish-white glazed plates were recovered from context (5197), their profiles suggesting a date in the early 18th century. Also from this context were two rim sherds from a shallow dish decorated internally with a trellis border along the rim edge and a bird sat on a flowering branch on the flange and well, the underside decorated with a ring running beneath the flange, all in shades of cobalt blue, probably dates to the early 18th century or perhaps the very late 17th. The bird design on this particular vessel is closely reminiscent of the 'bird on rock' motif used in conjunction with *Wan Li* decoration on products of the Pickleherring pothouse, Southwark, during this period (*See* Tyler, Betts & Stephenson 2008, 44–5, figs. 42–45). Two rim sherds from a single plate decorated internally with powdered blue grounds with reserved white panels containing what would be alternating leaf and flower sprays, all in blue, and with a reddish brown rim, were recovered from (5179). The sherds from this plate compare very closely in all details with an example illustrated in Archer (1997, 142, fig. B.55), attributed to the Vauxhall (Lambeth) pothouse under Mary Chilwell, dated to *c* 1745–50.

A sherd from the base, including foot-rim, of a bowl decorated internally with a sponged tree and brush-stroke sky, all in blue, completes the pieces of note in the tin-glazed ware category.

Unglazed earthenware

A handful of sherds of unglazed red earthenware were present throughout the overall assemblage. Of particular note however was the presence of fragments from sugar cone moulds. Two sherds were recovered that can be positively identified as belonging to such moulds. Context (5193) produced a rim from a mould with a profile not too dissimilar to an example from Bridge Street, Chester (Edwards 2008, 222, no. 70), although slightly less pronounced. A base sherd was recovered from (5196) that is very similar indeed to another example from the same Chester site (Edwards 2008, 222, no. 71). Both of the Delamere Street sugar cone mould sherds share the same hard, red fabric type and both have been shaved and smoothed on the interior surface. A third sherd, a body sherd from (5197) with similar smoothed interior surface, is also likely to be from a sugar cone mould.

Westerwald stoneware

Sherds of this distinctive cobalt blue decorated grey German stoneware were recovered from three contexts. The forms represented, straight-sided mugs and chamber pots, formed the staple Westerwald exports for the English market over the period *c* 1725–75 (Gaimster 1997, 94–5).

Context (5109) produced sherds, one the rim, from two straight-sided mugs, both with cordons around the upper body and both decorated with incised foliage pointed in blue. These mugs probably date to the first couple of decades of the 18th century; their decoration is similar to an example in Gaimster (1997, 267, fig 126 (L)) dated to *c* 1702–14. A third sherd recovered from the same context was from a chamber pot decorated with an incised zigzag medallion, bordered in blue, and containing a stylised flower with six petals pointed in blue. This form and decoration of this vessel compares very closely with an example recovered during building work in Norwich (Jennings 1981, 124, fig. 51 no. 848).

Several body sherds from a single rounded bottle or jug were recovered from (5179). These sherds were again decorated with incised foliage pointed in blue. A single body sherd, probably from a rounded jug, was recovered from (5214). This vessel was decorated with a cordon highlighted in blue.

White salt-glazed stoneware

White salt-glazed stoneware was particularly well represented in the pottery assemblage from (5109) accounting for just under 30% of the total sherd count and a little over 36% of the total base EVEs from this particular deposit. A range of both utilitarian, tea and tableware forms were positively identified including, although not exclusively, cups, mugs, teapots, bowls, dishes and plates; including plain-rimmed plates, octagonal plates with notched rims, and wavy-edged plates decorated with the moulded basketweave and mosaic pattern commonly in use in the 1750s. A single sherd from a cup or small bowl decorated with a scratch-blue floral design was also recovered from (5109), this was the only sherd decorated in this manner within the entire assemblage of white salt-glazed stoneware from the Delamere Street site.

Context (5193) produced a group of sherds that were probably from a single jug, decorated with turned bands. The spout on this jug is rather more rounded and subtle than the very angular spouts often encountered on many published examples (*See for example* Mountford 1971, figs. 114 & 115).

Two vessels were recovered from (5196). The first was a thrown straight-sided mug with reeded strap handle and turned bands; the handle similar to excavated examples from the Town Road site, Hanley, Stoke-on-Trent (*See* Edwards & Hampson 2005, 222, fig. 226). The second vessel from (5196) comprised a number of sherds, including the rim, from a thrown round-bodied teapot. The surface of this vessel was covered in numerous small raised lumps caused by trapped air bubbles in the clay expanding, and some bursting, outwards during firing. The vessel may, therefore, have been a waster or was perhaps sold as a second. Fragments of salt-glazed wasters and saggars recently found amongst rubbish dumped in fields at Heronbridge, south of Chester, have been interpreted as waste probably originating from the Chester White Ware Manufactory, Paper Mill Lane, Chester, operating from at least the later 1750s.

The white salt-glazed stoneware from Delamere Street is of the mass-produced plain white variety. With the exception of the moulded basketweave and mosaic decoration to the plate rims and the single scratch-blue decorated sherd, referred to above, decoration was generally limited to simple turned bands and a rouletted band encountered on a probable bowl from (5210). No sherds bearing polychrome enamel decoration were present amongst the overall assemblage.

Yellow ware

Yellow ware forms included small ointment or drug jars from (5179), (5210) and (5214), and the base from a small cup from (5210). The base from a large, handled- cup or posset pot was recovered from (5179), and a body sherd from a jar or mug from (5109).

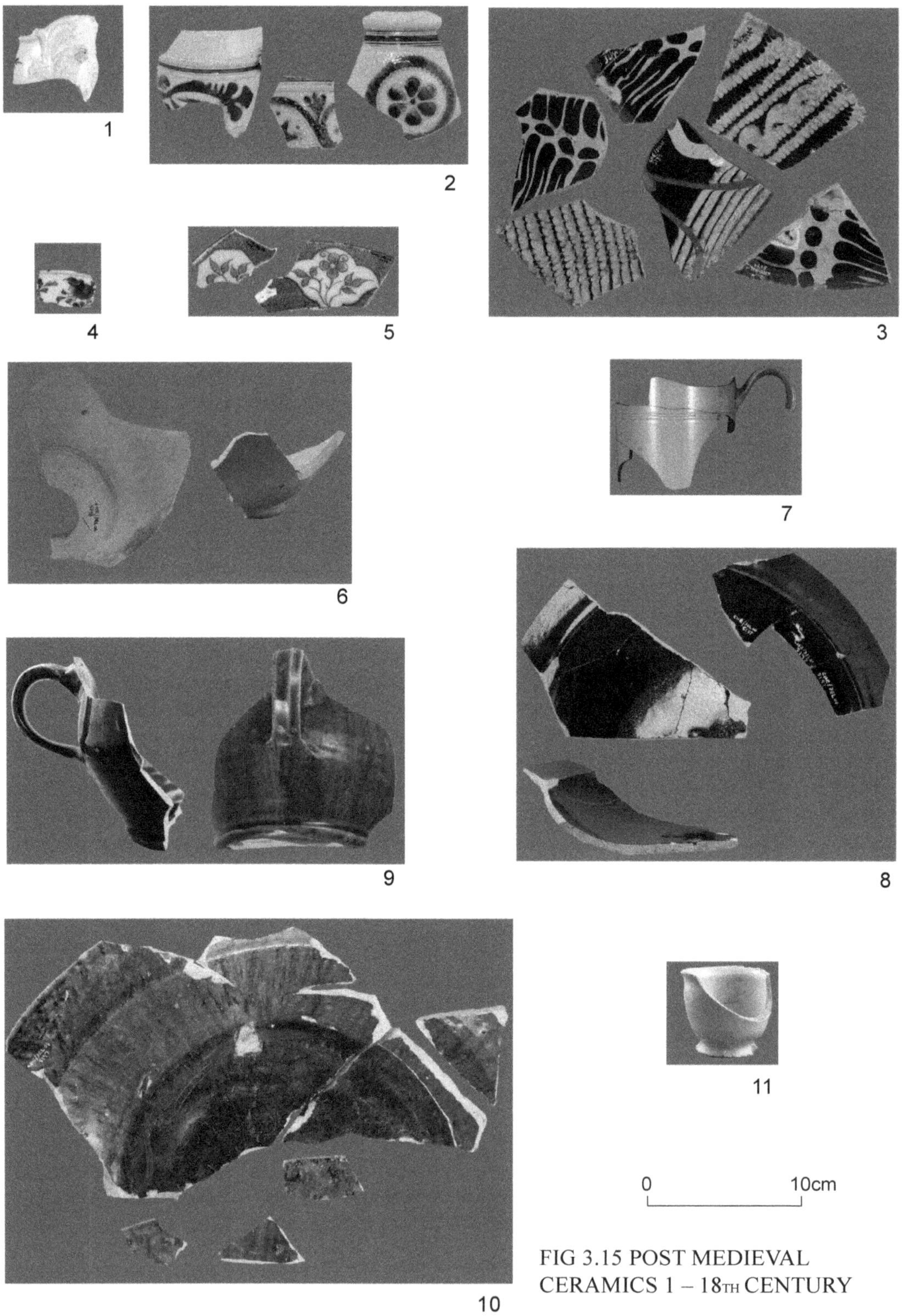

FIG 3.15 POST MEDIEVAL CERAMICS 1 – 18TH CENTURY

18th century Pottery

Context (5179)

TABLE 3.1. CONTEXT (5179) POST-MEDIEVAL POTTERY QUANTIFIED BY WARE, EVES, NO OF SHERDS AND WEIGHT

Ware	EVEs rim	*EVEs base*	*Sherd no*	*Wt (g)*
Agate ware	18	0	2	8
Blackware	224	210	35	4978
Midlands Purple	0	0	1	96
Mottled ware	0	289	30	820
Porcelain	18	0	1	4
Slipware	99	15	26	930
Slip-coated buffware	22	0	16	310
Speckled ware	25	43	3	500
TGW	13	75	7	184
Westerwald	0	0	5	76
Yellow	10	31	2	81

TABLE 3.2. CONTEXT (5179) POST-MEDIEVAL POTTERY AS A PERCENTAGE OF THE ASSEMBLAGE, QUANTIFIED BY WARE, EVES, NO OF SHERDS AND WEIGHT

Ware	EVEs rim %	*EVEs base %*	*Sherd no %*	*Wt (g) %*
Agate ware	4.2	0	1.6	0.1
Blackware	52.2	31.7	27.3	62.3
Midlands Purple	0	0	0.8	1.2
Mottled ware	0	43.6	23.4	10.3
Porcelain	4.2	0	0.8	0.05
Slipware	23.1	2.3	20.3	11.6
Slip-coated buffware	5.1	0	12.5	3.9
Speckled ware	5.8	6.5	2.3	6.3
TGW	3	11.3	5.5	2.3
Westerwald	0	0	3.9	0.9
Yellow	2.3	4.7	1.6	1

Context (5109)

TABLE 3.3. CONTEXT (5109) POST-MEDIEVAL POTTERY QUANTIFIED BY WARE, EVES, NO OF SHERDS AND WEIGHT

Ware	EVEs rim	*EVEs base*	*Sherd no*	*Wt (g)*
Agate ware	13	17	2	86
Blackware	375	485	40	7138
Brown stoneware	30	0	4	55
Creamware	3	0	1	23
Mottled ware	10	241	9	499
North Devon Gravel	0	38	1	158
Porcelain	0	0	1	18
Rhenish stoneware	100	0	1	117
Slipware	95	0	23	1070
Slip-coated buffware	15	25	3	110
Speckled ware	35	30	3	504
TGW	9	17	2	23
Unglazed earthen	31	0	3	146
Westerwald	50	0	3	122
White salt-glazed	86	483	40	1002
Yellow	0	0	1	5

Table 3.4. Context (5109) post-medieval pottery as a percentage of the assemblage, quantified by ware, EVEs, no of sherds and weight

Ware	EVEs rim %	*EVEs base %*	*Sherd no %*	*Wt (g) %*
Agate ware	1.5	1.3	1.5	0.8
Blackware	44	36.3	29.2	64.5
Brown stoneware	3.5	0	2.9	0.5
Creamware	0.4	0	0.7	0.2
Mottled ware	1.2	18	6.6	4.5
North Devon Gravel	0	2.8	0.7	1.4
Porcelain	0	0	0.7	0.2
Rhenish stoneware	11.7	0	0.7	1.1
Slipware	11.2	0	16.8	9.7
Slip-coated buffware	1.8	1.9	2.2	1.0
Speckled ware	4.1	2.2	2.2	4.6
TGW	1.1	1.3	1.5	0.2
Unglazed earthen	3.6	0	2.2	1.3
Westerwald	5.9	0	2.2	1.1
White salt-glazed	10.1	36.2	29.2	9.1
Yellow	0	0	0.7	0.05

Context (5210)

Table 3.5. Context (5210) post-medieval pottery quantified by ware, EVEs, no of sherds and weight

Ware	EVEs rim	*EVEs base*	*Sherd no*	*Wt (g)*
Agate ware	6	0	1	23
Blackware	149	310	29	5208
Brown stoneware	0	30	4	105
Mottled ware	28	198	4	340
Refined redware	7	0	1	6
Slipware	91	24	25	841
Slip-coated buffware	0	25	2	61
Speckled ware	8	78	9	616
TGW	0	12	2	85
White salt-glazed	0	100	1	96
Yellow	0	62	2	46

Table 3.6. Context (5210) post-medieval pottery as a percentage of the assemblage, quantified by ware, EVEs, no of sherds and weight

Ware	EVEs rim %	*EVEs base %*	*Sherd no %*	*Wt (g) %*
Agate ware	2.1	0	1.2	0.3
Blackware	51.6	36.9	36.3	70.1
Brown stoneware	0	3.6	5	1.4
Mottled ware	9.7	23.6	5	4.6
Refined redware	2.3	0	1.2	0.2
Slipware	31.5	2.9	31.3	11.3
Slip-coated buffware	0	3	2.5	0.8
Speckled ware	2.8	9.3	11.3	8.3
TGW	0	1.4	2.5	1.1
White salt-glazed	0	11.9	1.2	1.3
Yellow	0	7.4	2.5	0.6

Context (5196)

Table 3.7. Context (5196) post-medieval pottery quantified by ware, EVEs, no of sherds and weight

Ware	EVEs rim	*EVEs base*	*Sherd no*	*Wt (g)*
Blackware	38	43	6	706
Brown stoneware	0	0	1	81
Mottled ware	0	82	7	99
Slipware	24	0	8	270
Speckled ware	0	0	1	50
TGW	0	52	2	40
Unglazed earthen	0	50	1	125
White salt-glazed	28	0	8	69

Table 3.8. Context (5196) post-medieval pottery as a percentage of the assemblage, quantified by ware, EVEs, no of sherds and weight

Ware	EVEs rim %	*EVEs base %*	*Sherd no %*	*Wt (g) %*
Blackware	42.2	18.9	17.6	49
Brown stoneware	0	0	2.9	5.6
Mottled ware	0	36.2	20.6	6.8
Slipware	26.7	0	23.6	18.8
Speckled ware	0	0	2.9	3.5
TGW	0	22.9	5.9	2.8
Unglazed earthen	0	22	2.9	8.7
White salt-glazed	31.1	0	23.6	4.8

Context (5197)

Table 3.9. Context (5197) post-medieval pottery quantified by ware, EVEs, no of sherds and weight

Ware	EVEs rim	*EVEs base*	*Sherd no*	*Wt (g)*
Blackware	63	238	22	3883
Mottled ware	13	172	12	703
Slipware	50	18	18	573
Slip-coated buffware	25	97	9	599
Speckled ware	60	118	15	1500
TGW	81	112	8	220
Unglazed earthen	0	0	1	33
White salt-glazed	0	101	2	82

Table 3.10. Context (5197) post-medieval pottery as a percentage of the assemblage, quantified by ware, EVEs, no of sherds and weight

Ware	EVEs rim %	*EVEs base %*	*Sherd no %*	*Wt (g) %*
Blackware	21.6	27.8	25.3	51.1
Mottled ware	4.5	20.1	13.8	9.3
Slipware	17.1	2.1	20.7	7.5
Slip-coated buffware	8.6	11.3	10.3	7.9
Speckled ware	20.5	13.8	17.2	19.8
TGW	27.7	13.1	9.2	2.9
Unglazed earthen	0	0	1.2	0.4
White salt-glazed	0	11.8	2.3	1.1

Context (5214)

TABLE 3.11. CONTEXT (5214) POST-MEDIEVAL POTTERY QUANTIFIED BY WARE, EVES, NO OF SHERDS AND WEIGHT

Ware	**EVEs rim**	***EVEs base***	***Sherd no***	***Wt (g)***
Blackware	53	62	11	1774
Midland Purple	0	30	3	324
Mottled ware	41	161	9	588
Slipware	44	23	12	799
Speckled ware	0	85	5	535
TGW	0	32	1	24
Westerwald	0	0	1	15
Yellow	0	10	1	13

TABLE 3.12. CONTEXT (5214) POST-MEDIEVAL POTTERY AS A PERCENTAGE OF THE ASSEMBLAGE, QUANTIFIED BY WARE, EVES, NO OF SHERDS AND WEIGHT

Ware	**EVEs rim %**	***EVEs base %***	***Sherd no %***	***Wt (g) %***
Blackware	38.4	15.4	25.6	43.6
Midland Purple	0	7.4	7	8
Mottled ware	29.7	40	20.9	14.4
Slipware	31.9	5.7	27.9	19.6
Speckled ware	0	21.1	11.6	13.1
TGW	0	7.9	2.3	0.6
Westerwald	0	0	2.3	0.4
Yellow	0	2.5	2.3	0.3

Context (5193)

TABLE 3.13. CONTEXT (5193) POST-MEDIEVAL POTTERY QUANTIFIED BY WARE, EVES, NO OF SHERDS AND WEIGHT

Ware	**EVEs rim**	***EVEs base***	***Sherd no***	***Wt (g)***
Blackware	83	13	15	2115
Mottled ware	0	68	6	157
Porcelain	0	0	1	8
Slipware	80	15	15	555
Slip-coated buffware	33	38	3	97
Speckled ware	12	35	4	226
TGW	0	0	1	22
Unglazed earthen	7	0	1	85
White salt-glazed	22	0	4	62

TABLE 3.14. CONTEXT (5193) POST-MEDIEVAL POTTERY AS A PERCENTAGE OF THE ASSEMBLAGE, QUANTIFIED BY WARE, EVES, NO OF SHERDS AND WEIGHT

Ware	**EVEs rim %**	***EVEs base %***	***Sherd no %***	***Wt (g) %***
Blackware	35	7.7	30	63.6
Mottled ware	0	40.2	12	4.7
Porcelain	0	0	2	0.2
Slipware	33.8	8.9	30	16.7
Slip-coated buffware	13.9	22.5	6	2.9
Speckled ware	5	20.7	8	6.8
TGW	0	0	2	0.7
Unglazed earthen	3	0	2	2.6
White salt-glazed	9.3	0	8	1.9

The Nineteenth–Century Pottery

All contexts and associated pottery discussed in this section are accompanied by Tables 3.15-3.28.

Blackware and Brown-glazed wares

Black and brown-glazed red-bodied coarseware vessels were largey limited to large cylindrical storage jars with horizontally applied handles. Also present in the blackware catagory were cisterns, complete with hole for a bung or tap towards the base of the vessel, jugs and bowls. Brown-glazed vessels were again limited primarily to jars, but dishes and cups, although rare, were also identified.

Blue-bodied ware

Earthenwares with a solid blue body coated in a lead glaze were introduced during the mid 19th century; the Delamere Street assemblage produced a single vessel, a dish, recovered from (5139; Tables 3.15 and 3.16).

Blue transfer-printed ware

White-bodied vessels decorated with blue under-glaze transfer prints were common throughout the majority of the 19th century contexts; in the case of context (5115; Tables 3.17 and 3.18) this ware category accounted for 30% of the overall sherd count. By far the most prolific blue transfer-printed design encountered across the entire 19th century pottery assemblage was the common and well-known '*Willow*' pattern. This design was encountered on a variety of tableware vessels including circular and oblong plates and dishes, bowls, jugs, sauceboats, teapots, cups, mugs, lids, and tureens, as well as on more utilitarian type vessels such as jars and chamber pots. Occasionally vessels were encountered that were printed internally with the full Willow pattern design, but externally with restricted elements of it; such as the temple. An oval dish from (5136; Tables 3.19 and 3.20), for example, was printed in this manner. Other prints were encountered, mostly with floral designs or European figures, but were often too fragmentary for the designs to be identified with any degree of certainty.

A small number of blue transfer-printed ware sherds were marked on the underside of the vessel with printed slogans such as '*Stone China*' or '*Opaque China*', names used by many potteries during the period *c* 1830–80 to describe their ironstone-type whitewares despite the fact that many were simply standard white earthenware. Isolated numerals were occasionally encountered, printed on the underside of vessels; these probably represent batch numbers.

Context (5115) produced two Willow pattern plates with printed maker's marks on the underside; the first of these was marked: 'BMB-LANE END', and the second: 'STONEWARE W.E & C°'. The first mark may be that of Bradbury, Mason & Bradbury, recorded as operating in Longton during the early 1850s (Henrywood 2002, 83), whilst the latter may be that of William Emberton & Co. of the Highgate Pottery, Burslem, who were also operating during the early 1850s (Henrywood 2002, 124).

Factory-made slipware

This low-status yet popular decorated 19th century whiteware was surprisingly sparse amongst the overall assemblage. The standard range of forms including bowls, jugs and mugs and a colander were, however, represented. Decoration was limited to the banded variety, utilising blue, grey, and brown bands of slip.

Flow blue transfer-printed ware

Transfer-printed whiteware with heavy, dark 'flow blue' patterns were produced primarily for the American market during the 1820s–40s. Forms recovered from the Delamere Street site included cups, dishes, jars, jugs, plates and bowls. The base from a bowl from (5136) was printed with a diamond-shaped Design Registration Mark dating precisely the manufacture of this particular piece of pottery to the 7th May 1844.

Lustre ware

A single sherd, the rim from a mug, of copper-coloured lustre ware that was also sprig-decorated, was recovered from (5136).

Other transfer-printed wares

A small number of printed whitewares using colours other than blue were present throughout the overall assemblage. Forms identified printed with grey or blue-grey floral decorations included plates, oval plates, and chamber pots. A sherd from a plate from (5115) was marked on the underside with the pattern name AURORA in a bordered cartouche. A pattern named aurora is known in a flow blue pattern of flowers and scrolls on a vessel also marked J & A Bradley, who were retailers at 47 Pall Mall, London in around 1845–48 (Coysh & Henrywood 1989, 23, 32).

Pearlware

Plain white pearlware vessel forms present included bowls, chamber pots, jugs, mugs and ointment jars. Oval dishes and plates with blue rim edges were also recovered along with plates with under-glaze blue painted floral designs. A plate from (5116; Tables 3.21 and 3.22) was recovered with an over-glaze red and green painted floral design.

Porcelain

Amongst the few sherds of 19th century porcelain recovered were a mug and a plate, both from (5115), decorated with a blue 'Willow' type print, and a dish decorated with a dark blue floral design from the same context.

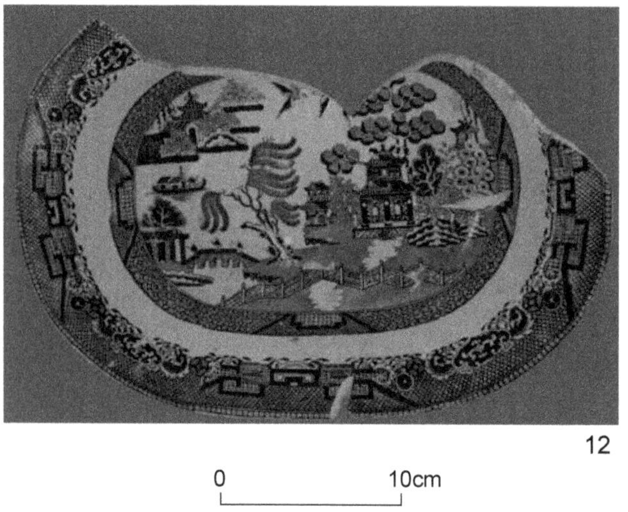

FIG 3.16 POST MEDIEVAL CERAMICS 2 – 19TH CENTURY

Rockingham glaze ware

A single sherd from a buff-bodied teapot coated in a thick, lustrous, treacly brown glaze, often referred to as 'Rockingham glaze', was recovered from (5136).

Saggar

The base from a thick-walled vessel, probably a saggar or other industrial vessel, in a coarse, buff fabric similar in appearance to that commonly used for such vessels was recovered from (5116). The presence of this vessel is more likely to be incidental rather than to indicate any nearby industrial activity.

Slipware

Although wheel-thrown trailed slip decorated dishes were recovered, the majority of the early to mid 19th century slipware vessels recovered were of the pink/red-bodied press-moulded variety. These dishes were simply decorated with trailed patterns in cream-coloured slip on dark brown backgrounds and are typical products of potteries in the Midlands, North of England and North Wales during this period.

Stonewares

Overall, light, and dark-glazed stonewares dominate the 19th century pottery assemblage; in the case of context (5136), for example, stonewares were shown to account for over 70% of the overall sherd count and over 63% of the overall sherd weight. Forms are almost exclusively limited to bottles, and of these bottles for ginger beer or other beverages are far and away the most prolific. Large bottles with a single handle used for storing and dispensing beverages were also common finds along with bottles commonly used to contain inks or blackening.

The stonewares have been sub-divided according to the glaze types used. The light-glazed stoneware vessels have been dipped into a clear vitreous glaze of the type developed and perfected by William Powell at his Temple Gate pottery in Bristol by *c* 1835, this glaze technique later to be widely accepted as 'Bristol ware' (Askey 1998, 126). Numerous bottles stamped close to the base with the wording: 'POWELL BRISTOL' or 'POWELL POTTRS BRISTOL', were recovered during the Delamere Street quarry excavation; fourteen bottles so marked with the former were recovered from context (5136) alone. Bottles from another famous Bristol stoneware manufacturer, J & C Price, stamped 'PRICE & SON MANUFACTURERS' or 'PRICE & SONS BRISTOL', were also recovered. Context (5136) produced a large Price & Son marked bottle that had been made for the Chester wine merchant *W. Walley.* A second large light-glazed bottle, marked for the merchant *Huxley & Co, Eastgate, Chester*, was also recovered from this context. Stoneware pharmaceutical vessels were represented by sherds from mortars recovered from (5116) and (5137; Tables 3.23 and 3.24).

Dark-glazed stoneware bottles were coated in ferruginous salt-glazes. These bottles were often found with the manufactures mark of *Joseph Bourne* of the Denby and Codnor Park pottery, Derbyshire. Bottles produced by this manufacturer were marked with Askey's type T6 stamp, dating the bottles to the period 1833–41 (Askey 1998, 63); several bottles marked this way were recovered from (5116). Dark-glazed bottles stamped with the '*EX*' excise duty mark were recovered from (5115) and (5136). These bottles date to the period 1817–34 as the requirement for bottles to be marked in this way, originally imposed in 1812 but not brought in until 1817, was finally abolished on the 14th August 1834 (Askey 1998, 62).

Unglazed earthenware

Unglazed red earthenware vessels were limited to a few sherds from cylindrical storage jars and horticultural vessels (flowerpots).

19th century Pottery

Context (5139)

Table 3.15. Context (5139) post-medieval pottery quantified by ware, EVEs, no of sherds and weight

Ware	EVEs rim	*EVEs base*	*Sherd no*	*Wt (g)*
Blackware	31	0	3	464
Blue-bodied ware	5	10	1	16
Blue TPW	0	20	1	11
Factory-made slip	0	63	1	200
Other TPW	0	0	1	26
Pearlware	0	27	1	149
Slipware	0	45	1	226
Stoneware (light glz)	0	0	2	201
Stoneware (dark glz)	237	190	13	1882
Whiteware	14	0	1	43

Table 3.16. Context (5139) post-medieval pottery as a percentage of the assemblage, quantified by ware, EVEs, no of sherds and weight

Ware	EVEs rim %	*EVEs base %*	*Sherd no %*	*Wt (g) %*
Blackware	10.8	0	12	14.4
Blue-bodied ware	1.7	2.8	4	0.5
Blue TPW	0	5.6	4	0.3
Factory-made slip	0	17.7	4	6.2
Other TPW	0	0	4	0.8
Pearlware	0	7.6	4	4.6
Slipware	0	12.8	4	7
Stoneware (light glz)	0	0	8	6.2
Stoneware (dark glz)	82.6	53.5	52	58.5
Whiteware	4.9	0	4	1.3

Context (5115)

Table 3.17. Context (5115) post-medieval pottery quantified by ware, EVEs, no of sherds and weight

Ware	EVEs rim	*EVEs base*	*Sherd no*	*Wt (g)*
Blackware	47	35	7	1372
Blue TPW	58	58	59	2686
Creamware	18	20	4	111
Factory-made slipw	20	45	2	85
Flow blue TPW	50	13	3	54
Other TPW	20	18	6	316
Pearlware	53	213	23	1135
Porcelain	53	65	7	107
Slipware	29	20	5	794
Stoneware (light glz)	400	734	53	5451
Stoneware (dark glz)	342	491	24	3791
TGW	0	0	1	2
Unglazed earthen	0	75	2	156
White salt-glaze	0	0	1	58

TABLE 3.18. CONTEXT (5115) POST-MEDIEVAL POTTERY AS A PERCENTAGE OF THE ASSEMBLAGE, QUANTIFIED BY WARE, EVES, NO OF SHERDS AND WEIGHT

Ware	**EVEs rim %**	***EVEs base %***	***Sherd no %***	***Wt (g) %***
Blackware	4.3	2	3.6	8.5
Blue TPW	5.3	3.2	30	16.7
Creamware	1.7	1.2	2	0.7
Factory-made slipw	1.8	2.5	1	0.5
Flow blue TPW	4.6	0.7	1.5	0.3
Other TPW	1.8	1	3	2
Pearlware	4.9	11.9	11.7	7
Porcelain	4.9	3.6	3.6	0.7
Slipware	2.7	1.2	2.5	4.9
Stoneware (light glz)	36.7	41	27	33.8
Stoneware (dark glz)	31.3	27.5	12.1	23.5
TGW	0	0	0.5	0.01
Unglazed earthen	0	4.2	1	1
White salt-glaze	0	0	0.5	0.4

Context (5136)

TABLE 3.19. CONTEXT (5136) POST-MEDIEVAL POTTERY QUANTIFIED BY WARE, EVES, NO OF SHERDS AND WEIGHT

Ware	**EVEs rim**	***EVEs base***	***Sherd no***	***Wt (g)***
Blackware	96	80	15	5335
Blue TPW	126	140	14	883
Brown glz	0	25	3	2552
Flow blue TPW	0	100	3	93
Lustre ware	3	0	1	12
Other TPW	0	22	1	37
Pearlware	23	90	5	264
Porcelain	0	150	2	74
Rockingham glz	0	0	1	53
Slipware	61	0	8	685
Stoneware (light glz)	1212	1605	98	13067
Stoneware (dark glz)	400	1058	39	5209
Unglazed earthen	0	38	1	659

TABLE 3.20. CONTEXT (5136) POST-MEDIEVAL POTTERY AS A PERCENTAGE OF THE ASSEMBLAGE, QUANTIFIED BY WARE, EVES, NO OF SHERDS AND WEIGHT

Ware	**EVEs rim %**	***EVEs base %***	***Sherd no %***	***Wt (g) %***
Blackware	5	2.4	7.9	18.4
Blue TPW	6.6	4.2	7.3	3.1
Brown glz	0	0.8	1.6	8.8
Flow blue TPW	0	3	1.6	0.3
Lustre ware	0.2	0	0.5	0.04
Other TPW	0	0.7	0.5	0.1
Pearlware	1.2	2.7	2.6	0.9
Porcelain	0	4.5	1	0.3
Rockingham glz	0	0	0.5	0.2
Slipware	3.2	0	4.2	2.4
Stoneware (light glz)	63	48.5	51.4	45.2
Stoneware (dark glz)	20.8	32	20.4	18
Unglazed earthen	0	1.2	0.5	2.3

Context (5116)

TABLE 3.21. CONTEXT (5116) POST-MEDIEVAL POTTERY QUANTIFIED BY WARE, EVES, NO OF SHERDS AND WEIGHT

Ware	EVEs rim	*EVEs base*	*Sherd no*	*Wt (g)*
Blackware	49	273	14	4735
Brown glz	82	0	5	1248
Blue TPW	116	266	31	1211
Factory-made slipw	51	123	8	471
Flow blue TPW	20	140	8	400
Other TPW	15	30	6	563
Pearlware	76	346	14	605
Porcelain	13	0	1	12
Saggar?	0	28	1	726
Slipware	39	53	12	1610
Stoneware (light glz)	260	772	43	8560
Stoneware (dark glz)	908	1175	64	7835

TABLE 3.22. CONTEXT (5116) POST-MEDIEVAL POTTERY AS A PERCENTAGE OF THE ASSEMBLAGE, QUANTIFIED BY WARE, EVES, NO OF SHERDS AND WEIGHT

Ware	EVEs rim %	*EVEs base %*	*Sherd no %*	*Wt (g) %*
Blackware	3	8.5	6.8	16.9
Brown glz	5	0	2.4	4.5
Blue TPW	7.1	8.3	15	4.3
Factory-made slipw	3.1	3.8	3.8	1.7
Flow blue TPW	1.2	4.4	3.8	1.4
Other TPW	1	0.9	2.9	2
Pearlware	4.7	10.8	6.8	2.2
Porcelain	0.8	0	0.5	0.04
Saggar?	0	0.9	0.5	2.6
Slipware	2.4	1.7	5.8	5.8
Stoneware (light glz)	16	24	20.8	30.6
Stoneware (dark glz)	55.7	36.7	30.9	28

Context (5137)

TABLE 3.23. CONTEXT (5137) POST-MEDIEVAL POTTERY QUANTIFIED BY WARE, EVES, NO OF SHERDS AND WEIGHT

Ware	EVEs rim	*EVEs base*	*Sherd no*	*Wt (g)*
Agate ware	12	0	1	40
Blackware	49	57	6	1853
Blue TPW	17	13	11	570
Flow blue TPW	4	128	4	138
Factory-made slip	15	53	5	120
Pearlware	52	357	8	1425
Slipware	63	9	5	625
Stoneware (light glz)	200	456	16	3871
Stoneware (dark glz)	0	0	34	5052

TABLE 3.24. CONTEXT (5137) POST-MEDIEVAL POTTERY AS A PERCENTAGE OF THE ASSEMBLAGE, QUANTIFIED BY WARE, EVEs, NO OF SHERDS AND WEIGHT

Ware	**EVEs rim %**	***EVEs base %***	***Sherd no %***	***Wt (g) %***
Agate ware	2.9	0	1	0.3
Blackware	11.9	2.9	6.7	13.5
Blue TPW	4.1	0.7	12.2	4.2
Flow blue TPW	1	6.6	4.4	1
Factory-made slip	2.7	3.6	5.6	0.9
Pearlware	12.6	18.3	8.9	10.4
Slipware	15.3	0.5	5.6	4.6
Stoneware (light glz)	48.5	23.4	17.8	28.3
Stoneware (dark glz)	0	45	37.8	36.8

Context (5122)

TABLE 3.25. CONTEXT (5122) POST-MEDIEVAL POTTERY QUANTIFIED BY WARE, EVEs, NO OF SHERDS AND WEIGHT

Ware	**EVEs rim**	***EVEs base***	***Sherd no***	***Wt (g)***
Blackware	9	0	1	32
Blue TPW	0	27	6	340
Other TPW	13	0	1	36
Pearlware	39	108	7	403
Porcelain	0	200	5	252
Slipware	0	40	2	750
Stoneware (light glz)	250	250	12	1462
Stoneware (dark glz)	0	50	1	307

TABLE 3.26. CONTEXT (5122) POST-MEDIEVAL POTTERY AS A PERCENTAGE OF THE ASSEMBLAGE, QUANTIFIED BY WARE, EVEs, NO OF SHERDS AND WEIGHT

Ware	**EVEs rim %**	***EVEs base %***	***Sherd no %***	***Wt (g) %***
Blackware	2.9	0	2.6	0.8
Blue TPW	0	4	25.6	13.7
Other TPW	4.2	0	2.6	1
Pearlware	12.5	16	17.9	10.7
Porcelain	0	29.6	12.8	6.7
Slipware	0	5.9	5.1	20
Stoneware (light glz)	80.4	37	30.8	38.9
Stoneware (dark glz)	0	7.5	2.6	8.2

Context (5138)

TABLE 3.27. CONTEXT (5138) POST-MEDIEVAL POTTERY QUANTIFIED BY WARE, EVEs, NO OF SHERDS AND WEIGHT

Ware	**EVEs rim**	***EVEs base***	***Sherd no***	***Wt (g)***
Blackware	34	0	2	1458
Stoneware (dark glz)	100	100	1	324

TABLE 3.28. CONTEXT (5138) POST-MEDIEVAL POTTERY AS A PERCENTAGE OF THE ASSEMBLAGE, QUANTIFIED BY WARE, EVEs, NO OF SHERDS AND WEIGHT

Ware	**EVEs rim %**	***EVEs base %***	***Sherd no %***	***Wt (g) %***
Blackware	25.4	0	66.7	81.8
Stoneware (dark glz)	74.6	100	33.3	18.2

The Pottery from the 2006 Excavation

The main body of the excavation, undertaken during 2006, produced a little under 800 sherds, weighing 22524g, of late post-medieval and early 20th century pottery. These sherds were recovered from a number of heavily truncated and disturbed features and deposits dating primarily to the later 19th and early 20th centuries. An assessment of the pottery from the excavation was undertaken (Dodd 2008a) which concluded that the assemblage overall held very little potential for further study as much of the very fragmentary pottery was almost certainly incidental rubbish survival with much re-deposited material, and furthermore, very few securely stratified groups were identified. What follows, therefore, is a discussion on some of the more interesting groups identified during the assessment of this material.

Discussion

With the exception of residual Roman and medieval sherds, the earliest datable pottery recovered from the 2006 excavation comprised a handful of sherds from a minimum of three yellow ware vessels, comprising mugs and jars, recovered from (2205) the fill of a pit feature. These vessels, which probably date to the middle of the 17th century, were all produced in a fine buff-coloured fabric coated with a clear yellow lead glaze. Pronounced ribbing or rounded cordons in bands of two or more are a feature of these vessels; characteristics noted on similar vessels from recent excavations within the city centre (*See* Edwards 2008, 202), whilst similar vessel forms were also recovered from Civil War period deposits at Beeston Castle, Cheshire (Noake 1993, 199, fig. 137, nos. 114–21).

Context (4552) was identified as being the uppermost backfill deposit to the post-medieval quarry encountered during this phase of the excavations. This context produced a handful of mid 19th century sherds comprising a range of mass-produced wares and forms similar to those outlined in detail above. Of particular note from this group was an almost complete flow blue facetted cup, minus handle, decorated with oriental inspired temples and floral transfer-printed designs.

Marked piece were scarce. An oval blue transfer-printed serving dish with '*Willow*' pattern design from (4543), interpreted as a rake-out deposit from a hearth, was transfer marked on the underside of the rim flange with a warranted Staffordshire makers mark for 'J.M & s'; the small 's' representing the term '*and son*'. These initials and suffix could relate to any number of 19th century Staffordshire potters with these particular initials (See, for example, those listed in Henrywood 2002). This deposit produced sherds from a number of plates and dishes decorated with the popular '*Willow*' pattern design.

A demolition deposit (2002) within Area A, located in the area formerly occupied by a Methodist chapel and associated Sunday school demolished during the 1960s, produced a sherd of white sanitary ware with the black printed mark for Thomas Twyfords Ltd, Hanley. This fragment, probably from a water-closet, is also marked 'VITROMANT' and 'MADE IN ENGLAND', the latter a term not used on ceramics until the 20th century.

SMALL AND OTHER FINDS

POST-MEDIEVAL GLASS by *Leigh Dodd*

The quarry backfill excavation produced a small assemblage of vessel glass from a handful of contexts dating to the first-half of the 18th century. The small size of the assemblage is surprising considering the large quantities of ceramics recovered from the same range of contexts. The range of vessel glass is limited to bottles, of onion (Fig. 3.17) and mallet forms, and a single phial, all in green glass. There is a total absence of drinking glasses and only two small fragments of window glazing were recovered. Surface condition of the glass varies with some pieces exhibiting no surface weathering and others with heavy weathering causing decay and flaking of the surfaces. Despite the small size of the overall assemblage remarkably several bottles were recovered in either complete or nearly complete condition.

Catalogue

1. Phial; one fragment comprising the complete pushed-in base from a small ?squat phial; clear green glass with slight weathering; base diam 47mm; first half 18th cent (17g). Context (5109), Phase 3.
2. Bottle; one fragment of base from an onion-shaped wine bottle; amber-green glass with heavy weathering; late 17th – early 18th cent (141g). Context (5109), Phase 3.
3. Bottle; one fragment of rim, neck and upper body from an onion-shaped wine bottle; green glass with heavy weathering; late 17th – early 18th cent (75g). Context (5109), Phase 3.
4. Bottle; one fragment of base and lower body from a mallet-shaped wine bottle; green glass with slight weathering; *c* 1720–60 (329g). Context (5109), Phase 3.
5. Bottle; complete squat mallet-shaped wine bottle; green glass with no weathering; rim diam 28mm, base diam 115mm, height 165mm; *c* 1720–40 (713g). Context (5179), Phase 3.
6. Bottle; complete onion-shaped wine bottle; green glass with no weathering; rim diam 26mm, base diam 126mm, height 170mm; *c* 1700–20 (872g). Context (5179), Phase 3.
7. Bottle; nearly complete squat mallet-shaped wine bottle missing rim and part of neck; green glass with slight weathering; base diam 108mm; *c* 1720–40 (669g). Context (5179), Phase 3.
8. Bottle; one fragment of complete base and lower body from a mallet-shaped wine bottle; green glass with heavy weathering; base diam 121mm; *c* 1720–40 (634g).
9. Bottle; one fragment of base from an onion-shaped wine bottle; green glass with heavy weathering; late 17th – early 18th cent (151g). Context (5179), Phase 3.
10. Bottle; one body fragment from an onion-shaped wine bottle; green glass with no weathering; late 17th – early 18th cent (71g). Context (5179), Phase 3.
11. Bottle; one fragment of base from a flat octagonal bottle; green glass with moderate weathering; *c* 1740 (239g). Context (5179), Phase 3.
12. Bottle; one fragment of rim and neck from a squat onion-shaped wine bottle; green glass with no weathering; rim diam 29mm; late 17th – early 18th cent (66g). Context (5179), Phase 3.
13. Bottle; one fragment of rim and neck from an onion or mallet-shaped wine bottle; green glass with slight weathering; rim diam 29mm; *c* 1700–40 (129g). Context (5193), Phase 3.
14. Bottle; nearly complete base from an onion-shaped wine bottle with high-kicked base; green glass with moderate weathering; base diam 118mm; late 17th – early 18th cent (329g). Context (5193), Phase 3.
15. Bottle; nearly complete onion-shaped wine bottle, minus rim; green glass with moderate weathering; base diam 126mm; *c* 1700–20 (978g). Context (5197), Phase 3.
16. Bottle; nearly complete small squat mallet-shaped wine bottle, missing part of rim and neck; green glass with no weathering; base diam 92mm, height 135mm; *c* 1720-40 (504g). Context (5197), Phase 3.
17. Bottle; nearly complete base from an onion-shaped wine bottle; green glass with heavy weathering; base diam 137mm; *c* 1700-20 (676g). Context (5197), Phase 3.
18. Bottle; one fragment of neck and upper body from an onion-shaped wine bottle; green glass with no weathering; rim diam 28mm; late 17th – early 18th cent (113g). Context (5210), Phase 3.
19. Bottle; one fragment of rim and neck from an onion-shaped wine bottle; green glass with slight weathering; rim diam 25mm; late 17th – early 18th cent (64g). Context (5214), Phase 3.

FIG 3.17 POST MEDIEVAL GLASS

THE CLAY PIPE *By Nigel Melton*

An assemblage of 74 of clay tobacco pipes comprising 43 bowls and bowl fragments, 25 stem fragments and 6 stem/bowl junctions was recovered. The pipes were recovered from Quadrant C, (5109, 5193, 5196 and 5210), and Quadrant E, (5179, 5197 and 5214) (Fig.3.18).

A study of the Chester pipe makers and their products was published by Rutter and Davey in 1980. This was based on pipes recovered from 15 excavations within the city and included a typology of the pipes that were produced in Chester, illustrations of makers' marks and the styles of decoration used, and a detailed chronology. All of the pipes recovered from the Delamere Street watching brief are Chester types and identifications and dating of the pipes are based on Rutter and Davey (1980).

There are large numbers of pipe makers recorded working at Chester from the 1640s through to the early 20th century. The industry peaked in the middle decades of the 18th century when, from the 1720s to 1780s, there were more than 30 makers in the city (Rutter and Davey 1980, 55). The pipes recovered from the Delamere Street watching brief provide a good cross section of the products of these makers. With the exception of the two, possibly residual, mid-17th century bowls from contexts (5197) and (5214) in Quadrant E, the Delamere Street pipes form a relatively tightly dated assemblage, the details of which may be summarized in Table 3.29.

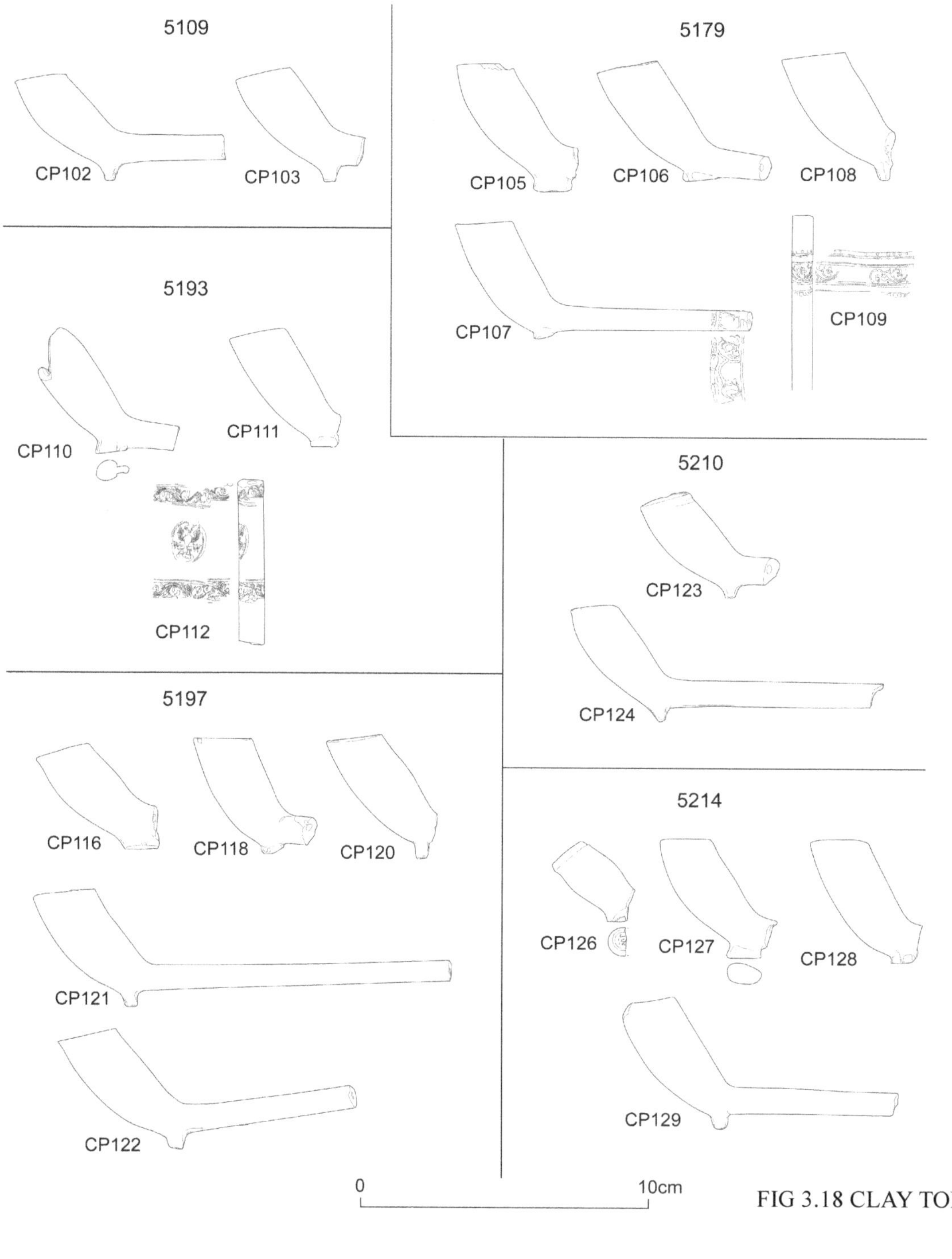

FIG 3.18 CLAY TOBACCO PIPE

TABLE 3.29: DATE SPREADS FOR CLAY TOBACCO PIPE ASSEMBLAGE

	Context		Pipe typology dates
Quadrant C	5109		Predominantly 1st quarter of the 18th century
	5193		Predominantly 2nd quarter of the 18th century
	5196		1st half of the 18th century
	5210		Predominantly 1st quarter of the 18th century
Quadrant E	5179		Predominantly 2nd quarter of the 18th century
	5197		Predominantly 2nd quarter of the 18th century
	5214		Predominantly 1st quarter of the 18th century

Catalogue

1. 7 bowl and bowl fragments, 3 stem fragments and 1 stem/bowl junction., Quadrant C, 5109, Phase 3a.

Bowls: ***Heel types***:

Chester type 69 (1690-1715), unmarked, stem bore $^{6}/_{64}$". ID 101.

Spur types:

Chester type 78 (1710-1720), unmarked, stem bore $^{5}/_{64}$ - $^{7}/_{64}$". 4 examples. ID 102.

Chester type 89 (1720-1750), unmarked, stem bore $^{5}/_{64}$" ID 103.

Bowl fragment, thin-walled, probably c. 1720-1750

Stem/bowl junction: Spur type, early 18th century, bore $^{7}/_{64}$" ID 104.

Stems: 50mm long, bore $^{7}/_{64}$"; 75mm long, bore $^{7}/_{64}$"; 66mm long, bore $^{5}/_{64}$".

2. 7 bowls and 5 stem fragments, Quadrant E, 5179, Phase 3a.

Bowls: ***Heel types***: Chester type 84 (1720-1750), unmarked, stem bore $^{5}/_{64}$" ID 105.

Chester type 85 (1720-1750), unmarked, stem bore $^{5}/_{64}$" ID 106.

Chester type 85? (1720-1750), unmarked, stem bore $^{5}/_{64}$"

Chester type 90 (1720-1750), stem bore $^{5}/_{64}$". Stem decorated with rolled-stamped border, weakly impressed, but a lattice type possibly similar to Rutter and Davey (1980), fig.58, no. 28. ID 107.

Spur types: Chester type 89 (1720-1750), unmarked, stem bore $^{5}/_{64}$"

2 fragments of thin-walled bowls, possibly from the same pipe, probably c.1720-1750 ID 108.

Stems: 32mm long, bore $^{5}/_{64}$"; 32mm long, bore $^{5}/_{64}$"; 79mm long, bore $^{5}/_{64}$"; 95mm long, bore $^{6}/_{64}$"; 100mm, bore $^{5}/_{64}$", decorated with rolled-stamped tulip and tendril border, possibly Rutter and Davey (1980), fig.60, no.65 (1720-1760). ID 109.

3. 4 bowls, 2 stem fragments and 1 stem/bowl junction, Quadrant C, 5193, Phase 3a.

Bowls: ***Small heel types***:

Chester type 79 (1710-1720), unmarked, stem bore $^{5}/_{64}$" ID 110.

Chester type 86 (1720-1750), unmarked, stem bore $^{5}/_{64}$. 2 examples. , ID 111.

Chester type 90 (1720-1750), unmarked, stem bore $^{5}/_{64}$" ID 112.

Stem/bowl junction: Heel type, possibly Chester type 84, early 18th century, bore $^{5}/_{64}$"

Stems: 60mm long, bore $^{7}/_{64}$", cut mouthpiece; 57mm long, bore $^{5}/_{64}$", rolled-stamped with oval stamp depicting eagle displayed and initials WW. Rutter and Davey (1980, fig. 54, no.10) illustrate this mark, but not the associated flower and tendril bands (Figure 1). No WW Chester maker was identified, but a date of 1700-1720 was suggested on stylistic grounds. FID 113.

4. 2 stem fragments and 1 stem/bowl junction, Quadrant C, 5196, Phase 3a.

Stem/bowl junction: Heel type, Chester type 83 (1720-1750), bore $^{5}/_{64}$"

Stems: 57mm long, bore $^{6}/_{64}$"; 90mm long, bore $^{5}/_{64}$", rolled-stamped, burnished. The fragment of surviving decoration consists of a border, possibly a fleur-de-lys type similar to Rutter and Davey (1980), fig. 59 no.50, dating to c. 1700-1720.

5. 18 bowls and 10 stem fragments, Quadrant E, Context 5197, Phase 3a.

Bowls: ***Heel types***: Chester type 22 (1640-1680), unmarked, stem bore $^{5}/_{64}$" , ID 114.

Chester type 68 (1690-1715), unmarked, stem bore $^{5}/_{64}$". 2 examples (1 damaged) ID 115.

Chester type 69 (1690-1715), unmarked, stem bore $^{7}/_{64}$" ID 116.

Chester type 83 (1720-1750), unmarked, stem bore $^{5}/_{64}$" ID 117.

Chester type 90 (1720-1750), stem bore $^{4}/_{64}$" - $^{5}/_{64}$" ID 118.

Chester type 92 (1720-1750), unmarked, stem bore $^{5}/_{64}$". 2 examples. XX, ID 119.

Spur types: Chester type 87 (1720-1750), unmarked, stem bore $^{5}/_{64}$". ID 120.

Chester type 88 (1720-1750), unmarked, stem bore $^{5}/_{64}$". 6 examples. ID 121.

Chester type 89 (1720-1750), unmarked, stem bore $^5/_{64}$". 3 examples. ID 122.

Stems: 46mm long, bore $^4/_{64}$"; 56mm long, bore $^5/_{64}$"; 58mm long, bore $^5/_{64}$"; 79mm long, bore $^6/_{64}$"; 82mm long, bore $^5/_{64}$"; 85mm long, bore $^5/_{64}$"; 100mm long, bore $^5/_{64}$"; 107mm long, bore $^5/_{64}$"; 132mm long, bore $^4/_{64}$".

6. 2 bowls, 3 stem fragments and 3 stem/bowl junctions, Quadrant C, 5210, Phase 3a.

Bowls: ***Spur types***: Chester type 66 (1690-1715), unmarked, stem bore $^7/_{64}$" ID 123.

Chester type 72 (1690-1715), unmarked, stem bore

Chester type 84 (1720-1750), unmarked, stem bore $^5/_{64}$" ID 124.

Stem/bowl junctions: Spur types, c.1690-1715, bore $^6/_{64}$ - $^7/_{64}$". 3 examples. ID 125.

Stems: 58mm long, burnished, bore $^7/_{64}$"; 50mm long, bore $^6/_{64}$"; 82mm long, bore $^6/_{64}$"

7. 5 bowls, Quadrant E, 5214, Phase 3a.

Bowls: ***Heel types***:

Chester type 27 (1640-1680), heel stamped NT, stem bore $^5/_{64}$". Bowl form and stamp identical to Rutter and Davey (1980), fig.45, no. 93. Probably a product of Nathaniel Thornley, recorded in Chester 1656-57 (*ibid*, 258) ID 126

Chester type 77 (1710-1720), unmarked, stem bore $^5/_{64}$". ID 127.

Chester type 79? (1710-1720), unmarked, stem bore $^5/_{64}$" (Heel missing)

Chester type 92 (1720-1750), stem bore $^5/_{64}$" ID 128.

Spur types:

Chester type 88 (1720-1750), unmarked, stem bore $^5/_{64}$" ID 129.

THE SMALL FINDS *By Chris Hewitson, and Erica Macey-Bracken. Species identification by Matilda Holmes*

Key Contexts

Only one of the six small finds recovered from the site was from a key context. This item was an incomplete copper alloy pin with a globular head, recovered from context 5179 (Phase 3b).

The copper alloy pin industry had existed since medieval times in England, with reference to the trade of pinners existing for the major centres of London, York and Coventry from the 14th to 15th century (Capel 2006, 135). By the post-medieval period the centres of industry had moved to areas where brass was available around Bristol in Gloucester, Stroud and Bristol itself (ibid, 136) By the 18th century the trade had begun to spread to new manufacturing centres where labour was cheaper such as Birmingham after 1750 (Cox 2005, 4-18). Localised development of the Pin Industry did not occur until the establishment of Works by Stubs Wood and Company at Warrington in 1823 (Dutton and Jones 1983, 183). Given the pin was recovered from an 18th century context this suggests that the pin was either a regional import from another manufacturing centre or locally produced from imported wire.

Catalogue

1. Part of copper alloy pin with globular head. Length 20mm; Quadrant E; 5179, SF 6, Phase 3b.

Non-Key Contexts

All of the small finds recovered from non-key contexts were worked bone and/or ferrous material, including domestic implements such as brush and implement handles and a knife. The most complete item recovered was a bone scale-handled tanged knife, with only the end part of the blade missing (5096, Phase 4a). The handle of the knife was incised with diagonal lines, and attached to the tang by iron rivets. Stylistically it was robust and it would appear to pre-date its 19th century context. Scale tanged knives were common from the medieval period but the style and size of the knife suggests it dates to the 17th or 18th century.

Two other handles were recovered that related to cutlery items. One was a tanged table knife with surviving squared handle (5136, Phase 4a). Stylistically this type of knife was typical of the 1850s (Brown 2001, 130-1). A further knife/ fork bone handle was recovered from layer 5116 (Phase 4b). It came from a scale-tanged knife handle of bone, with surviving rivets, stylistically 19th century in date (ibid, 129).

Three handles were recovered from 19th century contexts. A single handle from layer 5096 (Phase 4a) has the words EXTRA FIN [E] stamped on one side. The purpose of the handle was unclear but morphologically it looked to be a further brush handle. Two brush handles (5090, Phase 5; 5116, Phase 4b) had both broken at the point where the sockets for the brush bristles started. The handles all appear to relate to small brushes such as tooth brushes.

Catalogue

1. Curved handle of a narrow-bodied brush, tapering to a rounded point at one end, and broken off at the other. Size and form suggests a toothbrush. Made from a single piece of bone. Large mammal, rib/ pelvis. Dimensions 110mm, 13mm, 4mm; 5090, Quadrant A, Strip 3, SF3, Phase 5.
2. Near-complete bone-handled knife. Handle is incised with diagonal lines. Handle is composed of two scales of bone riveted together through the tang of the knife blade. Large mammal, rib. Handle: 95mm x 21mm x 14mm; Blade: 65mm x 27 – 16mm x 2mm. 5096, Quadrant D, Strip 2, SF1, Phase 4a.
3. Handle of an implement, possibly a piece of cutlery. Handle is rounded off at one end, tapering to a narrow neck at the other end, where it is broken. Handle is stamped with the words EXTRA FIN[E]. Unidentified mammal, rib/ pelvis. 100mm x 14mm x 4mm. 5096, Quadrant B, Strip 2, SF2, Phase 4a.
4. Plate from a handle with three copper alloy rivets spaced along the length. Item is stained blue-green around the site of two of the rivets. Large mammal, unknown large bone fragment. 99mm x 22mm – 14mm x 7mm. 5116, Area F, Strip 4, SF4, Phase 4b.
5. Curved handle of a narrow-bodied brush. Handle is made from a single piece of bone, is roughly circular in shape and highly polished with a rounded end. Unidentified mammal, unknown large bone fragment. 124mm x 10mm x 9mm. 5116, Area F, Strip 4, SF5, Phase 4b.
6. Knife handle, possibly from a piece of cutlery. The end has been squared off and the other end tapers to a point where it joins the iron tool. The shape of the knife is obscured by corrosion products. Unidentified mammal, large fragment. Handle: 83mm x 15 – 11mm x 7mm; Tool: 52mm x c.a. 10mm x c.a. 6mm. 5136, Phase 4a.

A MEDIEVAL SMALL FIND *By Hilary Cool*

In addition to the Roman finds uncovered during the 2006 excavation there is one medieval finds (No. 105, Fig.3.19) that is worth comment. There is a certain amount of debate over what these very distinctive bone implements with iron points were used for. It is generally agreed they were used in writing because of their association with wax tablets, inkwells etc at sites on the continent. Different authors, however, use the same evidence to argue that they were either used as parchment prickers for marking out writing guidelines (Biddle and Brown 1990, 733-5), or as styli for the wax tablets (Riddler, 1998, 272). What is certain though, is that they are a late medieval type, common in the 13th and 14th centuries. Though the context it came from forms part of the Phase 1 fill of pit [4041], it must clearly be intrusive. Implements such as these are regularly found on monastic and school sites suggesting there might have been such an establishment in the vicinity.

Catalogue

1. Writing implement. Bone shank tapering slight to both terminals; head is an expanded rounded cone divided into two ribs; flat base with small iron rod inserted. Length 50mm, head and maximum shank section 4.5mm. Area D, 4015, Group 11, ID 179, Phase 1.

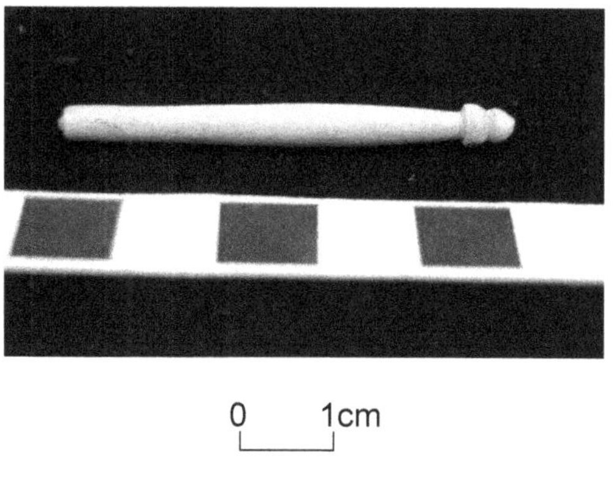

FIG 3.19 A MEDIEVAL SMALL FIND

THE COINS *By Roger White*

A small assemblage of modern coins was uncovered during the 2006 excavations on the site.

Catalogue

1. Farthing of George V, 1921; Area B, 2018, SF18, Phase 5.
2. Indian silver rupee of George VI, 1940, Area B, 2063, SF13, Phase 5.

CERAMIC BUIDLING MATERIAL AND ARCHITECTURAL STONE

THE CERAMIC BUILDING MATERIAL *by Jennie Stopford*

The tile assemblage consists of ten whole or part floor tiles, one ornamental brick and 12 fragmentary roof tiles.

The floor tile

There are four identifiable floor tile types as follows (for details of the recording methodology, see Stopford 1990):

Type 1 (5179, Area E)

One incomplete medieval floor tile decorated with a fairly common design consisting of a repeating eight-petalled flower design, with each flower set in a square and the squares set in a diagonal grid. The tile was probably made using a wooden stamp cut so that the design stood proud on the stamp, leaving an impression on the surface of the clay quarry. The fired tile measures 116mm across and 26mm deep.

The decorative technique used is generally described as 'line-impressed' because the tile was glazed to a single colour (dark green in this case) with the linear design impressed on the tile body. As the linear impressions are quite broad (*c.* 3mm) and have moulded (rather than straight) edges, the design is sometimes also described as in 'counter-relief'. Many versions of this design are known but a distinct feature in this case is that the two diagonal lines that form the grid do not meet in the centre; one of the lines is broken either side of the other in the centre of the tile.

The sides of the tile are steeply bevelled so that the upper surface would have fitted immediately next to its adjacent counterparts in a pavement, with the design running on as a continuous repeating pattern. Any mortar would have been held in the space left by the bevel. The tile was fired to produce a hard quarry that is oxidized throughout. The fabric is coarse, containing some pebbles of up to 10mm diameter, grog up to 5mm diameter and quartz sand up to 1mm, but had been well mixed. The base is uneven and coated with sand of up to 1mm diameter. There are no keys in the tile base and no nail-holes in the upper surface.

The tile is worn, with the glaze only surviving in the depressions made by the design stamp. It must have been in use in a floor for a lengthy period. No mortar remains on the sides or base.

Comparison with other material

Similar versions of this design are known from sites across the north of England, in the West Midlands and in Wales as well as elsewhere in Chester, including the site of the Roman amphitheatre, St John's Church and 11 Watergate Street (Lewis 1999, 752-4; Rutter 1990, 119/188; Stopford 2005, figs 22.1, 23.36 and 23.1, 24.33; Thompson 1976, fig 43, 1, 19, 20). However, apparently the same version as that from Delamare Street, with the grid line broken at the crossing, is known from Basingwerk Abbey, on the Dee estuary in north Wales (Lewis 1999, 93 and design no. 753). The Delamere Street tile differs slightly in manufacture from those of John Lewis' Group 53, being slightly larger, with steeply bevelled sides and without white specks in the glaze. It was not over-fired. The tiles from Basingwerk Abbey were assigned to the 15th century by John Lewis. In the north of England, tiles of similar design were thought to date from the second half of the 15th and early 16th centuries (Stopford 2005, 235 and 245).

Type 2 (5115; 5214)

The two tiles of this type are both triangles, one whole but worn with traces of green glaze and the other a fragment, glazed yellow over white clay with a line-impressed design that is previously unpublished. The design is a continuous repeating pattern that had a central fleur-de-lis enclosed in a circular band, with smaller fleur-de-lis in each quadrant also in circular bands. The line-impression is finer (1-2mm diameter) than on the Type 1 tile above and, at 22-22mm, the tiles were significantly less deep. However, when complete, their upper surface dimensions may have been similar to the Type 1 tile and some other aspects of their manufacture are similar. It is not impossible that they were in fact products of the same workshop. Green and yellow glazed tiles were often used together in late medieval pavements, to give an overall chequered effect.

The sides of the tiles were steeply bevelled and, while both were fired to form a hard quarry, the line impressed example was particularly highly fired. The quarry of the worn tile was oxidized throughout while the core of the line impressed tile was partly reduced to a dark grey. The fabric includes pebbles of up to 10mm diameter, grog up to 3mm diameter and quartz sand up to 1mm. There are a few voids suggesting some organics in the mix. The bases of both tiles are uneven and coated with sand of up to 1mm diameter. There are no keys in the bases and no nail-holes in the upper surfaces.

FIG 3.20 CERAMIC TILE AND BRICK

Both tiles had been scored and split from square tiles to make two triangles. The scoring was done before firing, with the tiles broken in two after firing. The scar from the break suggests that a wire may have been used for scoring (rather than a knife) since the scoring has a curved profile and is deeper at the corners than in the centre.

The yellow line-impressed tile is completely unworn, while the other tile is completely worn. The lack of wear on the yellow example does not necessarily mean it was never used; it could have been laid in an area where it was not walked on. No mortar remains on the yellow tile but the mortar on the worn example probably accreted after its original use.

The layer of white clay on the upper surface of the decorated tile is uneven (over 1mm in places) and must have been applied in a fairly plastic state (rather than as a thin slip). The white clay has not fully coated the quarry surface and consequently there are patches and spots that have fired dark brown.

Comparison with other material

The design on the yellow example is unknown elsewhere but line impressed tiles of similar type are well known from Chester and the surrounding area. Despite some differences from the Type 1 tile above, it is possible that they were the products of the same workshop with some broad similarities in manufacture, fabric and firing. The tiles could have been made in either the 14th or 15th centuries. The *fleur-de-lys* is one of the commonest motifs of the medieval period but the relative elaboration of the design and some aspects of manufacture suggest a later rather than earlier date.

Type 3 (5179; 5197; 5214)

This group consists of six tiles, one triangular with a green glaze remaining on the upper surface, four square and completely worn and one worn fragment. The worn tiles and fragment have either green or brown glaze on their sides and bases. The brown glaze may have been applied over a white slip and fired yellow.

The tiles are solidly made with vertical or only slightly bevelled sides, measuring 115-110mm across and 30-31mm deep. The fabric is oxidized throughout on all examples having been fairly well prepared with a fine sand (less than 1mm) but occasional grog and pebbles of up to 10mm. The fabric has cracked in two cases, with the upper surface of one tile beginning to sheer away. Some of the tile bases have mortar adhering to them. Where visible, the bases have a fine sandy coating and there are no keys. The triangular tile had been scored and split from a square tile using a blade that cut to an equal depth across the diagonal.

The nail-holes, close to the corners (5-10mm from the tile edges) on the upper surface of some of the tiles, are a distinctive feature. In the clearest case there are two nail-holes of 2-3mm diameter and up to 4mm deep in diagonally opposed corners. These holes are generally associated with a late-medieval tile-making technique where the quarries were either cut out or moved about using a wooden board with nails knocked through it.

Comparison with other material

It is not known if any of the worn examples were once decorated but the occurrence of nail-holes on medieval tiles is often associated with Plain-glazed tiles, which were either coated with a slip and glazed yellow, or glazed dark green or brown over the body fabric. Nail-holes are also often associated with tiles imported from the Netherlands in the later 14th to 16th centuries. In a study of all the tiles in the north of England the plain-glazed tiles thought to be imports had five nail-holes (one in each corner and one in the centre). However, from the mid 15th century some decorated types made in England were found to also have this feature, but often with two or four nail-holes rather than five. Nail-holes, for various reasons, are only ever visible on a proportion of examples (for example due to the frequent damage to tile corners). If the two diagonally opposed nail-holes identified here are a genuine characteristic of the group it might suggest that these are late medieval tiles that were made locally rather than being imported. They date to the fifteenth or early sixteenth century.

Type 4 (5197)

One corner fragment in two pieces, made entirely from white clay with a very low iron content that has fired to a pinkish white. The quarry is of fine sandy clay with up to 10mm white clay grog and a few 2mm pieces of red clay grog. The tile has been badly abraded but traces of green glaze remain on the upper surface with more on the sides and base. The tile depth is 25mm and the sides are slightly bevelled. The base has no added sand and there are no keys. Little can be said about this tile. The use of a white clay quarry is unusual in England and it may be a continental import. It is either medieval or early post medieval.

Ornamental brick

The assemblage includes a single large fragment of decorated brick (Fig. 3.20) with incomplete measurements of 148mm in length, 122mm in width and a complete depth of *c* 45. The upper surface has a raised rectangular panel running lengthwise on which, in further relief, there are three fleur-de-lis one above the other. The raised panel with the relief decoration has been glazed and fired brown. The glaze has been roughly applied and covers some of the adjacent areas but much of the lower, undecorated panel, was never glazed.

The blackening of part of the upper surface shows that the piece has been exposed to burning and, if this occurred when in its original location, it might suggest use as a stove tile or part of a fire surround. However it is thick for a stove tile and there are no signs of the flanges found on some examples.

The orange fabric is coarse including pebbles of 10mm, voids resulting from organic inclusions and a piece of wood or other cellular material that has not been destroyed in firing. The fabric is friable with much cracking. This is dated to the late medieval or early post-medieval.

The roof tile

The small assemblage of 12 tiles includes seven types of roof tile as follows:

Type 1 (5115)

1. One collared fragment of unglazed curved tile. Possibly Roman. Orange homogenous fabric.

Type 2 (5197)

2. One curved fragment with splashes of glaze on the upper surface, depth 14mm. Fabric partly reduced with a sandied underside.

Type 3 (5179, 5136)

3. One slightly curved fragment with a slight lip along the only complete edge, glazed upper surface, depth 17mm.
4. One flat fragment with no complete edges, glazed upper surface, depth 14mm.

Orange-brown glaze, oxidized fabric.

Type 4 (5090, 5214)

5. One fragment flat tile, upper side glazed. Depth 10mm.
6. One incomplete tile in two pieces, slightly curved, upper side glazed. Depth 18mm. The glaze is a shiny dark brown that has fused poorly with the body fabric, flaking off in places. The body fabric is distinctive, including a large proportion of fired and unfired white clay.

Type 5 (5090, 5214)

7. One fragment flanged, but otherwise flat or tapered, with a green slip on the upper side. Depth variable c.18mm.
8. One fragment flat tile with thin patchy yellow or black glaze. Depth 11mm.

Fabric low in iron and reduced to a light grey.

Type 6 (5210)

9. One incomplete tile in two fragments, flat, upper side glazed. Depth 18mm. Two incomplete tiles, one in two fragments, slightly curved, upper side glazed. Depth 18mm.
10. The red-brown glaze on both pieces shows brush marks. The glaze has not fused well with the body, flaking off in places. The orange fabric includes the occasional pebble and some large pieces of grog that have contributed to the tiles cracking and breaking up.

Type 7 (5179, 5193)

11. One incomplete salt-glazed tile in four fragments, flat or slightly curved, glazed patchy yellow and brown both sides. Incomplete length 143mm, incomplete width 95mm, depth 15mm.

THE ARCHITECTURAL STONE *By Michael Lobb*

Sandstone walls were analysed *in situ* and smaller stone samples were retrieved for closer inspection. The apsidal and square room (2092), located at the southeast corner of Area A, comprised a back-filled cellar. The walls were built of a combination of brick, random rubble, squared rubble and re-used masonry. Within the east wall of the room were two notable pieces of re-used masonry, laid side by side, within context 2090 (Fig. 3.7) These were retained after the removal of the wall, and cleaned of the thick white lime mortar used to construct the wall.

Both elements are composed of pale grey-white limestone, and have been worked into a similar shape, although their dimensions are slightly different. The exterior faces of the stone consist of two smooth faced surfaces set at right angles to each other; the top of each stone has also been worked to a smooth finish. The interior curve of each piece has been roughly worked with a flat chisel to create keying-in marks for fixing to other masonry. The underside of each piece also remains unfinished for keying in.

Whilst re-used, both elements retain some evidence of their original mortar, which was a dark orange-yellow, unlike the creamy-white mortar of wall 2092. This is principally found around the ends of each curve and around the underside. The second piece has evidence of a groove around the junction of the curve and the top face, which retains some of the original mortar.

From the finish applied to the various faces of the masonry elements, it appears that only three faces of the masonry were faced for display, being the top and two sides set at a right angle to each other (Fig. 3.21). This would appear to indicate that the elements formed the base of a masonry column, approximately 0.5m in diameter. In terms of date there is little that is diagnostic about the masonry, an approximate date may be reached by the relative dating of context (2092), though as the piece is clearly re-used, this will only provide the latest date of production.

Catalogue

1. Two pieces of ashlar faced pale grey-white limestone column base. Mortar adhering to two phases, reused in later wall construction; Size 400mm, 360mm, 200mm, Diameter *c* 500mm; Size 350mm, 260mm, 180mm, Area B, S20, 2092, Phase 4b, Ill. XX

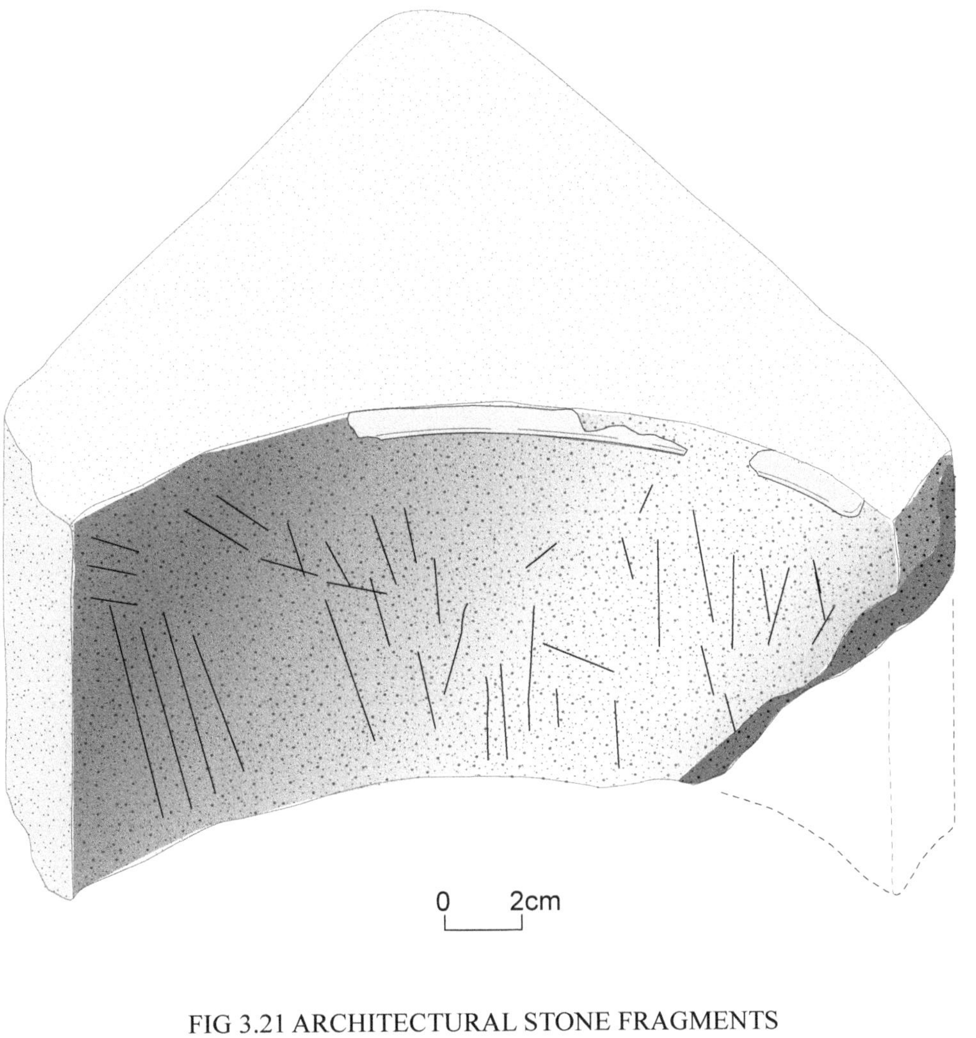

FIG 3.21 ARCHITECTURAL STONE FRAGMENTS

ANIMAL BONE *By Matilda Holmes*

Animal bone was recovered from Quadrants C, C/E, E and D of the post-medieval quarry, which was backfilled with deposits containing a small quantity of horse, cattle and other bones in the post-medieval and modern phases. Although the assemblage is small, it provides a useful insight into the deposition of waste in Chester. Horse and cattle bones predominated in both phases, although sheep/goat (of which sheep were positively identified, but no goat bones), dog and hare were also recovered (Table 3.30). Such a large quantity of horse bones is unusual in general domestic waste normally found in urban assemblages of this date, and may indicate that this was an area used for the deposition of refuse from different activities. This report will investigate the possible origins of the deposit.

Methodology

All fragments were recorded using the authors reference collection where necessary. Due to anatomical similarities between sheep and goat, bones of this type were assigned to the category 'sheep/goat', unless a definite identification using guidelines from Prummel and Frisch (1986) or Payne (1985) could be made. Bones that could not be identified to species were, where possible, categorised according to the relative size of the animal represented (small – rodent /rabbit sized, medium – sheep / pig / dog sized, or large – cattle / horse size). Ribs were not identified to species, and only the maxilla, zygomatic arch and occipital areas were identified from skull fragments.

Tooth height, wear and eruption were noted where appropriate using guidelines from Grant (1982), Levine (1982) and Silver (1969), as were bone fusion (Amorosi, 1989; Silver, 1969), metrical data (Albarella and Payne, 2005; Davis, 1992; von den Driesch, 1976), anatomy, side, zone (Serjeantson, 1996), pathology, butchery (Lauwerier, 1988; Sykes, 2007), bone working and condition (Lyman, 1994).

Condition and Taphonomy

The bones were in excellent condition although rather friable, as a number exhibited signs of fresh breakage, and 16 fragments could be conjoined to make 6 larger fragments. 8% of bones from post medieval and 3% of those from modern deposits exhibited signs of canine gnawing, this is not a large proportion of the assemblage, and it is likely that the bones were buried fairly quickly. There was no evidence for burning, and between 6% and 12% of bones had been butchered, suggesting that the bones had undergone minimal processing.

Butchery and Fragment Representation

Butchery took the form of chop, cut and saw marks on cattle and horse limb bones. The majority were transverse chops or cuts to the ends or mid shaft, but some had been split longitudinally, and others bore shaving marks on the shaft. One cattle horn core had been sawn through. Cattle bones were more often butchered and more fragmented than horse. Table 3.31 shows the proportion of bones more than half complete, of which the edible domestic species (cattle and sheep) were more fragmented than horse, dog and hare bones. This suggests that cattle and sheep carcasses underwent more processing before deposition. The parts of the carcass present in the assemblage for cattle and horse is shown in Table 3.32. All parts of the skeleton were recovered from horses in both phases, whereas cattle were best represented by horn cores and upper limb bones, indicating that only selected parts of the cattle carcass were present.

The Assemblage

Cattle

All ageing data came from bone fusion, of which the all bones were fused with the exception of one cattle calcaneum that would have come from an animal less than 42 months of age. No horn cores were complete enough to measure the greatest length, but all were curved, nearly all included part of the frontal bone, and those that could be allocated to a category were nearly all from males (9:1), following guidelines by Sykes and Symmons (2007).

Horse

All horse bones were fused, and the fusion of vertebrae and presence of third molars both indicate that these animals were over 5 years old. Further examination of the teeth heights suggest that animals were very old when they died – four of the five post medieval specimens and the only modern sample being over 20 years of age (the remaining post medieval mandible was from an animal 11-12 years old), which suggests that the horses had a long working life.

Given the high proportion of complete bones in the assemblage, a number of wither heights were available, as shown in Figure 3.22 (using indices from Kiesewalter, 1888), indicating that the animals disposed of at this site were between 1.48 and 1.63mtrs tall (14.3hh and 16.1hh). These are larger than those from many medieval sites, which were more in the range of 12.2 and 15 hands high (Rackham, 1995), and may indicate the presence of larger breeds that were introduced for draught, driving or traction in the 16th to 17th centuries (Langdon, 1986; Smil, 2000). The number of measurements available from such a small assemblage was too small to use for the calculation of log ratios, although they do fall within the range of measurements taken from post medieval horses at Bridge street, Dorchester (Bullock, n.d.) and Staff College, Bracknell (Holmes forthcoming). A full list of metrical data is included in the original database within the site archive.

For such a small assemblage, a number of pathologies were recorded, all from the post medieval phase. Many of them were exostoses or lipping to the bone associated with trauma or simply age related degeneration. Examples of this came from a lumber vertebra, 1st cervical vertebra (axis) articular surface of a scapula and a proximal metacarpal. A lateral metacarpal was fused to the metacarpal itself, also displaying a bony growth which is commonly known as a 'splint', and can occur in young horses, but in a position such as this would not have impeded the movement of the animal.

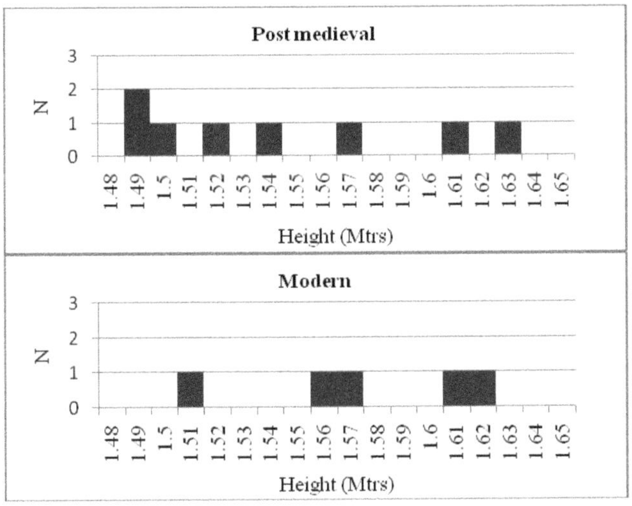

FIG 3.22 ANIMAL BONE 1

TABLE 3.30: SPECIES REPRESENTATION (FRAGMENT COUNT)

Species	Post Medieval		Modern
Cattle	37		10
Sheep / Goat	2		2
Sheep	1		1
Horse	61		29
Dog	2		2
Hare	1		
Total Identified	104		44
Unidentified Mammal	41		1
Large Mammal	6		9
Total	151		54

TABLE 3.31: % PROPORTION OF BONES OVER 50% COMPLETE

	Post Medieval		Modern
Cattle	68		57
Horse	89		86
Sheep / Goat	33		100
Dog	100		100
Hare	100		
n	75		31

TABLE 3.32: FRAGMENT REPRESENTATION (EPIPHYSIS COUNT)

	Post Medieval		Modern	
Anatomy	Cattle	Horse	Cattle	Horse
Mandible **		6		1
Horn Core***	6		4	
Skull	1	1		
Atlas		1		
Axis				
Sacrum				
Cervical		2		3
Thoracic	5	3		
Lumber		1		
Scapula D	2	3		1
Humerus P		4		1
Humerus D		4	1	1
Radius P	2	4		4
Radius D		2		2
Pelvis	1	2		1
Femur P	1			
Femur D	2			1
Tibia P	1	1		3
Tibia D		1		4
Calcaneum	2			
Metacarpal P		3		
Metatarsal P		4		1
Metacarpal D		1		
Metatarsal D		4		1
1st phalange *		1		
2nd phalange *				
3rd phalange *				
Total	23	48	5	24
MNI	5	5	4	4
* count adjusted for frequency bias				
** mandibles with molars				
*** horn cores >50% complete				

Other animals

A dog tibia, the proximal end of which was unfused, indicated an individual less than 18 months old, the rest of the dog bones were fused. Shoulder heights were available from a modern dog humerus which would have stood 0.5mtrs high (Harcourt, 1974). One pathological bone was recovered from the modern phase, as lipping on the medial and lateral aspects of the humeral trochlea, which may have been caused by trauma to the bone, or degeneration as a result of old age. Sheep mandibles gave tooth wear evidence to suggest animals were between 2 and 4 years of age.

Discussion

A predominance of cattle horn cores is indicative of tanning or horn working waste (Prummel, 1978; Serjeantson, 1989; Yeomans, 2008). Coupled with this is the presence of largely complete horse carcasses that had undergone minimal butchery, a trend that has also been noted from tanning sites (Shaw, 1996). Other deposits of post medieval horse bones have been interpreted as knacker waste, the meat being given to dogs, but these are always more heavily butchered (Holmes Forthcoming; Thomas, 2000), which is not the case at Delamere Street. It is therefore likely that this deposit originated as waste from hide processing, whereby cattle hides were sent to the tanner from the butcher complete with horns (later discarded by the tanner). Horses, however, were not eaten at this time, the whole carcass being acquired the tawyer or tanner, who would skin them and then dispose of the bodies with other waste (Albarella, 2003). A contemporary site from City Road, Chester (Sykes *et al.*, n.d.) has produced an assemblage of cattle horn cores and equid remains, similar to that of Delamere Street, which was also interpreted as tanning debris.

Traditionally, the hide processing areas of Chester were situated in the east of the town, around the east gate and St John's areas (Laughton, 2008), indeed the City Road site is also in this area, and contained many of the features of a tannery. However, the Delamere Street assemblage is in the north of the town, so the deposition of what is most likely waste from large, skinned animals, could reflect an opportune dumping of offensive waste in a newly disused quarry. A larger assemblage was recovered from Kingston-Upon-Thames, Surrey (Serjeantson *et al*, 1992), from a minimum number of 11 horse carcasses disposed of along with 80 cattle horn cores dated to the 14th century. It is possible, therefore, that this was a continuation of a medieval practice for disposing of old, worn out horses.

THE HUMAN BONE *By Samantha Hepburn*

Introduction

All of the remains from the post medieval quarry were retrieved from along the southern edge (Quarry Face 3) of the quarry, with a large part of the assemblage coming from a single deposit (5214) within Quadrant E. The remains represent a minimum number of seven adult individuals. This was achieved using guidelines set out in McKinley (2004) with the aid of skeletal recording sheets.

Preservation

A total of 81 fragments/ complete bones were recovered during the watching brief, see Table 3.33 for a breakdown of fragment representation. All of the remains were disarticulated. Using the system of bone surface preservation recommended by McKinley (2004) the bone from the two sites was recorded as Grade 1-2. The edges and ends of many of the long bones were damaged. The fragmentary nature of the bone has greatly reduced the amount of metrical information that could be derived from the remains. It has been possible to fit several of the fragments together, and as a result reduce the minimum number of individuals represented by the assemblage.

Age at Death and Sex Determination

Due to the nature of the bones present it was not possible to offer any detailed estimation as to the age the individuals present. The majority of the bone is robust; and where the proximal and distal ends are present, epiphyseal fusion is evident, which implies that the bone present belongs to adult individuals above the age of twenty based on the ages at which epiphyseal fusion occurs in the humerus, ulna, and femur (Gray 1918).

The only bones that were able to be sexed came from the layer (5214); the remains were sexed using dimorphic aspects stated in Bass (1995) and consisted of, part of a female left pelvic innominate, and the right half of a mandible which displayed male characteristics. The skull fragments from this context form part of the cranial vault, four of the fragments fitted together to form a near complete frontal bone, that had slightly pronounced brow ridges, which might suggest it was male. Although due to a lack of any other sexual diamorphic attributes on the skull fragments it is impossible to say definitively what the sex of the individual was.

Pathology/Trauma

New bone growth on the cortical bone is present on several femur and humerus fragments across the entire assemblage, in particular along the linea aspera of the femurs. Not enough bone survives to be able to draw any firm conclusions as to the cause; it could be a result of disease, physical action or age. Early signs of osteo-arthritis are present on the proximal head of an ulna (5214) and the distal end of a femur (5096) both display new bone grow and a slight porosity (two of the signs recommended by Waldron and Rogers (1991) for possible diagnosis if eburnation is not present).

The near complete frontal bone from (5214) shows signs of weapon trauma above the left orbit, there was a faint suggestion of a 30mm lateral cut 1mm deep. There are no associated radiating fractures, which suggests, that the injury was made with a sharp blade that scored the bone as it cut through the flesh. The wound shows signs of healing over a long period of time, so it is unlikely to have caused the death of the individual.

TABLE 3.33 HUMAN BONE FRAGMENT REPRESENTATION

	2006 Excavation	*2008 Excavation (5214)*			**2008 Excavation Other Contexts**
		Left side	**Right Side**	**Number of Fragments**	
Skull	-	-	-	26	-
Humerus	1 (Left)	5	-	6	-
Ulna	-	-	4	4	-
Ribs and Sternum	-	-	-	2	1
Pelvis	-	-	-	1	-
Femur	2 (Right side)	4	2	10	3 (left side)
Tibia	-	-	2	5	-
Fibula	-	-	-	1	-
Unidentifiable small fragments	-	-	-	20	-
Total no. of fragments	3	77			4
	84				

The femoral head recovered from (1008) displays clean tool marks on the joint surface, which is consistent with new damage caused during the excavation process. Tool marks are also present on fragments of a femur and tibia from (5214) and a femoral head from (5228), unlike the other marks these are not fresh cut marks made during retrieval. It is most likely that they made when then burials were removed from their original resting place, prior to deposition within the quarry.

Radiocarbon Dates

Radiocarbon dating was recommended to date the remains retrieved from the post-medieval quarry. The potential for Roman inhumations north of the town walls had been highlighted by previous work (Margary 1955, 35; Jermy 1961; Thompson 1965, 48). Two samples were taken from different contexts and quadrants of the quarry in order to establish a potential spread of dates for the human remains and minimise the potential for a single dump affecting the results. The first was from a large quarry fill in Quadrant D adjacent to Quarry Face 2 (5096, Phase 3b), and the second from a smaller deposit (5214, Phase 3a) within Quadrant E adjacent to Quarry Face 3 which contained the main deposit of human remains. In each case part of the shaft from a femur was used to extract the collagen needed to produce the date. The radiocarbon dating was carried out by Beta Analytic Inc, an ISO/ISE 17025:2005 accredited laboratory.

Discussion

The human bone from the post medieval quarry at Delamere Street came from a varying series of deposits along the SE quarry face, with the greatest concentration of remains coming from (5214, Phase 3a) in Quadrant E, an area which contained the earliest sequence of deposits. The radiocarbon date for this assemblage taken from a fragment of a single adult femur came back as AD 1160 to 1280; this is at odds with the other finds retrieved from this deposit which date to the first half of the 18th Century. Cut marks on two of the bone fragments (a femur and a tibia) from this assemblage were made by hand tools; they were made in antiquity as opposed to during excavation. A smaller amount of human remains were also recovered from the quarry from three other contexts. A radiocarbon date taken from an adult femur from (5096, Phase 3b) a large quarry fill produced a date range of AD 1430-1520 to Cal AD 1590 to 1620. Again the date of the remains is inconsistent with other material recovered from this deposit.

The presence of these earlier remains within the quarry suggests a period of major redevelopment in the area. The later remains could be explained by the cutting of the Chester and Nantwich canal which opened in 1779 (Lewis and Thacker 2003), disturbed earlier burials which were subsequently deposited in the quarry backfills. The suggested date of deposition for layer (5214) of the early-18th century would suggest a separate date of deposition. The obvious candidate for the source of the earlier medieval burials and possibly the late-medieval to early post-medieval burials would be the Hospital of Little St John to the west of the current site. Founded in the 1190's it provided accommodation for 13 inmates, it survived the dissolution due to its hospital status, but was subsequently demolished in 1644 to prevent it from giving cover during the siege of Chester. The site was not rebuilt upon until 1717, when a chapel, charity school and six almshouses were constructed (Baggs *et al* 1980). The hospital was in use during the time periods that both sets of radio-carbon date's span. The evidence for the area around Northgate suggests development had been occurring throughout the quarries lifespan, and as a result it may have become a convenient dumping ground for the surrounding area.

Sample Data	Measured Radiocarbon Age	13C/12C Ratio	Conventional Radiocarbon Age (*)
BA1410-5096	340+/-40BP	-19.9 o/oo	420+/-40BP

Analysis: AMS-Standard Delivery
Material/Pre-treatment: (bone Collagen): collagen extraction: with Alkali
2 Sigma Calibration: Cal AD 1430 to 1520 (Cal BP 520 to 430) and Cal AD 1590 to 1620 (Cal BP 360 to 330)

Sample Data	Measured Radiocarbon Age	13C/12C Ratio	Conventional Radiocarbon Age (*)
BA1410-5214	710+/-40BP	-18.6 o/oo	810+/-40BP

Analysis: AMS-Standard Delivery
Material/Pre-treatment: (bone Collagen): collagen extraction: with Alkali
2 Sigma Calibration: Cal AD 1160 to 1280 (Cal BP 790 to 670)

SHELL *by Erica Macey-Bracken and Chris Hewitson*

Twenty three shells were recovered from the key contexts of the site. The shells were a mixture of oyster (*Ostrea edulis*), common cockle (*Cerastoderma edule*), and terrestrial snails of the genus (*Capaea hortensis*) (Table 3.34).

The oyster shells were mostly complete, and displayed a size range between 70 and 128mm in diameter, with the largest coming from layer 5179. Several of the oyster shells displayed parasitic alteration in the form of perforations on the exterior surfaces (5197, 5179, 5109). A single shell (5179) was infested with calcareous worm holes (*pomatocerus*) on its exterior face. None of the oyster shells had evidence for mortar adhering on their surfaces. The purpose of the shells is unclear but the presence of parasitic alterations may suggest they were empty when brought to site and were not therefore food items. This may point to their use in mortar production.

A single common cockle (*Cerastoderma edule;* 5210) was recovered. Together the collection of marine molluscs was small given the scale of the watching brief and is certainly insufficient for detailed statistical analysis or comparison. However, local sources have been suggested for both the oyster in Colwyn Bay and common cockle from any of the local estuary of the Wirral as well as Morecambe Bay, Bangor and Conway (Harrison 1990, 58). There is insufficient evidence to suggest a source from further afield.

A series of terrestrial snail shells were recovered from the interior of a bottle. All appeared to be a grey-white variety from the genus *capaea hortensis*. The snails probably accumulated prior to the deposition of the bottle.

TABLE 3.34 SHELL SPECIES

Context	***Oyster*** *Ostrea edulis*	***Snail*** *Capaea hortensis*	***Cockle*** *Cerastoderma edule*
5109	4		
5179	5	11	
5197	1		
5210	1		1
TOTAL	**11**	**11**	**1**

THE HISTORICAL DEVELOPMENT OF THE POST-MEDIEVAL QUARRY AT GORSE STACKS

The topography of the northeast of Chester in the late-medieval period

The site of Gorse Stacks lies to the northeast of the walled medieval city. The walls that surround the city date from the foundation of the Roman civic centre. The Roman fortress was succeeded in the early Middle Ages by a small fortified town on the same site. Two sides of the Roman walls were abandoned, probably in the 10th century and by the early 12th century the circuit of walls had reached its modern extent (VCH 2003, Vol 5.1, 1). The walls were built up in stone between the 13th and 14th centuries and the repair of the city wall demanded additional stone resources. Whereas on the western and southern sides the natural escarpment of the Dee valley acted as a defensive barrier, to the north and east the walls were supplemented by a large ditch known from medieval times as the Kale-Yard Ditch [le Calliord Diche]. These two elements dominated the topography of this part of the town until the cutting of the Chester Canal in 1771 (which adopted and subsequently destroyed the line of the ditch along the walled city's northern edge).

The streets that lay to the north of the city were laid out, predominantly in medieval times. Northgate, the principal route north of the town, was laid out along the line of the Via Decorum, the main north access route to the Roman civic centre. The suburb's main axis lay along Upper Northgate Street, the road north into Wirral, which from early times forked to form Bache Way (later Liverpool Road) running north, and Mollington Lane (later Parkgate Road) running north-west.

The area between Brook Street and Upper Northgate Street, northeast of the city, was wholly agricultural. The greater part comprised the Town Field (also called Chester Field), the western boundary of which formed Windmill or Besom Lane (later Victoria Road) in the 13th century.

The hospital of St. John precinct occupied a narrow strip of land extending westwards from the gate outside the town ditch, with most of the area belonging to the monks of St. Werburgh's (VCH 2003, Vol 5.1, 209-220). Between Windmill Lane and Upper Northgate Street lay crofts or gardens, mostly the property of Chester Abbey and including the Battle Croft, whose name implies that it was the site of judicial battle in the Middle Ages. (VCH 2003, Vol 5.1, 209-220). Bag Lane (later George Street) extends away from Northgate and follows the line of the city ditch directly to the east; it was laid out by the 16th century, but probably pre-dates this time.

Many town and cities in medieval England were dominated by ecclesiastical religious bodies that controlled the daily interaction of medieval life. Chester was no exception and by the late medieval period precincts associated with the Benedictines, Franciscans, Carmelites and Dominicans were distributed throughout the town, though within the confines of the city walls. As well as these, a number of lesser houses existed throughout the city. The north-eastern corner of the walled city from Northgate to the east was dominated by the precinct of the Benedictine Abbey. Unlike the other major orders that were destroyed by the dissolution in 1538, the abbey survived and was re-founded as the cathedral for the Diocese of Chester. It still stands today and dominates this area of the town (Ward 1990, 1).

The post-medieval development of Gorse Stacks

By the early 17th century the town was surrounded by suburbs to the northeast. The largest expansion lay outside the Eastgate along Foregate Street, which was built up as far as Boughton and in the vicinity of Cow Lane (later Frodsham Street) and the Gorse Stacks. Outside the Northgate, where over thirty houses and several bake-houses and barns were burnt down in 1564 or 1565, building extended mainly along Upper Northgate Street (VCH 2003, Vol. 5.1, 220-9).

The area between the North Gate and Cow Lane was known as the Gorse Stacks from at least the 16th century. The name is derived from the practice of stacking gorse, furse or faggots outside of the city gates, in order to minimize the fire risk associated with their component materials. Certainly, concern appears to have been raised about the storing of gorse within the city from at least the 16th century, since an order by the mayor of Oct 27th 1592 states:

> *No gorse to be stacked behind houses* (ref CCA ZA/F/4/30).

By the turn of the century (1599-1600) fines are also documented relating to offences concerning such materials;

> *William Alcock for keeping a great quantity of faggots and gorse in his backside* (6s.8d.).
>
> *James Broster for putting three or four loads of gorse in his backside and timber [in] the street* (6s.8d.) (Ref CCA ZM/B/28a).

Principal amongst the users of gorse were the Company of Bakers. It represented an easily accessible source of fuel for baker's ovens. Even the Company of Bakers could fall foul of the law as shown by an order of 1764 where;

> *the Company of Bakers were not to lay or stack faggots or gorse thorn or other fuel on any corporation waste land without special leave (ZA/B/4/(222v).*

During the Civil War in early 1643, the city's defenders demolished buildings at Boughton, which could provide shelter for attackers and this appears to have included the area around the Northgate (VCH 2003, Vol 5.1, 121). After the destruction caused by the Civil War siege of Chester the suburbs were rebuilt, albeit slowly, but do not appear to have expanded much beyond the mid 17th-century limits before 1745 (VCH 2003, Vol. 5.1, 220-9).

The area around the quarry appears to have suffered during the Civil War. Situated just to the east of the Northgate it would have held a point of strategic significance. William Wilson, tanner, related that he owned *'a parcel of ground without the Northgate, near the quarry, upon which several buildings stood until they were burnt down in the siege'* (dated 1660 ref CCA ZA/B/2/129-129v). The quarry was clearly being worked at this time as a dispute had arisen between William Wilson about the quarry that adjoined his property and William Bristow who '*intended to cut this rock down*'.

By the start of the 18th century, expansion had occurred beyond the city walls and with this came increased pressure on land and building materials. The Company of Bakers tried to remove the order preventing the storing of gorse or furse in the city, but this was rejected (ref CCA ZA/F/50e/63). The Gorse Stacks was confirmed as a suitable place for the storage of Gorse in 1713 (ref CCA ZA/F/49f/3) and further storage at Gallows Hill, Eccleston Lane, Hough Green and Watergate suggest demand for fuel and pressure on storage facilities had increased dramatically by this time.

Tanner's, Butcher's, Cow Lane and the Cattle Market

References to the tanning industry, butchers and the cattle markets exist from the late-medieval period throughout the extra-mural suburbs to the north and east of the town. Cow Lane had existed from the medieval period and it was here that the tanning industry and slaughterhouses appear to have been preferentially clustered. The earliest reference comes from the Sherrif's note-books (Hamon Godeman and John Bradfield), referring to the lane in 1506:

'*...obstructed the highway called le Cowlane with filth*' (CCA ZS/B/6 a).

In the later 18th and 19th century, Philip Chevers, a butcher was known to have owned property including 2 messuages, cow houses as well as the quarry (CAA CR111/1).

The cattle market was moved from Upper Northgate Street in the 19th century to a location about 200m to the east of the site at the junction between George Street and Cow Lane. Initial petitions were made for it be moved in 1818 (CCA ZA/B/5/f.207), but it was not actually established until 1849 (CCA ZC46C). By the late 19th century, slaughterhouses are known to have been concentrated around the cattle market. Sir George Greaves is known to have owned two slaughterhouses adjacent to the cattle market between 1827 and 1882 (CAA ZC46C/2).

Quarrying in Gorse Stacks

The earliest post-medieval references to the quarry at Gorse Stacks appear from the mid-16th century. The quarry does not appear on the Hollar plan of 1653, although there is the possibility that the 16th century quarry was not well established and that quarrying was not therefore noted.

The earliest references come from the Corporation lease books which refer to:

> *the demise of a lease to Thomas Barker, embroiderer of a vacant plot near the Gorse Stacks adjoining a close called the Quarell on the south (*dated 1574 *ref CAA ZCHB/3 f.16v).*
>
> *a lease to Arthur Chauntrell , Sherman of a garden place outside the Northgate between the Quarell near the Northgate and the lane leading to the Gorse Stacks (dated 1584 ref CCA ZCHB/3 f.50v.-51).*

The close is referred to as 'le Quarrel' in the counterpart of a fee farm (dated 1574; ref CCA ZCHD/7/5). The Chester Workhouse was established in a building adjacent to the quarry in 1575 (Utting 1969).

The quarry appears to have continued to be in existence throughout the 16th and 17th century, although its extent is not clear. At the beginning of the seventeenth century it was referred to in documents relating to a lease from the Corporation for a building on a garden, without Northgate, between "*quarrel neere the Northgate*" and the highway leading to Gorse Stacks (dated 20 May 1611 ref CAA ZCHD/7/1).

The copies of the minutes of the Corporation Assembly state that a fee farm (fixed yearly income) be granted to William Ince on condition that he surrenders his interest in the Northgate quarry (date 1711 ref CCA ZA/F/49c/20). Interestingly a Robert Ince, shoemaker, is referred to in the Corporation Lease Book (ref CCA ZCHB/3 f.8.) and the counterpart of a fee farm with reference to a plot of land adjoining '*le quarrel*' at Gorse Stacks (ref CCA ZCHD/7/5) in 1574.

Several petitions are put forward for the lease of waste land at Gorse Stacks and it would appear that the corporation had begun to allow small scale exploitation of some of the land at this time.

In 1715 John Jones, mason or stone getter, paid William Coker for part of the stone joining his quarry under a lease for 7 years. John Jones was to pay *'£30, to leave the walls in good repair and have liberty to carry his stone through the old quarry'* (ref CCA ZTCD/116). The fact that even at this time it was referred to as the old quarry suggests it had become partly disused or was perhaps not in continuous use.

A petition of 1720/ 1721 from James Hutchison sought permission to get stone from the ground owned by the hospital of St. John the Baptist, held by John Moreton and adjoining the city wall near the Northgate.

A detailed account of the work done at the quarry is dated 1728/ 9 (CCA ref ZM/UV/3/34) and it is around this time that we get the first accurate plan of the quarry at Gorse Stacks, which is referred to on the 'De Laveaux plan' of 1745 (Fig 3.4). The plan does not give us any indication of the depth of the quarry, but it would appear that it was well established. Furthermore, the Delamere Street quarry [the site] conforms to the L-shape outline of the quarry seen during excavations.

A further quarry is clearly visible to the north of the Delamere Street quarry. The likelihood is that this is associated with the quarry described in a conveyance from the Ecclesiastical Commissioners for England of 1867 as '*...a piece of land fronting Upper Northgate Street and Windmill Lane, together with 14 houses, 9 cottages, 2 joiners shops, a shop, a bakehouse, a quarry, and with gardens and yards adjoining...*' (ref CCA ZC133C/1). It is referred to as the Bishops Quarry in a declaration of 1880 (ref CCA ZC1C/4), though it is unclear when this quarry first came into existence. A George Eaton, stone cutter, petitioned to build a house on a 99 year lease in the area of the quarry in 1763-4, but the success of this petition is unclear (ref CCA ZA/B/4/(208)).

The expansion of the quarry did not occur until the late-19th century. One reason for this must surely lie with the cutting of the adjacent Chester Canal. The likelihood is much of the stone removed during the period of canal building (completed 1779) was used in construction within the town or associated with the canal structures. This may well have lessened the demand for an expansion of the quarry.

By the 19th century the quarry appears to have been in the possession of Phillip Chevers, a butcher from Chester. His will of June 1791, attested on the 13th September 1862, references property in Chester, including '*2 messuages, cow houses and a quarry contiguous to a road leading from Northgate to Gorse Stacks*' (ref CCA CR111/1).

The development of the quarry as seen in the cartographic sources

The historical development of the quarry suggests that initial excavation started in the 16th century, but in order to understand the function of the quarry as finally investigated it is important to understand its chronological development.

Survey work in 2008-9 as part of the 'Mapping Medieval Chester Project' suggests that around 1500, a series of

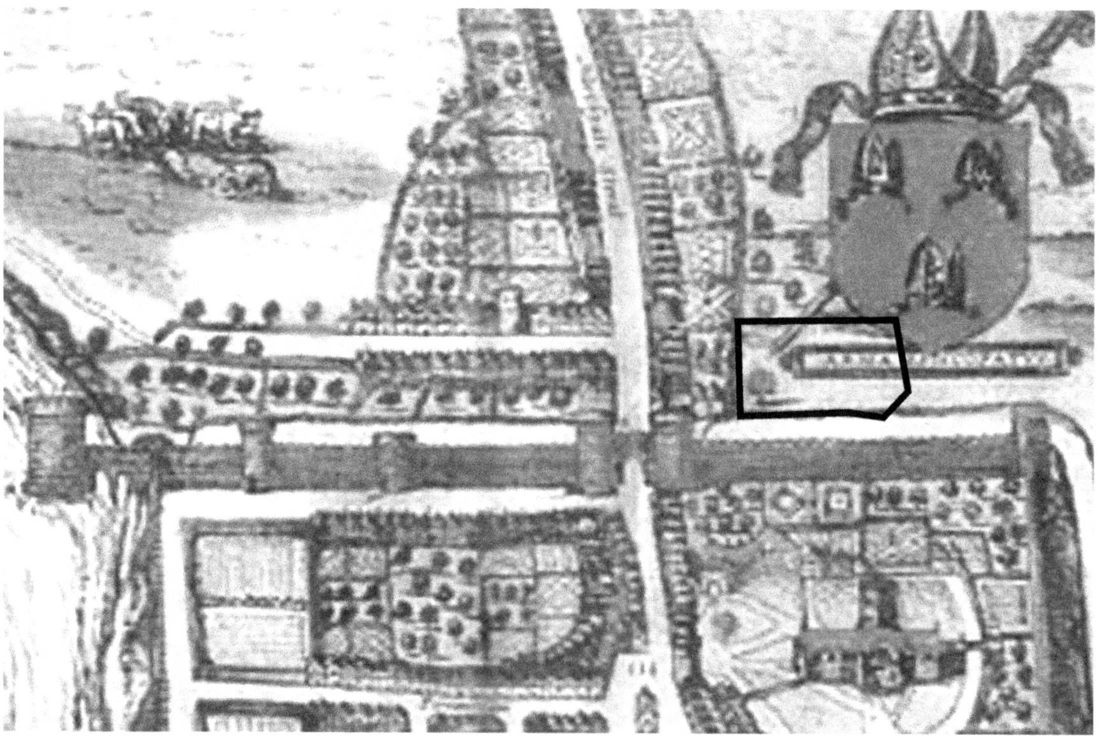

FIG. 3.23 BRAUN'S MAP OF CHESTER 1571

narrow plots were lining Upper Northgate, extending to the rear in the form of possible burgage plots, although these rear plots are poorly defined as a result of later quarrying activity (http://www.medievalchester.ac.uk).

The earliest plans of the city should not be seen as accurate geographical representations, but they do provide an indication of the city as it stood in the 16th and 17th centuries. The earliest map of the city is probably Braun's Map of 1571 (Fig 3.23). The development of the Northgate extramural settlement is clearly visible on the plan stretching either side of the road, but there is an indication that there is little or no development to the east of the main street and gardens.

Smith's pictorial map of 1581 is the first to delineate Bag Lane that later became George Street, leading to the horse market, which became the cattle market in the 19th century. It does not show the delineated line of Windmill Lane, but it does suggest that the first buildings were constructed along the northern line of Bag Lane.

John Speed's map of 1610 (Fig 3.24) reiterates the earlier plan of Smith's map. It does depict a short lane leading off Bag Lane to the north, into the gardens at the rear that may depict the initial elements of the quarry. It also depicts the line of Windmill Lane (Victoria Road) for the first time. Hemingway's engraving of 1645 (Fig 3.25), depicts the line of the Civil War ditch both in the primary defence of the city and the subsequent refortification of the city that resulted in the demolition of the buildings around the Northgate. Hollar's plan of 1653 still shows the buildings standing, but suggests a redrawn version of Speed's earlier map.

Nathaniel Buck's prospect of the city is taken from the riverside and does not depict the area of Northgate and the site. It is not until the 18th century that we can visualise the quarry as the De Lavaeux plan of 1745 represents the earliest precise image of the quarry. By this period it is clear that the centre of the quarry is well developed, which would correlate with quadrants A to D, but it did not extend as far (east) as the excavated quarry. Elsewhere, the house plots line the George Street frontage, with a gap in between for the quarry.

Stockdale's plan of 1795 does not clearly depict the quarry, but it does show the canal for the first time. Houses are depicted on the frontage as with De Lavauex's plan, but the land plot where the quarry once stood is now occupied by buildings and it may be that initial backfilling of the quarry had begun to occur at this time.

The Stockdale plan may be deceptive and not actually define the quarry, even though it may have been in operation. Bateman's plan of 1816 defines the land plot to the rear of the houses that closely replicates the boundary of the excavated quarry and although the quarry may not have been totally excavated, it may reveal the possible extent of land that could be excavated. This corresponds closely with the ownership of Phillip Chever between 1791 and 1862 and it may be that during this period that the greatest expansion of the quarry occurred. The quarry is certainly still visible on the 1870s Ordnance Survey map.

The quarry is also visible on John McGeegan's highly detailed image of the city, painted from a balloon in 1855. It lies to the rear of a series of houses backing onto George Street, Victoria Street (formerly Windmill Lane) and the newly aligned Delamere Street respectively. The

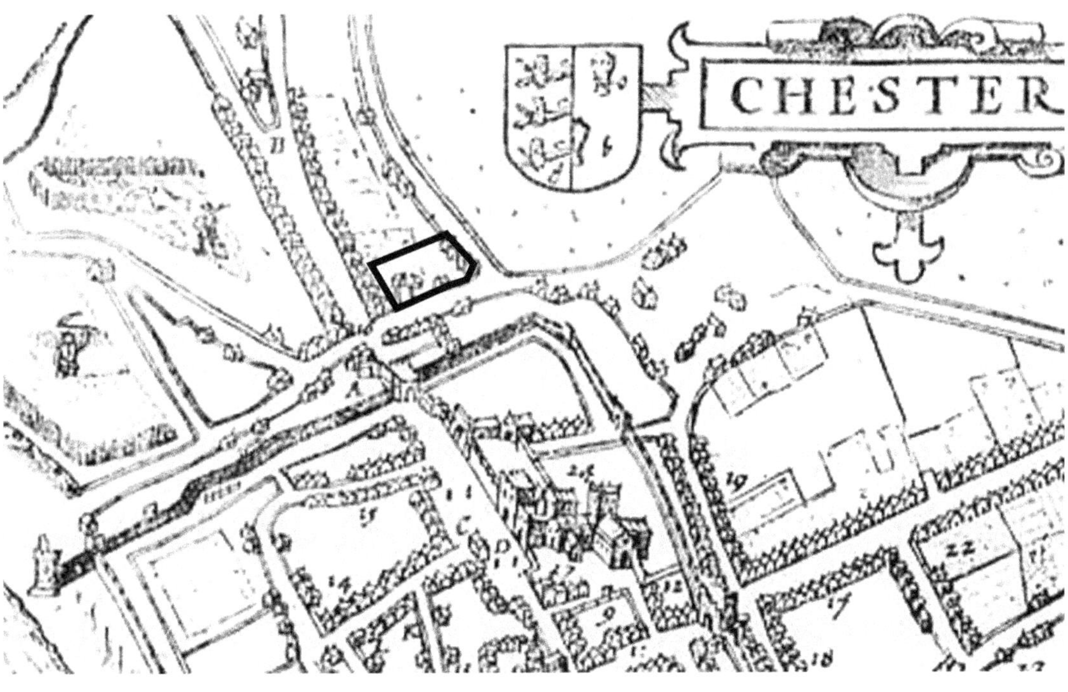

FIG. 3.24 SPEED'S MAP OF 1610

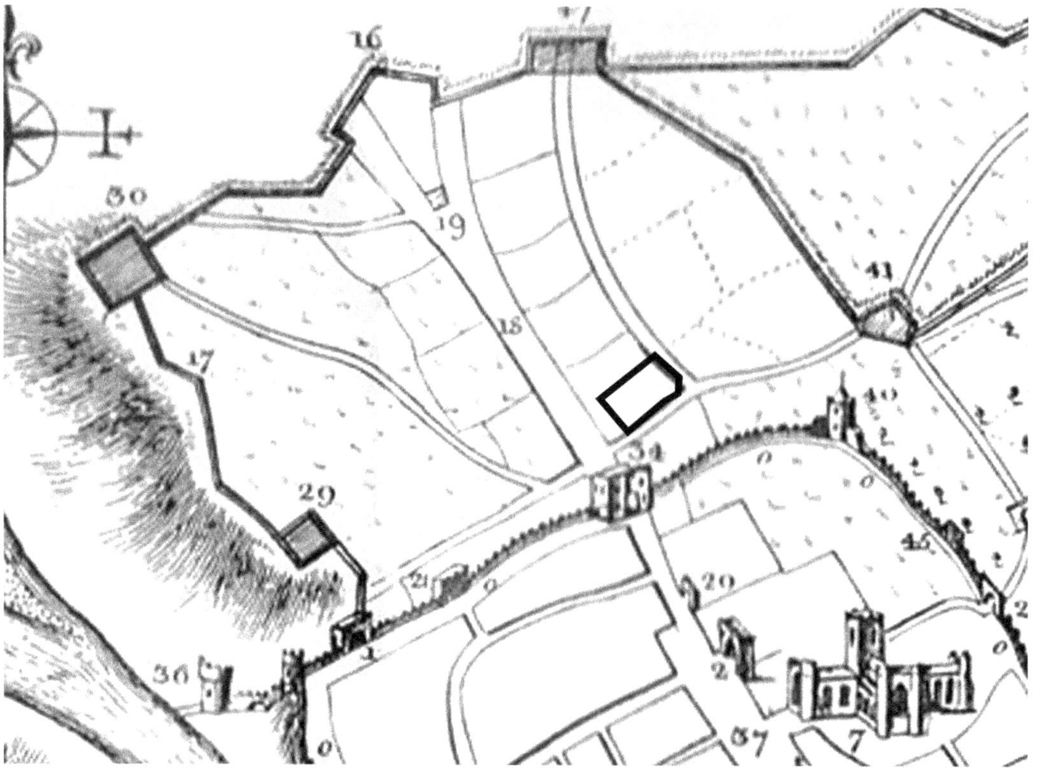

FIG. 3.25 HEMMINGWAY'S ENCRAVING OF 1645 (PRODUCED 1881)

FIG. 3.26 JOHN MCGEEGAN'S IMAGE 1855

quarry does not appear to be pronounced or deep, but it does appear to be lined by houses. It has not expanded to the east, which suggests that the final stages of extraction occurred in the decades of the 1850s and 1860s and make up the majority of the eastern side of the quarry.

By the 1st Edition 1:500, 1873-4 Ordnance Survey map, much of the quarry has been built over and this appears to depict the final phase of quarry development (Fig 3.9). The new Methodist Chapel (Plate 3.5) was built on the corner of George Street in 1888 (Bounds 1975:3). The chapel appears on the 1911 Ordnance Survey map of the city as 'Methodist chapel and Sunday school'. The chapel served as an amalgamation of several smaller institutions, which were no longer able to cope with the number of parishioners. The whole area was extensively redeveloped in the 1960s and 1970s as part of the new Bus Station. Buildings within the site began to be demolished in the late 1960's culminating in the demolition of the Methodist chapel fronting onto George Street in 1970.

DISCUSSION

The Quarry Features

The solid geology of Chester comprises Lower Triassic Sandstones of the Sherwood Sandstone Group, which consist largely of red, yellow, and brown mottled, pebbly sandstones (www.naturalengland.org.uk). Triassic sandstones and other inter-bedded lithologies have been worked for building material across much of their outcrop, which extends from the Devon coastline, through the English Midlands, into Northern England and south-west Scotland. A number of Triassic red sandstones were quarried for building material in the Cheshire area particularly from 1850s onwards. The main building sandstone crops out on the ridges around the margins of the Cheshire Plain, with many 19th century quarries at Runcorn, Rainhill, Storeton and Woolton amongst other places; sadly, many of these features are now buried beneath the suburban sprawl of Liverpool (www.englishstone.org.uk).

The Sherwood Sandstone Group like most sandstone is useful for building since it is sufficiently soft for working and is laid down in geological units (beds), conducive to quarrying (Ashurst and Dimes 1977, 10-13). Within the Delamare quarry the rock was strong enough to allow excavation of a vertical quarry face and would have produced stone suitable for both building (roughly faced blocks and ashlar) and rubble for use as 'hard core'.

The site of the quarry on the plateau of the ridge is perhaps a reflection of the localised demand for stone as opposed to the ease with which the stone could be extracted. The quarry was small and one of several in the area; all were rectangular or sub-rectangular in form. In the example of the excavated quarry the evidence for the vertical jointing of the rock was sparse, but appeared to be perpendicular to the quarry faces (see Plate 3.7, 3.8). Evidence from the Hambleton Hills, North Yorkshire, points to plateau quarries being sub-square in formation, aligned along the available vertical jointing patterns (Cooper 1977, 166).

The removal of overburden above the quarry is difficult to assess as none of this original material survived. The shallow depths of the natural sandstone beneath the surface may have dictated the desirability of the Gorse Stacks site for quarrying and the evidence of the present street level above the surviving sandstone suggests that less than 1m of overburden would have been present originally. Again the examples noted above from the Hambleton Hills suggest that plateau-top quarries are generally worked with a depth of around 1.6m of overburden including the topsoil, subsoil and frost-shattered rock termed 'rag' (Cooper 1977, 167).

The method of extraction within the quarry is critical to the understanding of the use of the stone. The longevity of the quarry suggests that the earliest materials would have been excavated by hand along a series of natural bedding planes (Greenwell 1913, 139-60). There is no evidence for blasting in the quarry.

Methods of cleaving the rock would have varied. Wedge pits may have been used and would have involved boring a series of vertical holes along the line of the desired block size and then wedging it apart using 'splits' and 'feathers' (Greenwell 1913, 151). An alternative technique, which was applied since Roman times, involved the excavation of a series of slots vertically through the bedding plane and inserting wedges or crowbars to loosen the horizontal sheets. The distinctive cross-hachured pick marks visible within the majority of the quarry faces would appear to indicate that the later technique was adopted. This technique was still being used by hand at Ham Hill quarry, Somerset, as late as the 1920s and the quarry faces still display a distinctive pattern of tool marks comparable with Delamere quarry (Stanier 2000, 76-81, fig.45).

Excavation of a test-pit in the base of the quarry suggested around 2m of sandstone waste was present and that little waste material had been removed. The depth of waste material suggests that stone may have been dressed on the quarry floor prior to removal and this was the most common approach before the 19th century (Stanier 2000, 28). Although no pieces of dressed stone were identified within the quarry itself, reused dressed stone (*see Lobb Chapter 15*) was found in the cellar construction of 2092 (Group 107/ 108) dating to the 19th century. This would appear to suggest that dressing of the stone was occurring locally if not within the quarry itself.

Quarries should normally be stepped for safety purposes, but it is clear in the case of this quarry that although stepping may have occurred initially, it was abandoned in order to maximise the output of the quarry, resulting in vertical faces of more than 6m.

Excavating a quarry vertically into the ground would have presented a number of problems. The principal one of these would have been access and it is presumed that this would have been achieved by a series of ramps, though the confined space of the quarry would have made these difficult to operate. It is possible that carts were accommodated in the quarry and the presence of horse bones within the faunal assemblage (see Holmes above) may point to the use of animals for haulage.

The lifespan of the quarry may have allowed steam driven cranes by the later period, but no evidence was recovered during excavations to this affect. The presence of stone pudlocks within the quarry face appears to be associated with scaffolding. A number of examples of these pudlocks of varying size can be found (see Plate 3.6). The insertion of scaffold points throughout the quarry may suggest cranes were used to remove stone vertically from the quarry and it is known more generally that heavy timber cranes were introduced from the 19th century. These cranes were generally either of tripod construction or in the case of vertically sided quarries, a single vertical timber supported by a series of ties (Stanier 2000, 26-8).

Dating

The ceramic building material recovered could be conclusively dated to before the Civil War. The presence of glazed tiles, both plain and decorated, dated to the period of the 15th and 16th century. They are clear indicators that activity from this date was occurring in the area, and it was of a status sufficiently high to warrant decorative ornamentation of the houses. However, caution should be applied since ceramic building material is often a poor indicator of age, since it generally has a long life-span, usually within the fabric of buildings, and particularly in the case of tile, it is reused extensively.

In the case of the tile, two sources are possible. Although separated by the city walls the Benedictine Abbey lay to the south of the site and it may be possible that material from the Abbey found its way as far as this quarry. A greater possibility though, is the hospital of St. John's located just outside the Northgate. The hospital was founded in *c* 1190 and was cleared as a result of the Civil War in 1644. It was rebuilt after the war and became incorporated into the Blue Coat School in 1715-7 (VCH 2005, Vol. 5, 2, 156-9). Such a building was of sufficiently high status to warrant floor tiles and decorated brick; stone from the quarry was certainly used in its reconstruction (*see above*) and it is possible that some of the cleared material may have found its way into the quarry when stone was extracted.

The presence of human bone within the quarry is enigmatic. Radiocarbon dates taken from the larger bones are contradictory, giving dates of Cal AD 1160 to 1280 and Cal AD 1430 to 1520. This suggests that at least some of the material has been re-deposited after burial elsewhere. The absence of material predating the Civil War strongly suggests that activity including quarrying was small-scale during the medieval period.

The destruction wrought during the Civil War in the Northgate area (the area of Gorse Stacks lay between the primary defences constructed *c.* 1642 and the secondary later defences 1645), led to widespread demolition of houses. Evidence for these were relatively absent from the archaeological record. A single pit in Area A [2321] contained ceramics dating to the Civil War period (*see Dodds Chapter 13*). However, the general absence of material may support the theory that large scale clearances occurred in the area and it was not occupied during this period. The clearances may also provide a reason for the absence of pre-existing medieval deposits.

The historical evidence (*see Chapter 17*) suggests increased activity for the period 1700-50 when the land around the quarry began to be more heavily occupied, as seen by the increase in the number of petitions made to the City Corporation to allow the construction of houses. This could explain why earlier deposits became incorporated into the quarry. In particular it is interesting to note the clear concentration in the material to the period 1700--1750. Of the sealed 18th century contexts it is clear this period was one when deposition of material was occurring in the quarry. Dodd points to the absence of Creamware sherds within the assemblage as indicating that deposition occurred prior to the 1760/ 70s. The lifespan of the quarry was extensive, historically known to last from the 16th to the 19th centuries, but still open well into the 20th century. The dating of the assemblage may suggest there were periods when the quarry was actively operating and periods when there was inactivity, with a geographic shift from one area of the quarry to another.

The majority of clay tobacco pipes recovered dated from the period 1690-1750, with only a single bowl pre-dating the Civil War. The majority of the surviving glass assemblage came from this period, with the recognisable forms being predominantly onion form wine bottles. It would appear clear that there was a distinct early to mid-18th century episode of deposition in the quarry. The glass assemblage had no medicine or perfume bottles and this may suggest something of the status of this area of the city. Although medicine and perfume bottles predominate more uniformly in 19th century assemblage, there absence in the 18th century assemblage conforms to the extramural nature of settlement in this part of the town, where luxury goods did not predominate.

A third possibility for the deposition of the material in the quarry exists. The construction of the Chester Canal was undertaken in 1771 and followed the line of the former city ditch. The ditch was reused during the Civil War as the primary defence for the area in 1642 and as such would undoubtedly have been dug out and cleaned, with the result that much of the material pre-dating the mid-17th century would have been removed. Subsequent deposition in the ditch would date from the late-17th century through to around the period 1750-1760 when the construction of the canal would have resulted in the removal of all these deposits. The material from the canal would have to have been deposited somewhere. The stone itself would undoubtedly have been a useful building resource; if not used within the town, it would have probably been reused during construction of the canal itself. However, the fills of the ditch and upper subsoil would have needed to be removed elsewhere and probably to the most convenient dumping spot. Since the age range of the superficial deposits in the quarry correlate with those of the former city ditch deposits, this may provide a suitable answer.

Economy and Trade

The presence of butchered waste from both cattle and horse remains has been suggested to be associated with the tanning industry (*see Holmes Chapter 16*). The presence of horn cores has been repeatedly used as an indicator of the tanning industry in widespread locations (e.g. Northampton, Shaw 1996; Birmingham, Rátkai and Patrick 2009). Chester's position as a port and trading centre with markets to Ireland encouraged the growth of leather trades from the 16th to 18th centuries (Clarkson 1965, 27-34). The Irish Campaigns that followed the Civil War opened up new markets, but also meant that from the mid-17th century there was an influx of imported material from Ireland that passed through Chester and hides made up an important element of this trade. (VCH 2003, Vol. 5, 1).

The presence of horse bones, largely unbutchered, may suggest that the site was exposed to skinning waste. There was a tendency towards older animals with trauma marks that may suggest these were draft animals that had reached the end of their natural lives. Their sale and subsequent butchery as suggested by Holmes would be associated with the economic value of their hides as opposed to their meat. Another possibility should be examined however; the quarry was actively operating throughout this period and there is no evidence of mechanized power. Therefore, horses and carts may have been used to remove dressed stone via a series of ramps; alternatively timber cranes may have relied on animal power (see Stanier, 2000, 37).

Canine bones were recovered from the assemblage and have been associated with the tanning industry. Historical evidence from the 19th-century Rheadr Tannery in Wales, now moved to the Welsh Folk Museum, revealed evidence for mastiffs being used in tanneries, not only to provide protection, but also to clean hides (Jenkins 1973). Dogs would remove blood and flesh from the hide without causing undue damage to them. Residual blood left on hides would lead to staining of the hides during the tanning process and reduce their value. The height of the dogs has been estimated at 0.5m at the withers and points to very large animals similar to mastiffs. Comparable examples have been found in Birmingham at the site of Gibb Street (Baxter in Hewitson *et al* forthcoming).

The presence of sheep/ goat bones does not suggest tanning waste was exclusive within the assemblage. However, the assemblage was large and distributed over a large geographic spread within the quarry. No firm evidence could tie the assemblage to skinning processes. This may reflect the collection policies, as the bones associated with the skinning industry, the metacarpals and metatarsals, usually left connected to the skins, are small and indistinct. A visual collection policy for the deposits would have led to a bias towards larger bones as mentioned previously and therefore it cannot be assumed that they were absent from the archaeological record.

The noted presence of animal hair and leather fragments from a later 19th century pit within the quarry may suggest tanning, but no further work was undertaken on samples. The pit, which was stone lined, was isolated and probably a waste or cess pit associated with quarry buildings that were constructed in the 19th century. Recently Rátkai (*forthcoming* with reference to Birmingham tanneries) has questioned the validity of using single sources of evidence for indicating tanning. Instead she suggests that a suite of structural and environmental/ ecofactual evidence is required and alternative reasons for the presence of horncores may include butchery or bone-working.

The area to the west of Gorse Stacks has long been associated with the cattle industry (*see above*). The historic location of the cattle market was approximately 200m to the east of the quarry site. William Wilson, a tanner, is known to have owned land around the quarry in 1660, although at this time his buildings were burnt down (see above). By the late-18th century Philip Chevers, a butcher, owned the quarry and he may be responsible for the animal bone waste deposited in the area. However, any one of these three may have used the quarry as a useful dump for waste.

The evidence from the ceramics is typical of the late-post medieval to early modern period. The development of a fully integrated market at this time meant that whereas the medieval period was dominated by local trade, established links began to develop that were nationwide by the 18th century. Unsurprisingly given the dominance displayed by the Staffordshire potteries during the 18th and 19th centuries, much of the ceramics come from this region (Barker 2004). Certainly the 18th century deposits are dominated by the remains of these potteries with Agate Ware, Redwares and Porcelain found within the assemblage. Chester's importance as a port meant that both coastal and continental trade would bring products to the area and this is reflected in the presence of Rhenish Stonewares, Westerwald wares and North Devon Gravel-tempered Wares. Indeed the dominance of the Staffordshire Potteries may have led to a reduction in the range of wares within the assemblage by the 19th century with far more local input. The continuation of porcelain forms probably from the Longton Potteries provides continuity between the two centuries. Interestingly the presence of Stonewares from Bristol may suggest a change in trade patterns brought about by the development of the canal network. Although a coastal route existed, the presence of Stonewares associated with the 'Bristol ware' glaze may suggest links up the River Severn and via the Staffordshire & Worcester/ Shropshire Union canal network. The range and forms recovered from the quarry are typical for the 17th and 18th century and have been recorded elsewhere in Chester. It is unfortunate that the earliest deposits did not reveal assemblages from the 16th and 17th centuries, which would have allowed for a greater chronological understanding of Chester's pottery industry.

The excavations at Delamere Street have shown a pattern of activity that dates to the Roman period and is repeated almost exactly in the post-medieval period, stone removal followed by extensive refuse deposition. Due to the reductive nature of quarrying there are large gaps in the chronology of the site, buildings and structures from the Anglo-Saxon and medieval periods are entirely absent. It is tempting to think that there may be some survival of this activity beneath George and Delamere Street. A few fragments of tile and pottery are all that is left of this activity so its extent and form can only be guessed at.

The assemblage recovered from both the Roman and post-medieval phases of this site are some of the largest ever recorded in Chester by a single excavation. The impact of this assemblage for the interpretation of material recovered in the future cannot be underestimated. It is hoped that by the publication of this material new understandings of the development of Roman and post-medieval Chester can be reached.

ACKNOWLEDGEMENTS

The test pitting was undertaken in 2002 by Earthworks Archaeology and the team of excavators was supervised by Leigh Dodd with assistance from V Clarke, G Luke, G Jones, L Bognar, A Boganr, A Byard, and managed by Will Walker. The work was commissioned by Rufus Estates Limited and London and Amsterdam Developments Limited.

The main body of the project was commissioned by Watkins Jones and the thanks must go for their co-operation and assistance throughout the production of this report. Thanks are due to Nick Bumby, John Thomas and all the staff of Watkin Jones for their assistance throughout the project. The project was monitored by Mike Morris for Chester City Council and I would like to extend thanks on behalf of the University of Birmingham for his assistance, co-operation and input into the production of this report. In addition to this the assistance of the Chester Archaeology Service would like to be acknowledged for all their co-operation during the production of the specialists reports. Individual specialists have reiterated the assistance of members of staff within the process below.

Work on the excavation of the site in 2006 was undertaken by Bob Burrows, Kristina Krawiec and Phil Mann, assisted by Anthony Aston, Emily Betts, Lara Bishop, Liz Bishop, Paul Breeze, Dave Brown, Mark Charles, Paul Collins, Gary Coop, Sean Daly, Emily Hamilton, Cat Hopwood, Dan Jackson, Mark Middleton, Hailey Richardson, Emma Sautejeau, Elle Vincent and Nick Witchell. Work involved in the watching brief on the quarry in 2008 was undertaken by David Brown, Ellie Buttery, Sam Hepburn, Shane Kelleher, Kristina Krawiec, Michael Lobb, and Phil Mann.

Thanks are also due to a number of specialists for their input into the project. These include for the Roman chapter Iain Baxter, Hilary Cool, Alison Heke, Pam Grinter, Samantha Hepburn, Michael Lobb, Erica Macey-Bracken, Rod MacKenzie, Rebecca Nicholson, Jane Timby, Felicity Wild and Roger White. For the post-medieval chapter these include Megan Brickley, Leigh Dodd, Alison Heke, Malcolm Hislop, Matilda Holmes, Michael Lobb, Nigel Melton and Jennie Stopford.

In particular Jane Timby would like to thank the Chester Archaeological Service for allowing access to their fabric series and in particular to Alison Heke for her assistance in providing background information on pottery work in Chester. Iain Baxter would like to thank Gill Jones and Peta Sadler for sharing their research on the significance of the CEJ in cattle mandibles; Gill Jones for references relating to cattle teeth; Sheila Hamilton-Dyer for her identifications of the wild bird and fish remains; Nadja Pöllath for references relating to spavin in cattle; Simon Davis for discussion and comparison of data on cattle metapodials; Umberto Albarella for providing statistical data in electronic form on the Elms Farm assemblages; and, last but by no means least, Peter Carrington, Ian Smith and Mike Morris of Chester Archaeology for access to the unpublished work of Judy Cartledge on the animal bones from the fortress and helpful discussions of this and other Chester assemblages and archaeology. Hilary Cool would like to thank Fraser Hunter for allowing her to see his work on beaded torcs and to Roger White for information about the Rocester torc. Both of them have very kindly allowed her to cite their work within this publication in advance of their own.

The Roman chapter was written by Kristina Krawiec and Chris Hewitson and the post-medieval chapter was written by Samantha Hepburn and Chris Hewitson. The report was edited by Kristina Krawiec and Chris Hewitson with further comments from Alex Jones, Simon Esmonde-Cleary, Ben Gearey and Roger White.

Illustrations were principally produced by Nigel Dodds and Helen Moulden. The survey drawings were prepared by Kristina Krawiec, Phil Mann and Chris Hewitson. Initial rough out drawings for the Roman pottery were prepared by

BIBLIOGRAPHY

Albarella, U. 2003. Tawyers, tanners, horn trade and the mystery of the missing goat. In Murphy, P. and Wiltshire, P. (eds.) *The environmental archaeology of industry* 20, 71-86.Oxford: Oxbow.

Albarella, U. and Davis, SJM. 1994. *The Saxon and Medieval animal bones excavated 1985-1989 from West Cotton, Northamptonshire.* London: English Heritage, AML Report 17/94.

Albarella, U. and Davis, SJM. 1996. Mammals and birds from Launceston Castle, Cornwall: decline in status and the rise of agriculture. *Circaea* 12 (1), 1-156.

Albarella, U. and Payne, S. 2005. Neolithic pigs from Durrington Walls, Wiltshire, England: a biometrical database. *Journal of Archaeological Science.* 32, 589-599.

Allason-Jones, L. 1989. *Earrings in Roman Britain*, BAR British Series 201, Oxford.

Allason-Jones, L. and Bishop, MC. 1988. *Excavations at Corbridge: the Hoard*, English Heritage Archaeological Report 7, London.

Allason-Jones, L. and Miket, R. 1984. *The Catalogue of Small Finds from South Shields Roman Fort*, Soc. Ant. Newcastle upon Tyne Monograph 2, Newcastle upon Tyne.

Amorosi, T. 1989. *A Postcranial Guide to Domestic Neo-Natal and Juvenile Mammals: The Identifying and Ageing of Old World Species*.: BAR International Series 533. Oxford.

Anderson, AC, and Anderson, AS, (eds). 1981. *Roman pottery research in Britain and North-west Europe. Papers presented to Graham Webster*, BAR S123, Oxford.

Andrews, AH and Noddle, BA. 1975. Absence of Premolar Teeth from Ruminant Mandibles found at Archaeological Sites. *Journal of Archaeological Science* 2,137-144.

Archer, M. 1997. *Delftware: The Tin-Glazed Earthenware of the British Isles*. London: The Stationary Office.

Armitage, PL. 1982a. 'A System for ageing and sexing the Horn Cores of Cattle from British post-medieval sites (17th to early 18th century) with special reference to unimproved British Longhorn Cattle'. In Wilson, R., Grigson, C. and Payne, S. (eds.) *Ageing and Sexing Animal Bones from Archaeological Sites.* BAR British Series 109, 37-54. Oxford.

Armitage, PL. 1982b. 'Studies on the remains of domestic livestock from Roman, medieval, and early modern London: objectives and methods'. In: Hall, A.R. and Kenward, H.K. (eds). *Environmental Archaeology in the Urban Context.* CBA Research Report 43, 94-106. London.

Armitage, P.L. and Clutton-Brock, J. 1976. A System for Classification and Description of the Horn Cores of Cattle from Archaeological Sites. *Journal of Archaeological Science* 3, 329-348.

Arthur, P and Marsh, G. (eds). 1978. *Early fine wares in Roman Britain*, BAR British Series 57. Oxford.

Ashurst, J and Dimes FG. 1977. *Stone in Building: Its Potential and Uses Today*, London Architectural Press.

Askey, D. 1998. *Stoneware Bottles, from Bellarmines to Ginger Beers 1500–1949*. (2nd edn). BBR Publishing.

Atkinson, D. 1914. A hoard of samian ware from Pompeii, *J Roman Stud* **4**, 27-64.

Baggs, AP, Kettle, AJ, Lander, SJ, Thacker, AT, Wardle, D, 1980. 'Hospitals: St John the Baptist, Chester', *A History of the County of Chester, Vol 3*, 180-183. http://www.british-history.ac.uk/report.aspx?compid=39986

Barker, D and Cole, S. 1998. 'William Littler at Longton Hall'. in Barker, D & Cole, S (eds). *Digging for Early Porcelain: The Archaeology of Six 18th century British Porcelain Factories.* City Museum & Art Gallery, Stoke-on-Trent. 4–21.

Barker, D and Halfpenny, P. 1990. *Unearthing Staffordshire.* City of Stoke-on-Trent Museum and Art Gallery.

Barker, D 1997 'Post-Medieval Pottery'. in Wilson, D, Bagnall, A & Taylor, B *Excavation of a Medieval Moated Site at Northwoods Farm , Dodcott-cum-Wilkesley, Cheshire.* Staffordshire Archaeological Studies 9. City Museum & Art Gallery, Stoke-on-Trent. 34–46.

Barker, D. 2004. 'The industrialization of the Staffordshire potteries', in Barker, D and Cranstone, D (eds) *The Archaeology of Industrialization*, 203–22. Leeds: Maney,

Barker, D. 2007. Creamware in Context. *In:* Walford, T & Massey, R (eds). *Creamware and Pearlware Re-examined.* English Ceramic Circle.

Barone, R. 1980. *Anatomia Comparata dei Mammiferi Domestici. Vol. III Splancnologia*. Bologna.

Bartosiewicz, L., Van Neer, W. and Lentacker, A. 1997. *Draught Cattle: their osteological identification and history.* Koninklijk Museum voor Midden-Afrika, Tervuren, België, Annalen Zoölogische Wetenschappen/Annales Sciences Zoologiques, Musée Royale de l'Afrique Central, Tervuren, Belgique.

Bass, WM. 1995. *Human Osteology: A Laboratory and Field Manual.* Columbia: Missouri Archaeological Society. Vol 2.

Baxter, IL. 1998. Species identification of equids from Western European archaeological deposits: methodologies, techniques and problems. In Anderson, S. (ed.) *Current and Recent Research in*

Osteoarchaeology, 3-17. Proceedings of the third meeting of the Osteoarchaeological Research Group. Oxford: Oxbow.

Baxter, IL. 2002. 'Occipital Perforations in a Late Neolithic Probable Aurochs (*Bos primigenius* Bojanus) Cranium from Letchworth, Hertfordshire, U.K'. *International Journal of Osteoarchaeology* 12, 142-3.

Baxter, IL. 2006. 'A Dwarf Hound Skeleton from a Romano-British Grave at York Road, Leicester, England, U.K., with a discussion of other Roman small dog types and speculation regarding their respective aetiologies'. in Snyder, L.M. and Moore, E.A. (eds). *Dogs and People in Social, Working, Economic or Symbolic Interaction* (Proceedings of the 9th Congress of the International Council for Archaeozoology 2002), 12-23.Oxford: Oxbow Books.

Bayley, J. and Budd, P. 1998. 'The clay moulds', in Cool and Philo, 195-222.

Bayley, J. and Butcher, S. 2004. *Roman Brooches in Britain: a Technological and Typological study based on the Richborough Collection*, Rep. Res. Com. Soc. Ant. London 68. London.

Biddle, M. and Brown, D. 1990. 'Writing equipment' in Biddle, M. 1990. *Object and Economy in Medieval Winchester*, 729-53.Winchester Studies 7.ii, Oxford.

Birks, H.J.B and Birks, H.H. 1980. *Quaternary Palaeoecology*. Edward Arnold London.

Bishop, MC and Coulston, JCN. 2006. *Roman Military Equipment from the Punic Wars to the Fall of Rome*, (Second edition, Oxford).

Bishop, MC. 1998. 'Military equipment', in Cool and Philo, 61-81.

Bishop, MC., 2002. *Lorica Segmentata. Volume 1: A handbook of articulated Roman plate armour,* JRMES Monograph 1 (Duns).

Black, EW. 1996. 'Box flue-tiles in Britannia: the spread of Roman bathing in the first and second centuries'. *Archaeol J* 153, 60-78

Blagg, TFC. 1976. 'Tools and Techniques of the Roman Stonemason', *Britannia* 7, 152–72.

Blockley, M. 1989. 'The Tile'. in Blockley, K. *Prestatyn 1984-5: An Iron Age farmstead and Romano-British industrial settlement in North Wales*. BAR British Series 210, 164. Oxford.

Boessneck, J. 1969. 'Osteological Differences between Sheep (*Ovis aries Linne*) and Goat (*Capra hircus Linne*)'. in Brothwell, D.R. and Higgs, E. (eds.) *Science in Archaeology*, 331-359. London: Thames and Hudson.

Boessneck, J. 1969. Osteological Differences between Sheep (*Ovis aries Linne*) and Goat (*Capra hircus Linne*). In: Brothwell, D.R. and Higgs, E. (eds.) *Science in Archaeology*, pp. 331-359. London: Thames and Hudson.

Boessneck, J. von den Driesch, A. Meyer-Lemppenau, U. and Wechsler-von Ohlen, E. 1971. *Die Tierknochenfunde aus den Oppidum von Manching*. Wiesbaden (Die Ausgrabungen in Manching, 6).

Bounds,K 1975. *Reminiscences of Chester Methodism.*

Brailsford, JW. 1962. *Hod Hill Volume One. Antiquities from Hod Hill in the Durden Collection*, London.

Brain, CK. Some principles in the interpretation of bone accumulations associated with man. In: Isaac, G. and McCown, E. (eds.) *Human origins*, 97-116. California: Benjamin.

Brodribb, G. 1987. *Roman brick and tile*, Gloucester: Alan Sutton Publishing.

Brown, P. 2001. *British Cutlery: An Illustrated History of Design Evolution and Use*. York: Philip Wilson.

Burnett, A. 1987. *Coinage in the Roman World,* London: Seaby.

Capel, C. 2006. *Objects: Reluctant Witnesses to the Past*, Abingdon: Routledge.

Carringon, P. 1981. Some second century mortaria from Chester, in A.C. Anderson and A. S. Anderson, 465-9.

Carrington, P, 1977. 'Severn Valley ware and its place in the Roman pottery supply at Chester: a preliminary assessment', in Dore, J. and Greene, K. 147-55.

Carrington, P. 2006. *Towards a model of Roman society in Cheshire: First to third centuries AD*, Chester City Council: http://www.chesterarchaeolsoc.org.uk/R_cheshire_bibl.html accessed 02-03-2010

Carrington, P, and Ward, TE. 1982. *Roman kiln wasters from the Chester vicus*, unpub archive report, Chester Archaeology Unit

Cartledge, J.A.1991.*Chester Animal Bone* (WYASR26): Chester Roman Sites. Unpublished report prepared for West Yorkshire Archaeology Service.

Casey, PJ. and Hoffmann, B. 1995. 'Excavations on the Corbridge Bypass 1974', *Archaeologia Aeliana Series 5* 23, 17-45.

Castle Lyons. Y Cymmrodor 41. London: Society of Cymmrodorion

Clark, KM. 1995. The later prehistoric and protohistoric dog: the emergence of canine diversity. *Archaeozoologia* 7 (2): 9-32.

Clarkson, LA. 1965. 'The organisation of the English *leather industry* in the late sixteenth and seventeenth centuries', *Economic History Review* 13, 245-256.

Clay, C. *2004*. Iconoclasm in Roman *Chester*: the significance of the mutilated *tombstones* from the north wall. *J Brit Archaeol Assoc* 157, 1-16.

Constable, C and Parsons, G. 2001. *Archaeological evaluation of Land at the Bars, Russell Street, Chester*. L-P Arcaheology unpub rep.

Cool, H.E.M. and Philo, C. (eds.) 1998. *Roman Castleford Excavations 1974-85. Volume I: the Small Finds,* Yorkshire Archaeology 4, Wakefield.

Cool, H.E.M. 1983. *A Study of the Roman Personal Ornaments made of Metal, excluding Brooches, from southern Britain*, unpubl. PhD Thesis, Univ. Wales.

Cool, H.E.M., Lloyd-Morgan, G. and Hooley, A.D. 1995. *Finds from the Fortress*, Archaeology of York 17/10, York.

Cool, H.E.M. 2006. *Eating and Drinking in Roman Britain*, Cambridge.

Cool, H.E.M. 2007. 'Metal and glass', in Catchpole, 171-83. Catchpole, T. 'Excavations at the Sewage Treatment

Works, Dymock, Gloucestershire, 1995',*Transactions Bristol and Gloucestershire Archaeological Society* 125, 137-219.

Cool, H.E.M. 2008. 'Copper alloy', 'Bone, antler and ivory', in Garner, 307-17, 329-331.

Cool, H.E.M. and Baxter, M.J. 1999. 'Peeling the onion ; an approach to comparing vessel glass assemblages', *Journal of Roman Archaeology* 12, 72-100.

Cooper, N. 1999. 'The small finds', 239-82. in Connor, A. and Buckley, R. 1999. *Roman and Medieval Occupation in Causeway Lane, Leicester,* Leicester Archaeology Monographs 5, Leicester.

Cooper, R.G. 1968. *English Slipware Dishes 1650–1850.* London

Cooper, R.G. 1977. 'Quarrying in the Hambleton Hills, North Yorkshire: The Problem of Identifying Disused Workings', *Industrial Archaeology Review* Vol. 1/2.

Cox, N. 2005. 'Gloucester Folk Museum and the Mechanistaion of the Pin Industry', *Gloucester Society for Industrial Archaeology Journal*, 4-18.

Coysh, A.W and Henrywood, R.K. 1989. *The Dictionary of Blue and White Printed Pottery 1780–1880 Volume II.* Woodbridge: Antique Collector's Club.

Crabtree, P. 1989. West Stow, Suffolk: Early Anglo-Saxon Animal Husbandry. *East Anglian Archaeology* 47.

Crummy, N. 2004. 'The Lead Canister' in Brooks, H., Crummy, N. and Archibald, M. M. 'A Medieval Lead Canister from Colchester High Street: Hoard Container, or Floor Safe?' *Medieval Archaeology* 48, 131-42.

Crummy, N. 1983. *The Roman Small Finds from Excavations in Colchester 1971-9*, Colchester Archaeol. Rep. 2, Colchester.

Curzon, J.B, and Hanson, W.S. 1971. 'The pottery kiln', in Jones, GDB. 'Excavations at Northwich (Condate)', *Archaeol J* 128, 31-77.

Dannell, G.B. 1996. Petrecus Connected: Thirty Years on, in Bird, J. Hassall, M. and Sheldon, H. (eds.), *Interpreting Roman London: Papers in memory of Hugh Chapman*, Oxbow Monograph 58, 129-33.

Dannell, G.B, Dickinson, B.M, Hartley, B.R, Mees, A.W, Polak, M, Vernhet, A, Webster, P.V. 2003. *Gestempelte Südgallische Reliefsigillata (Drag. 29) aus dem Werkstätten von La Graufesenque*, Verlag des Römisch-Germanischen Zentralmuseums, Mainz

Dannell, G. 1971. 'The samian pottery' in B. Cunliffe, *Excavations at Fishbourne 1961-1969, Vol. II: The Finds*, Reports of the Res Comm Soc Antiq London 27, 260-316.

Dannell, G. 2006. The samian pottery mould-makers Lupus ii, Sabinus iii and their connections, in P. Ottaway (ed.), *A Victory Celebration. Papers on the archaeology of Colchester and Late Iron Age - Roman Britain presented to Philip Crummy*, Colchester, 83-104.

Dannell, G, Dickinson, B, and Vernhet, A. 1998. Ovolos on Dragendorff form 30 from the collections of Frédéric Hermet and Dieudonné Rey in J. Bird (ed.), *Form and Fabric: Studies in Rome's material past in honour of B.R. Hartley*, Oxbow Monograph 80, 69-109.

Davis, S.J.M. 1980. Late Pleistocene and Holocene equid remains from Israel. *Zoological Journal of the Linnean Society* 70 (3): 289-312.

Davis, S.J.M. 1992. *A rapid method for recording information about mammal bones from archaeological sites*, London: English Heritage AML Report 19/92.

Den Boesterd, M.H.P. 1956. *The Bronze Vessels in the Rijksmuseum G.M. Kam at Nijmegen* (Nijmegen).

Dickinson, B. 2005. 'P-14 unmasked, and what happened next', in GB. Dannell and PV. Irving (eds.), *J Roman Pottery Stud* **12**, *An Archaeological Miscellany: Papers in honour of K.F. Hartley,* 97-105.

Dickinson, B, and Hartley, B. 1993. 'Illustrated samian' in J. Monaghan, *Roman Pottery from the Fortress: 9 Blake Street,* The Archaeology of York **16**, Fasc. 7, York, 722-5, 745-69.

Dobney, K., Jaques, D. and Irving, B. 1996. *Of butchers and breeds. Report on vertebrate remains from various sites in the City of Lincoln.* Lincoln Archaeological Studies 5.

Dobson, B and Mann, J.C. 1973. 'The Roman Army in Britain and Britons in the Roman Army', *Britannia 4*, *191-205.*

Dodd, L and Clarke, V. 2002. *The Area of the Delamere Street Bus Station, Chester; An Archaeological Evaluation*, Earthworks Archaeology unpub rep.

Dodd, L J (2008b) *Excavation at Delamere Street, Chester, 2008: Post-Excavation Assessment of Post-Medieval Pottery and Glass from Watching Brief.* Unpublished report on behalf of Birmingham Archaeology.

Dore, J, and Greene, K. (eds) 1977. *Roman pottery studies in Britain and beyond*, BAR Supplementary series 30, Oxford.

Driesch, A. von den. 1976. *A guide to the measurement of animal bones from archaeological sites.* Peabody Museum Bulletin 1, Cambridge Mass., Harvard University.

Driver, J. 1982. 'Medullary bone as an indicator of sex in bird remains from archaeological sites'. in Wilson, R., Grigson, C. and Payne, S. (eds.) *Ageing and Sexing Animal Bones from Archaeological Sites*, 251-254. BAR British Series 109. Oxford.

Dutton, H.I and Jones, S.R.H. 1983. 'Invention and Innovation in the British Pin Industry, 1790-1850', *The Business History Review*, Vol. 57/2, 175-193.

Eckardt, H and Crummy, N. 2008. *Styling the Body in Late Iron Age and Roman Britain*, Monographies Instrumentum 36 (Montagnac).

Edwards, J. and Hampson, R. 2005. *White Salt-Glazed Stoneware of the British Isles.* Antique Collectors Club.

Edwards, J. 2008. 'Post-Medieval Pottery'. *In:* Garner, D. *Excavations at Chester, 25 Bridge Street 2001: Two Thousand Years of Urban Life in Microcosm.* Chester Archaeological Service Excavation & Survey Report 14, 190–242.

Egan, G. 1998. *The Medieval Household Daily Living c. 1150 - c. 1450*, Medieval Finds from Excavations in London 3. (2nd Edn), New York: Woodbridge and Rochester.

Eisenmann, V. 1981. 'Etude des dents jugales inferieures des *Equus* (Mammalia, Perissodactyla) actuels et fossiles'. *Palaeovertebrata* 10: 127-226.

Ellis, P and White, R. (eds) 2006. Wroxeter Archaeology: Excavations and research on the defences and in the town, 1968-1992. Shropshire Archaeological and Historical Society Transactions vol LXXVIII. English Heritage

English Heritage. 1991. *Management of Archaeological Projects* (2nd edn). London.

Esmonde-Cleary, S. *1987. Extra-mural areas of Romano-British towns, BAR* British Series 169. Oxford.

Evans, E.M 2000. *The Caerleon canabae: excavations in the civil settlement.* Britannia Monograph 16

Evans, D.R and Metcalf, V.M. 1992. *Roman Gates Caerleon*, Oxbow Monograph 15 (Oxford).

Exner, K., 1939. 'Die provinzialrömischen Emailfibeln der Rheinlande', *Bericht der Romisch-Germanischen Kommission* 29, 31-121.

Thompson, F.M. 1976, 'Excavations of Roman Amphitheatre at Chester' in *Archaeologia* 105, 219.

Faiers, J. 2000. 'The Period 2 assemblage', in Ellis, P. (ed) *The Roman baths and* macellum *at Wroxeter. Excavations by Graham Webster 1955-85*, English Heritage Archaeol Rep 9.

Feugère, M. 1985. *Les Fibules en Gaule Méridionale de la conquête à la fin du Ve siècle après J.-C.*, Revue Archéologique de Narbonnaise Supplément 12, Paris.

Fiches, J-L, Guy, M et Poncin, L. 1978. Un lot de vases sigillés des premières années du règne de Néron dans l'un des Ports de Narbonne', *Archaeonautica* 2, 185-219.

Forbes, R.J 1963. *Studies in ancient technology: Ancient geology, ancient mining and quarrying.* E.J Brill, Leiden, Netherlands

Fox, A. 1940. 'The legionary fortress at Caerleon Monmouthshire: excavations in Myrtle Cottage Orchard 1939', *Archaeologia Cambrensis* 95, 102-52.

Gaffney, V.L and White R.H. 2007. *Wroxeter, The Cornovii, and the Urban Process: Final Report on the Wroxeter Hinterland Project, 1994-1997, Volume 1: Researching The Hinterland*, Journal of Roman Archaeology sup series no. 68.

Gaimster, D. 1997. *German Stoneware 1200–1900.* London: British Museum Press

Garner, D. 2008. *Excavations at Chester: 25 Bridge Street 2001*, Archaeological Service Excavations and Survey Report 14, Chester.

Garner, D. 2009. 'Chester Ampitheatre' *in* Wilmott, T (ed) Roman Amphitheatres and Spectacula: a 21st-Century perspective: *Papers from an international conference held at Chester, 16th-18th February*, BAR International Series 1946.

Gillam, J. 1976. Coarse fumed ware in North Britain and beyond, *Glasgow Archaeol J* 4, 57-80

Godden, G.A. 2004. *Godden's Guide to English Blue and White Porcelain.* Woodbridge: Antique Collector's Club.

Grant, A. 1982. 'The use of toothwear as a guide to the age of domestic ungulates' in Wilson, B., Grigson, C. and Payne, S. (eds.) *Ageing and Sexing Animal Bones from Archaeological Sites.* Oxford: BAR British Series 109, 91-108.

Gray, H. 1918. *Anatomy of the Human Body 20th edition.* Philadelphia: Lea & Febiger.

Greenwell, A and Elsden, JV. 1913. *Practical Stone Quarrying*, London: Crosby Lockwood and Son.

Grigson, C. 1976. 'The Craniology and Relationships of Four Species of *Bos*. 3. Basic Craniology: *Bos taurus* L. Sagittal Profiles and other Non-Measurable Characters'. *Journal of Archaeological Science* 3: 115-136.

Grimes, W.F. 1930. *Holt, Denbighshire: the works-depot of the Twentieth Legion at Castle Lyons,* Y Cymmrodor 41, London.

Hall, R.A. 1997. *Excavations in the Praetentura: 9 Blake Street*, Archaeology of York 3/4 (York)

Harcourt, R. 1974. The dog in prehistoric and early historic Britain. *Journal of Archaeological Science.* 1, 151-175.

Harrison, L.A. 1990. 'Invertebrate Remains' in Matthews, K *Excavations at Chester: The Evolution of the Heart of the City*, Chester Archaeological Service Excavation and Survey reports 8, 55-9.

Hartley, B.R. 1972. 'The samian ware' in S. Frere, *Verulamium excavations Vol I*, Rep Res Comm Soc Antiq London 28, 216-62. Oxford.

Hartley, B.R. 2007. 'Samian ware' in W.S. Hanson, *Elginhaugh: a Flavian fort and its annexe*, Britannia Monograph 23, London.

Hartley, B.R, and Dickinson, BM. 2008. *Names on Terra Sigillata. An index of makers' stamps and signatures on Gallo-Roman Terra Sigillata (Samian ware)*, Vols 1-3, London.

Hartley, K.F. 1972, 'The mortarium stamps', in S. S. Frere, *Verulamium excavations*, Vol 1, Res Rep Soc Antiq London no 28, 371-81. Oxford.

Hartley, K.F. 1981. Painted fine wares made in the Raetian workshops near Wilderspool, Cheshire, in AC. Anderson and AS. Anderson, 471-8

Hartley, K.F and Webster P.V. 1973. The Romano-British kilns near Wilderspool, *Archaeol J* 130, 77-103.

Haverfield, F. 1890. 'Catalogue of Roman inscribed and sculptured stones in Grovesnor Museum, Vol 4

Hawkes, C.F.C. and Hull, M.R. 1947. *Camulodunum: First Report on the Excavations at Colchester 1930-1939*, Rep. Res. Comm. Soc. Antiq. Lond. 14, Oxford.

Hayes, L. 2005. Excavation at the former Rylands Garage, City Road, Chester, *J Chester Archaeol Soc* 80, 99-137.

Henig, M. 1974. *A Corpus of Roman Engraved Gemstones from British Sites*, BAR British Series 8, Oxford.

Henrywood, R.K. 2002. Staffordshire Potters 1781–1900. Woodbridge: Antique Collectors Club.

Hepburn, S. and Krawiec, K. 2009. Delamere Street, Chester, Archaeological Watching Brief on the

Excavation of the Quarry, 2008, Birmingham Archaeology unpub rep.

Hermet, F. 1934. *La Graufesenque (Condatomago)*, Paris.

Heyworth, M. 1993. 'The Metal Slag' in Casey, PJ. and Davies, JL. *Excavations at Segontium (Caernarfon) Roman Fort 1975 – 1979,* CBA Research Report 90, 211. London.

Hinchcliffe, J and Williams, J.H, with Williams, F. 1992. *Roman Warrington. Excavations at Wilderspool 1966-9 and 1976*, Brigantia monog 2.

Holmes, M. *forthcoming*. The Animal Bones. In Lowe, J. (ed.) Bronze Age and Iron Age occupation and post-medieval lime kilns at Staff College, Bracknell. *Berkshire Archaeological Journal*

Hunter, F. *forthcoming*. 'Changing objects in changing worlds - dragonesque brooches and beaded torcs', in Lewis *et al*.

Hunter-Mann, K. 2001. *Ceramic building materials. Minimum standards for recovery curation, analysis and publication.* Archaeological Ceramic Building Materials Group draft document.

Lewis. J.M. 1999. *The Medieval Tiles of Wales* (National Museum and Galleries of Wales, Cardiff).

Jacobs, J. 1913. Sigillatafunde aus einem römischen Keller zu Bregenz, *Jahrbuch für Altertumskunde* 6, 172-84.

Jenkins J.G. 1973. *The Rhaeadr Tannery*, National Museum of Wales: Welsh Folk Museum.

Jenkins, I. 1985. 'A group of silvered-bronze horse-trappings from Xanten (*Castra Vetera*), *Britannia* 141-64.

Jennings, S. 1981. Eighteenth Centuries of Pottery from Norwich. *East Anglian Archaeology Report* 13.

Jermy, K.E. 1961. 'The Roman road in Wirral' *Journal of the Chester Archaeological Society* 48.

Johnstone, C. and Albarella, U. 2002. *The Late Iron Age and Romano-British Mammal and Bird Bone Assemblage from Elms Farm, Heybridge, Essex (Site Code: HYEF93-95)*. English Heritage Centre for Archaeology Report 45/2002.

Jones, G. and Sadler, P. 2006. *Cattle mandibles - separating the adults from the old. The eruption of the cement-enamel-junction of the lower third molar as a way of defining old cattle.* Unpub rep.

Kiesewalter, L. 1888. Skelettmessungen an Pferden als Beitrag zur theoretischen Grundlage der Beurteilungslehre des Pferdes. Unpub dissertation. Leipzig.

King, A. 1978. A comparative survey of Bone Assemblages from Roman sites in Britain. *Institute of Archaeology Bulletin*, University of London. 15, 207-232.

Knorr, R. 1919. *Töpfer und Fabriken verzierter Terra Sigillata des ersten Jahrhunderts*, Stuttgart.

Koster, A. 1997. *Description of the Collections in the Provincial Museum G.M. Kam at Nijmegen 13: The Bronze Vessels 2. Acquisitions 1954-1996 (including vessels of pewter and iron.* Nijmegen.

Kratochvil, Z. 1969. Species criteria on the distal section of the tibia in *Ovis ammon* F. *aries* L. and *Capra aegagrus* F. *hircus* L. *Acta Veterinaria (Brno)* 38: 483-490.

Krawiec, K. 2008. Excavations at Delamere Street, Chester, 2006: Post Excavation Assessment and Research Design. Birmingham Archaeology unpub rep.

Langdon, J. 1986. *Horses, Oxen and Technological Innovation: The Use of Draught Animals in English Farming From 1066 to 1500.* Cambridge: Cambridge University Press.

Laughton, J. 2008. *Life in a Late Medieval City: Chester 1275-1520.* Oxford: Oxbow.

Lauwerier, R. 1988. Animals in Roman Times in the Dutch Eastern River Area. *Amersfoort: ROB Nederlandse Oudheden* 12.

Lawson, AK. 1995. 'A hoard of Roman harness fittings', 984-1000 in Blockley, K, Blockley, M, Blockley, P, Frere, S.S. and Stow, S. 1995. *Excavations in the Marlowe Car Park and surrounding areas*, Archaeology of Canterbury V, Canterbury.

Levine, M. 1982. 'The use of crown height measurements and eruption-wear sequences to age horse teeth' in Wilson, B., Grigson, C. and Payne, S. (eds.) *Ageing and Sexing Animal Bones from Archaeological Sites.* Oxford: Council for British Archaeology. Vol 109 pp. 223-243.

Lewis, C.P and Thacker, A.T (eds). 2003. *A History of the County of Chester*, *Victoria County History*.

Lewis, J.M. 1999. Lathom House: The Northern Court. *Journal of the British Archaeological Association* CLII, pp. 150-71.

Lewis, M, Worrell, S and Geake, H. (eds) forthcoming. *A Decade of Discovery: the Portable Antiquities Scheme*, BAR British Series, Oxford.

Lloyd-Morgan, G. 1988. 'The ivory buckle', in Ward, 122-3.

Luff, R. 1993. *Colchester Archaeological Report 12: Animal bones from excavations in Colchester, 1971-85.* Colchester Archaeological Trust.

Lyman, R.L. 1994. *Vertebrate Taphonomy.* Cambridge: Cambridge University Press.

MacGregor, A. 1985. *Bone, antler, ivory and horn. The technology of skeletal materials since the Roman period.* London: Croom Helm.

MacGregor, M. 1976. *Early Celtic Art in North Britain*, Leicester.

Mackreth, D. 1998. 'Copper-alloy brooches (Site 1, Site 2 and field-walking)' in Timby, 113-49.

Mackreth, D. 2000. 'The Romano-British brooches', in Ellis, 144-59.

Magregor, M, Mainman, A.J and Rogers, N.S.H. 1999. *Bone, Antler, Ivory and Horn from anglo-Scandinavian and Medieval York,* The Archaeology of York 17/12 (York).

Manaseryan, N.H., Dobney, K. and Ervynck, A. 1999. On the causes of perforations in archaeological domestic cattle skulls: new evidence. *International Journal of Osteoarchaeology* 9, 74-75.

Mann, J.C. 1983. *Legionary recruitment and veteran settlement during the principate.* Institute of Archaeology Occas Pap 7, London.

Manning, W.H. 1985. *Catalogue of the Romano-British Iron Tools, Fittings and Weapons in the British Museum*, London. 134-7.

Manning, W.H, Price, J and Webster, J. 1995. The *Roman Small Finds Report on the Excavations at Usk 1965 – 1976* Cardiff: University of Wales Press.

Margary, I.D. 1955. Roman roads in Britain: vol 1. London:Phoenix House

Marsh, G. 1978. 'Early second century fine wares in the London area', 119-223, in P. Arthur, P and Marsh, G. (eds), *Early fine wares in Roman Britain*, BAR British Series 57. Oxford.

Mason, D.J.P. 1978. 'The extra-mural area' in Strickland, TJ and Davey, PJ (eds) New Evidence for Roman Chester. Liverpool: Liverpool University Press.

Mason, D.J.P. 1980. *Excavations at Chester, 11--15 Castle Street and neighbouring sites 1974--8: a possible Roman posting house (Mansio).* Chester Archaeological Excavation and Survey Rep 2.

Mason, D.J.P. 1987. *Chester: The Canabae Legionis.* Brittania 18, 143-68.

Mason, D.J.P. 1988. *Prata legionis* in Britain. *Britannia* 19, 163-89.

Mason, D.J.P. 2001. *Roman Chester: City of the Eagles* Tempus, Stroud.

Mason, D.J.P. 2002. ' The Construction and Operation of a Legionary Fortress' in Carrington, P *Deva Victrix: Roman Chester Re-assessed*, Chester.

Mason, D.J.P. 2005. *Excavations at Chester: the Roman Fortress Baths : Excavation and Recording 1732-1998* Chester: Archaeological service excavation and survey report.

Mason, D.J.P. 2007. Mason, D.J.P., 'Excavations at Chester: The Roman Fortress Baths, Excavation and Recording 1732-1998', Britannia 38, 396-397.

Mason, D, & Frost, P. 2001. *Delamere Street Bus Station, Chester.* ***Chester.*** *Mason* Welland Archaeological Consultancy.

Matthews, K.J. 1995. Excavations at Chester : the evolution of the heart of the city; investigations at 3-15 Eastgate Street 1990/1. Chester Archaeological Service Excavation and Survey Report 14, Chester.

Matolcsi, J. 1970. Historische Erforschung der Körpergröße des Rindes auf Grund von ungarischem Knochenmaterial. *Zeitschr. f. Tierzüchtg. u. Züchtungsbiol., Hamburg* 87, 89-137.

Mattingley, H. and Sydenham, EH. 1926. *Roman Imperial Coinage vol. II: Vespasian to Hadrian,* London.

Maxfield, V.A. 1981. *The Military Decorations of the Roman Army*, London.

May, E. 1985. Widerristhöhe und Langknockenmasse bei Pferd – ein immer noch aktuelles Problem. *Zeitschrift fur Saugertierkunde* 50, 368-382.

McDonnell, G. 1983. 'Tap Slags and Hearth bottoms or, How to identify slags' *Current Archaeology* 86 VIII/3, 81-83.

McDonnell, G. 1991. A model for the formation of smithing slags. *Materialy Archeologiczne* 26, 23-6.

McKinley, J. 2004. *Compiling a skeletal inventory: disarticulated and co-mingled remains* in Guidelines to the Standards for Recording Human Remains, IFA Paper no 7.

McNeil, R. 1983. Excavation of an Eighteenth Century Pottery Kiln in Eccleston Street, Prescot (Site F). *Journal of the Merseyside Archaeological Society* **5** (1982–3), 49–94.

McPeake, J.C., Bulmer, M. and Rutter, J.A. 1980. 'Excavations in the garden of no. 1 Abbey Green Chester, 1975-77. Interim Report', *Journal Chester Archaeological Society* 63, 15-37.

Mees, A.W. 1995. *Modelsignierte Dekorationen auf südgallischer Terra Sigillata,* Forschungen und Berichte zur Vor- und Frühgeschichte in Baden-Württemberg, Band **54**, Stuttgart

Middleton, T. 1988. 'The Cliffs: Quarry and Caves, Little Haywood, Staffordshire' *Mining History: Bulletin of the Peak District Mines Historical Society* Vol. 10/3.

Miles, H. 1977. 'The Honeyditches Roman villa, Seaton, Devon', *Britannia* 8, 107-148.

Miller, S.N. 1922. *The Roman Fort at Balmuildy on the Antonine Wall*, Glasgow

Moser, F, and Tilhard, J –L. 1987. Un nouvel atelier de sigillée en Aquitaine, *Aquitania* 5.

Mould, Q. 2008. 'Iron' in Garner, 317-22.

Mountford, A. 1971. *The Illustrated Guide to Staffordshire Salt-Glazed Stoneware.* London: Barrie & Jenkins

Museums and Galleries Commission. 1992. *Standards in the museum care of archaeological collections.* London: Museums and Galleries Commission.

Newstead, R. 1901. 'A descriptive account of Roman and other objects recovered from various sites in Chester and district 1898-1901', *Journal Chester Archaeological Society* new series 8, 81-106.

Newstead, R. 1914. The Roman Cemetery in the Infirmary Field, Chester. Part I Annals of Archaeology and Anthropology 6, 121-67

Newstead, R. 1921. The Roman Cemetery in the Infirmary Field, Chester. Part II Annals of Archaeology and Anthropology 8, 49-60

Newstead, R. 1924. 'Report on the excavations on the site of the Roman camp at the Deanery Field, Chester', *Annals of Archaeology and Anthropology* 9, 59-86.

Newstead, R. 1928. 'Records of archaeological finds at Chester', *Journal Chester Archaeological Society* new series 27, 59-191.

Newstead, R. 1939. 'Records of archaeological finds at Chester ii-iv', *Journal Chester Archaeological Society* new series 33, 1-117.

Newstead, R. and Droop, J.P. 1932. 'The Roman amphitheatre at Chester', *Journal Chester Archaeological Society* new series 29, 1-40.

Newstead, R. and Droop, J.P. 1940. 'Excavations at Chester, 1939. The Princess Street Clearance area', *Journal Chester Archaeological Society* new series 34, 1-47.

Noake, P. 1993. The Post-Medieval Pottery. In: Ellis, P (ed). *Beeston Castle, Cheshire: a report on the excavations 1968–85.* English Heritage Archaeological Report 23, 191–210.London: HBMCE.

O'Connor, T.P. 1993. Birds and the scavenger niche. *Archaeofauna* 2, 155-162.

O'Connor, T.P. 1988. *Bones from the General Accident Site, Tanner Row.* The Archaeology of York: The Animal Bones 15/2. CBA.

Oldenstein, J. 1977. 'Zur Ausrüstung römischer Auxiliareinheiten. Studien zu Beschlägen und Zierat an der Ausrüstung der römischen Auxiliareinheiten des obergermanisch-raetischen Limesgebietes aus dem zweiten und dritten Jahrhundert n. Chr.', *Bericht der Römisch-Germanischen Kommission* 57 (1976), 49-366.

Orr, D.G. 2003. Samuel Malkin in Philadelphia: A Remarkable Slipware Assemblage. *In:* Hunter, R (ed). *Ceramics in America 2003,* 252–54.University Press of New England.

Oswald, F. 1936-7. *Index of figure types on Terra Sigillata,* Univ Liverpool Annals of Archaeol and Anthropol, Suppl.

Oswald, F. 1931. *Index of Potters' Stamps on Terra Sigillata 'Samian Ware',* Margidunum.

Oxford: Archaeopress (BAR Brit Ser **417**)

Payne, S. 1969. A metrical distinction between sheep and goat metacarpals. In: Ucko, P. and Dimbleby, G. (eds.) *The domestication and exploitation of plants and animals,* 295-305. London: Duckworth.

Payne, S. 1985. Morphological distinctions between the mandibular teeth of young sheep, *Ovis,* and goats, *Capra. Journal of Archaeological Science* 12, 139-147.

Payne, S and Bull, G. 1988. Components of variation in measurements of pig bones and teeth, and the use of measurements to distinguish wild from domestic pig remains. *Archaeozoologia* 2, 27-65.

Peacock, D.P.S. 1977. Roman amphorae: typology, fabric and origins, *Colln de l'École Française de Rome* 32, 261-78.

Peacock, D.P.S, and Williams, D.F. 1986. *Amphorae and the Roman economy: an introductory guide,* London.

Pearson, A. 2006. *The Work of Giants: Ston and Quarrying in Roman Britain,* Stroud: Tempus.

Philpott, R, Brennard, M, Carrington, P, Shottor, D and Stallibrass, S. 2005. *North West Regional Research Framework: Romano-British Period.*

Polak, M. 2000. *South Gaulish Terra Sigillata with potters' stamps from Vechten,* Rei Cretariae Fautorum Acta, Supplement 9, Nijmegen.

Poole, B 2004 *Archaeological Excavation Report of Land at the Bars: Area 2, Chester.* L-P Archaeology unpub rep.

Price, J. and Cottam, S. 1997. 'The Roman Glass', in Wenham and Heywood, 118-31.

Price, J. and Cottam, S. 1998. *Romano-British Glass Vessels: a Handbook,* CBA Practical Handbook in Archaeology 14, York.

Pringle, S. 2006. 'Early box flue tiles in London'. *London Archaeologist* 11/5, 124-9.

Prummel, W. 1978. Animal bones from tannery pits of 's-Hertogenbosch. *Berichten van de rijksdienst voor het oudheidkundig bodemonderzoek.* 28, 399-422.

Prummel, W. and Frisch, H. 1986. A guide for the distinction of species, sex and body side in bones of sheep and goat. *Journal of Archaeological Science.* 13, 567-577.

Rackham, J. 1995. 'Physical remains of medieval horses', 19-32, in Clark, J. (ed.) *The Medieval Horse and It's Equipment.* London: HMSO.

Rátkai, S. 1996. 'Post-medieval pottery' in Ellis, P. *Excavations at Chester: Chester Castle, The Seventeenth–Century Armoury and Mint.* Chester Archaeology, Excavation and Survey Report 10, 23–6.

Rátkai, S. forthcoming Tanning and Related Industries in Birmingham, 1200-1800.

Rátkai, S and Patrick, C. 2009. The Bull Ring Uncovered: Excavations at Edgbaston Street, Moor Street, Park Street and The Row, Birmingham City Centre 1997-2001. Oxbow.

Reece, R. 1987. *Coinage in Roman Britain* Seaby, London

RFG & FRG 1993 Roman Find Group and Finds Research Group AD 700-1700.

Richmond, I.A and Webster, G. 1950. Excavations in Goss Street, Chester, 1948-9, *Chester Archaeol Soc J* [offprint], 1-38.

Riddler, I. 1998. 'Wax tablets' in Egan, G (ed) *The medieval household : daily living c. 1150-1450,* London : Museum of London.

Riha, E. 1979. *Die römischen Fibeln aus Augst und Kaiseraugst,* Forschungen in Augst Band 3, (Augst)

Riley, M. 2009. '18th Century Chinese Coffee Cups' *Antique Collecting.* Vol 44/1, 37–41. Woodbridge: Antique Collectors Club.

Rogers, G.B. 1974. *Poteries sigillées de la Gaule Centrale I: les motifs non figurés,* Gallia Suppl 28.

Rogers, G.B. 1999. *Poteries sigillées de la Gaule Centrale II: les potiers,* premier Cahier du Centre Archéologique de Lezoux.

Romeuf, A–M. 2000. *Le quartier artisanale Gallo-Romain des Martres-de-Veyre (Puy-de-Dôme),* deuxième Cahier du Centre Archéologique de Lezoux.

Rutter, J.A. 1987. 'Finds from No. 1 Abbey Green' in Ward. S. *Excavations at Chester: The Civil War Siegeworks 1642–6.* Grosvenor Museum Archaeological Excavation and Survey Reports 4, 19–28. Chester City Council.

Rutter, J.A. 1990. Arrowcroft scheme 1977-9. In S.W. Ward, *Excavations at Chester, the lesser medieval religious houses, sites investigated 1964-1983,* 210-223. Grosvenor Museum archaeological and survey reports no.6. Chester City Council.

Rutter, J.A and Davey, PJ. 1980. Clay Pipes from Chester, in Davey, P J (ed) *The Archaeology of the Clay Tobacco Pipe III,* BAR British Series 78, 41-273, Oxford.

Rutter, J.A. 1990, 'Floor tile catalogue' in S W Ward *Excavations at Chester. The Lesser Medieval Religious*

Houses. Sites investigated 1964-1983 (Grosvenor Museum Archaeological Excavation and Survey reports No. 6).

Saller, RP and Shaw, BD. 1984. 'Tombstones and Roman Family Relations in the Principate: Civilians, Soldiers and Slaves,' *Journal of Roman Studies* 74, 124-56.

Schmid, E. 1972. *Atlas of Animal Bones for Prehistorians, Archaeologists and Quaternary Geologists.* Amsterdam, London & New York: Elsevier.

Schramm, Z. 1967. Long Bones and Height in Withers of Goat. *Roczniki Wyzszej Szkoly Rolniczei w Poznaniu, Posen* 36, 89-105.

Scruby, A. 2001. *Proposed Residential development on Land between Russell Street, Steam Mill Street and Canal Side, Chester; A Programme of archaeological evaluation.* Earthworks unpub rep.

Sedlmayer, H. 1999. *Die römischen Bronzefässe in Noricum*, Monographies instrumentum 10, Montagnac.

Sellye, I. 1939. *Les bronzes* émaillés *de la Pannonie Romaine*, Dissertations Pannonicae Ser. 2. Fasc. 8, Budapest.

Serjeantson, D. 1989. 'Animal remains and the tanning trade' in Serjeantson, D. and Waldron, T. (eds.) *Diet and Crafts in Towns: The Evidence of Animal Remains From the Roman to the Post-Medieval Periods.* Oxford: BAR British Series 199, 129-46.

Serjeantson, D. 1996. 'The animal bones' in Needham, S. and Spence, T. (eds.) *Refuse and Disposal at Area 16 East Runnymede: Runnymede Bridge Research Excavations.* London: British Museum Press 2, 194-223.

Serjeantson, D., Waldron, T. and Bracegirdle, M. 1992. 'Medieval horses from Kingston upon Thames'. *London Archaeologists.* 7(1), 9-12.

Shaw, M. 1996. The excavation of a late 15th- to 17th-century tanning complex at The Green, Northampton. *Post-Medieval Archaeology.* 30, 63-127.

Silver, I.A. 1969. 'The Ageing of Domestic Animals' in Brothwell, D. and Higgs, E. (eds). *Science in Archaeology.* London: Thames & Hudson, 283-302.

Smil, V. 2000. 'Horse power'. *Nature.* 405 (6783). p. 125.

Stanfield, J.A and Simpson, G. 1958. *Central Gaulish potters*, London.

Stanier, P. 1996. 'The Quarried Face: evidence from Dorset cliffstone quarries' *Mining History: The Bulletin of the Peak District Mines Historical Society*, 13/2.

Stanier, P. 2000. *Stone Quarry Landscapes: The Industrial Archaeology of Quarrying*, Stroud: Tempus.

Stephens, J. 2008. 'Ancient Roman hairdressing: on (hair) pins and needles', *Journal of Roman Archaeology* 21, 111-32.

Stopford, J. 1990, *Recording medieval floor tile* (CBA Practical Handbook 10).

Stopford, J. 2005, *Medieval floor tiles of northern England. Pattern and purpose: production between the 13th and 16th centuries* (English Heritage, Oxbow Books).

Strickland, T.J and Ward, S. 1978. *Chester Excavations, Northgate Brewery 1974-5: A Roman's Centurion's Quarters and Barracks*, Chester Archaeology Service Report 1.

Swan, V.G. 1984. *The pottery kilns of Roman Britain*, RCHM supp ser 5, London.

Swan, V.G. 2008. 'Builders, suppliers and supplies in the Tyne-Solway region and beyond' in Bidwell, P (ed) *Understanding Hadrian's Wall: Papers from a conference held at South shields, 3-5 November 2006, to mark the publication of the fourteenth edition of the Handbook to the Roman Wall [South Shields]: Arbeia Society*, 49-82.

Sykes, N. 2007. *The Norman Conquest: a Zooarchaeological Perspective.* BAR International Series 1656. Oxford.

Sykes, N and Symmons, R. 2007. Sexing cattle horn-cores: Problems and progress. *International Journal of Osteoarchaeology.* 17, 514-23.

Sykes, N, Wan, V, Robinson, J, Timney, B and Reynolds, R. (n.d.). Horn Cores, Horses and Hides: The Animal Remains from City Road, Chester. *Bioarchaeology Research Laboratory, University of Nottingham. Unpublished report.*

Tassinari, S. 1993. *Il Vasellame Bronzeo di Pompei,* Rome.

Taylor, I. 1985. 'The pottery', in Members of the Carlisle Regional Group, 'Fieldwalking in the Solway Plain', *Trans. Cumberland & Westmoreland Antiq. & Arch. Soc.* 85, 250-260.

Taylor, J. 1991. *The Roman pottery from Castle Street, Carlisle: excavations 1981-2*, Cumberland Westmorland Antiq Archaeol Soc Res ser no 5.

Teichert, M. 1975. Osteometrische Untersuchungen zur Berechnung der Widerristhöhe bei Schafen. In: Clason, A.T. (ed.) *Archaeozoological Studies*, 51-69. Oxford: Elsevier.

Terrisse, J –R. 1968. *Les céramiques sigillées Gallo-Romaines des Martres-de-Veyre*, Gallia Suppl. 19.

Thomas, R. 2000. Food for the dogs? The consumption of horseflesh at Dudley Castle in the eighteenth century. *Environmental Archaeology.* 5, 83-91.

Thompson, F.H. *1965. Roman Cheshire*, Chester.

Thompson, F.H. 1969. 'Excavations at Linenhall Street, Chester 1961-2', *Journal Chester Archaeological Society* new series 56, 1-21.

Thompson, F.H. 1976. 'The excavation of the Roman amphitheatre at Chester', *Archaeologia* 105, 127-239.

Thompson, F.H and Tobias, F.W. 1957. 'Excavations in Newgate Street, Chester 1955', *Journal Chester Archaeological Society* new series 44, 29-40.

Tilhard, J–L. 2009, *Les céramiques sigillées d'Espalion (Aveyron – France). Localisation de l'atelier, production, diffusion.* Périgueux.

Timby J.R. 1991. The Berkeley Street pottery kiln, Gloucester, *J Roman pottery studies* 4, 19-32.

Timby, J.R. 1998. *Excavations at Kingscote and Wycomb, Gloucestershire*, Cirencester.

Tomber, R and Dore, J. 1998. *The National Roman fabric reference collection: a handbook*, Museum of London / English Heritage/ British Museum.

Towle, A. Forthcoming. Tower Wharf Excavation. Earthworks Archaeology report.

Tyler, K, Betts, I and Stephenson, R. 2008. *London's Delftware Industry: The tin-glazed pottery industries of Southwark and Lambeth.* MoLAS Monograph **40**. London

Ulbricht, I. 1978. *Die Geweihverarbeitung in Haithabu.* Neumünster: Karl Wachholtz Verlag.

Utting, EJ. 1969. The working of the poor law in Chester in Tudor times, unpub B. Phil. Liverpool University.

Van der Voort, GF. 1999. *Metallography, Principles and Practice,* Ohio.

Von den Driesch, A. 1976. *A Guide to the Measurement of Animal Bones from Archaeological Sites.* Cambridge: Harvard University Press.

Waldron, T and Rogers, J. 1991. 'Inter-observer variation in coding osteoarthritis in human skeletal remains', *International Journal of Osteoarchaeology* 1, 49-56.

Walker, K. 1990. Guidelines for the preparation of excavation archives for long-term storage, Archaeology Section of the United Kingdom Institute for Conservation.

Ward, C. 1999. *Iron Age and Roman Piddington: The Roman Ceramic and Stone Building Materials 1979-1998.* Fascicule 4, The Upper Nene Archaeological Society.

Ward, M, and Carrington, P. 1981. 'A quantitative study of the pottery from a Roman extra-mural building at Chester', in AC. Anderson and AS. Anderson, 25-38.

Ward, S.W. 1987. *Excavations at Chester, the Civil War siegeworks 1642 – 1646.* Chester Archaeology Excavation and Survey Reports 4.

Ward, S.W. 1988. *Excavations at Chester. 12 Watergate Street 1985.* Chester Archaeology Excavation and Survey Reports 5, Chester.

Ward, S.W. 1990. *Excavations at Chester, the lesser medieval religious houses: sites investigated 1964-1983.* Chester Archaeology Excavation and Survey Reports 6, Chester.

Warry, P. 2006. '*Tegulae*: Manufacture, typology and use in Roman Britain'. BAR British Series 417. Oxford.

Watkinson, D and Neal, V. 1998. *First Aid for Finds (3rd edition)*, RESCUE and the Archaeology Section of the United Kingdom Institute for Conservation.

Webster, J. 1992. 'The objects of bronze', in Evans and Metcalf, 103-43.

Webster, P. 1989. 'The coarse ware' in J. Britnell, *Caersws vicus, Powys*, BAR British Series 205, 89-121. Oxford.

Webster, P. 1991. 'Pottery supply to the Roman north-west', *J Roman Pottery Stud* 4, 11-18.

Wenham, L.P and Heywood, B. 1997. *The 1968 to 1970 excavations in the vicus at Malton, North Yorkshire*, Yorkshire Archaeological Report 3, Leeds.

White, R. 2007. *Britannia Prima.* Tempus.

White, R. forthcoming. *'A British beaded torc from the Old Shops Site, Mill Street, Rocester'* in Ellis, P, Halstead, J and Burrows, R (eds). Excavations at Mill Street and Northfield Avenue, Rocester, Staffordshire, 2000-2002.Britiash Archaeological Report: Birmingham Archaeology Monograph series.

Wild, F. 1988. 'The samian ware' in BJN Edwards and PV Webster, *Ribchester Excavations 3: Excavations in the Civil Settlement 1968-1980. B. Pottery and Coins*, Cardiff, 9-50.

Wild, F. 1998. 'Three stamped decorated bowls from Gloucester' in J. Bird (ed.), *Form and Fabric: Studies in Rome's material past in honour of B.R. Hartley*, Oxbow Monograph 80, 111-3.

Wild, F. 2007. 'The samian ware' in RA Gregory, *Roman Manchester: The University of Manchester's Excavations within the Vicus 2001-5*, 56-168. Oxford.

Wild, J.P. 1970. 'Button-and-loop fasteners in the Roman provinces', *Britannia* 1, 137-55.

Williams, D.F. 1982. Petrological analysis of Roman pottery from Chester, unpub archive report.

Williams, D.F and Carreras C. 1995. 'North African amphorae in Roman Britain: a re-appraisal', *Britannia* **26**, 231-52.

Yeomans, L. 2008. Historical and zooarchaeological evidence of horn-working in post-medieval London. *Post-medieval Archaeology.* 42, 1-13.

Young, R and Humphrey, J. 1999. 'Flint use in England after the Bronze Age: Time for a re-evaluation?', *Proc. Prehistoric Society* 65, 231-42.

http://www.englishstone.org.uk/documents/dimension%20stone%208.html, accessed 2 December 2009.

http://www.naturalengland.org.uk/ourwork/conservation/geodiversity/englands/counties/area_ID5.aspx, accessed 2 December 2009.

http://www.medievalchester.ac.uk, accessed: 18 September 2009.

Cartographic Sources

Anon. 2001. *Printed Maps in the Cheshire Record Office* Chester.

Batenham, 1816. *Batenham's map of Chester.*

Braun, G. 1572-1618. *Civitates Orbis Terrarum.*

Buck, N. 1728. *Nathaniel Buck's Prospect of the City of Chester*

De Lavaux. 1745. Lavaux's Map of Chester (plus detailed enlargements).

Hemingway. 1836. *Hemingway's map of Chester* (published 1836 but based on a view of the city as it appeared in 1645).

Hollar. 1653. *Hollar's Plan of Chester.*

McGeegan, J. 1855. *John McGeegan's Image of City of Chester drawn from a Hot air balloon.*

Smith, W. 1588. *William Smith's plan of Chester* from his 'The Particuler Description of England. With the portraitures of certaine of the Cheiffest cities and townes' British Library, Harley MS 1046, fol. 173

Speed, J. 1610. *John Speed's map of Chester.*

Stockdale. 1795. *Stockdale's map of Chester.*

Ordnance Survey. 1874. *1st Edition Chester - Cheshire: 038/11', 1:2,500* (surveyed 1869-70).

Ordnance Survey. 1882. *England - Cheshire: 038, 1:10,560*

Primary Sources
All sources are from the Chester City Archives and noted CCA accordingly.

CR111/1 Attested copy of the will of Philip Chevers of Chester, butcher. *25 June 1791.*

ZA/B/2/129-129v William Wilson, tanner, related that he owned a parcel of ground without the Northgate, near the quarry, upon which several bays of building stood until they were burnt down in the siege. *10th August, 1660.*

ZA/B/4/(208) Upon reading the petition of George Eaton for a lease for 99 years of a piece of waste ground near the Gorsestacks in the lane leading from the Northgate to the Gorsestacks, 12 yards long by 8 and about 6 feet from the pavement of the lane, to build a house, an order was made for viewing and reporting. *1763-64.*

ZA/B/4/222v *[No Title].* Company of Bakers were not to lay or stack faggots of gorse thorns or other fuel on any Corporation waste land without special leave. *19th July 1764.*

ZA/B/5/f.207 Petition from Henry Potts, esq., and others for the cattle market held weekly in Upper Northgate Street, to the inconvenience of residents, and coaches passing to and from Liverpool and Parkgate, to be removed to waste ground in Gorse Stacks, which petitioners will fence at their own expence, accepted. *12 June 1818.*

ZA/F/4/30 *[No Title].* Mayors Orders, signed. *Oct. 27th, 1592.*

ZA/F/49c/20 Copy minutes of assembly, 19 January 1710/11, inscribed with report of committee of inquiry dated 2 March 1710/11 into petition from William Ince, [see ZA/F/49c/11], recommending the grant of a fee farm on condition the petitioner surrenders his interest in the Northgate quarry. *2 March 1710/11.*

ZA/F/49f/3 Copy minutes of assembly, 17 June 1712, inscribed with report of committee of inquiry appointed to decide places for stacking gorse. *30 October 1713.*

ZA/F/50e/63 Petition from the Aldermen, Stewards and Company of Bakers opposing an Assembly order of 13 May 1709.

ZC133C/1 Copy conveyance from the Ecclesiastical Commissioners for England to George Hodgkinson, auctioneer, of a piece of land fronting Upper Northgate Street and Windmill Lane, together with 14 houses, 9 cottages, 2 joiners shops, a shop, a bakehouse, a quarry, and with gardens and yards adjoining, as set out in a schedule. *14 Nov. 1867.*

ZC1C/4 Declaration by Charles Townsend of Chester, esq., before John Gamon, Commissioner of Oaths, relating to a house and shop on the E. side of Northgate Street in the occupation of Henry Sellers, and a piece of ground forming part of the Bishops Quarry in Northgate Street with the warehouse and outbuildings late in the occupation of R.D. Grindley, stated to be part of the estate of the Dean and Chapter of Chester, and from which he had collected the rents since 1844 as Clerk to the Dean and Chapter. *14 Feb. 1880.*

ZC46C Premises in the Co Gorse Stacks, [n.d.].

ZCHB/3 Corporation Lease Book. This book consists mainly of copies of grants, leases, bonds and documents setting up various charities in the city. *1508 – 1705.*

ZCHD/7/11 [no title] Also a building on a garden, without Northgate, between "quarrel neere the Northgate" and the highway leading to Gorse Stacks, now in occupation of 2. *20 May 1611.*

ZCHD/7/5[no title] One void plot of land near Gorse Stacks adjoining a certain close called 'le Quarell' on the south and a barn, now in the tenure of Randle Ince, on the east, 24 Royal virgates by 6 royal virgates*14 SEPTEMBER 1574.*

ZM/B/28a [*No Title]. Fines. 1599-1600.*

ZM/UV/3/34 Account of work done at the Quarry. *Feb., 1728[/9].*

ZS/B/6 a Hamon Godemon and John Bradfeld, Sheriffs, *1505-06.*

ZTCD/116 Conditions for leasing a large orchard without Northgate extending from Northgate Garden along the Walls to the Quarry westward and two small dwelling houses adjoining the orchard lately held by Mr Humphrey Read. *18th May 1766.*

www.ingramcontent.com/pod-product-compliance
Ingram Content Group UK Ltd.
Pitfield, Milton Keynes, MK11 3LW, UK
UKHW061847041026
15591UKWH00015B/953

9 781407 310015